SHEET METAL WORKERS HANDY BOOK

By Edwin P. Anderson

THEODORE AUDEL & CO.
a division of
HOWARD W. SAMS & CO., INC.
4300 West 62nd Street
Indianapolis, Indiana 46268

SECOND EDITION

FIRST PRINTING—1975

International Standard Book Number: 0-672-23235-9

Foreword

The purpose of this book is to present the fundamentals of sheet metal work layout in a clear and simple language.

The subject matter is divided into two parts. The first part concerns itself with *Practical Pattern Cutting,* followed by instructions in the use of various *Sheet Metal Machines, Welding of Sheet Metal* and *Boiler Plate Work.*

The second part conveys information about *Drafting, Plan Reading, Geometrical Problems* and *Mathematics* necessary for proficient workmanship in the modern sheet metal shop.

Due to the fact that a certain amount of geometrical knowledge is always required, especially in sheet metal patterns of a more intricate nature, the reader is advised to study Chapter 10, which gives a step-by-step instruction in 50 problems, many of which are actually encountered in laying out of patterns of sheet metal work. Proficiency in the solution of these problems will prove to be of inestimable value wherever expert workmanship is to be attained.

Contents

CHAPTER 1

Aircraft Sheet Metal Work

Sheet Metals Used.—A wide variety of sheet metal is used in aircraft work, including iron, steel, copper, brass, aluminum and various plated sheets. Aluminum alloy, however, on account of its lightness and strength together with numerous other desirable qualities, is perhaps used more extensively in this kind of work than any other metal.

Sizes.—Sheets of different metals are rolled in stock sizes and thicknesses. The length and breadth are almost universally given in inches and the thickness usually by gauge.

This use of gauges often lead to confusion, as different mills use different gauges and even the same mill may use different gauges on different metals.

For example a No. 10 sheet of copper, measured on "Stubbs" gauge is 0.134 in. thick; No. 10 sheet iron on the U.S. Standard gauge is 0.141 in. thick and No. 10 zinc is only 0.020 in. thick.

Characteristics of Materials—Aluminum.—The characteristics of aluminum which are the chief factor in determining its usefulness in aircraft work are: 1. Light weight, approximately ⅓ of that of steel. 2. High resistance to corrosion when properly treated. 3. The ease of fabrication (i.e. facility for bending without sacrifice of strength). It should be noted that the term *aluminum* as used here, refers to all of its alloys, because of the fact that all of them contain approximately 95% of pure aluminum. The word *duralumin*, a term widely used in trade circles refers to certain alloys manufactured by the Aluminum Company of America.

Commercially pure aluminum may contain upwards of 1% of other elements, chiefly iron and silicon. It is not particularly strong, but in its soft condition is easily formed.

The metals with which aluminum is commonly alloyed in various quantities to produce various characteristics are: Copper, silicon, manganese, magnesium, chromium, iron, zinc and nickel. These metals may be added singly or in combination to produce high strength and high resistance to corrosion.

Pure aluminum is used in aircraft work principally for non-structural parts, such as cowling, wings and any other part which has to be "*bumped*" to shape, also for welded gas and oil tanks.

Aluminum alloys are used for structural parts such as ribs, beams, wing skins, hulls, in engines and propellers, etc. As a matter of fact, it would hardly be possible to mention any part of the aircraft in which its use has been excluded.

Stainless Steel.—Another material widely used in aircraft construction is stainless steel. The prime factors contributing toward its use are: 1. Great strength. 2. Resistance to rust, corrosion and general deterioration. 3. Its adaptability to modern production methods.

Stainless steel contains approximately 18% of chromium with less than 0.70% of carbon and 8% nickel. It is used for wing ribs, control cables, fuel tanks and a great variety of other parts.

Copper.—This is an excellent conductor of electricity and heat, and in aircraft is used chiefly in the form of tubing for fuel and oil lines, and in most electrical parts. Unlike steel, it is annealed or softened by heating and sudden cooling, whereas vibration, hammering, rolling, or bending hardens it.

Copper surfaces, like those of all other metals must be tinned before soldering, because solder adheres to them poorly. Muriatic (hydrochloric) acid is usually employed as a cleaner, because it easily removes scale, rust, and other deposits that may adhere to copper. However, the acid, as well as the copper article, should be cold when cleaned, otherwise the copper will be attacked by the acid.

Brass.—This is an alloy of copper and zinc. Its use in aircraft is limited to various pipe fittings, tanks and valves.

Sheet Metal Work.—Since sheet metal arrives from the mills in standard lengths, widths and thicknesses it must be cut and shaped according to requirements in the shop. Thus each individual item is precisely laid out in the drafting room with dimensions and tolerances, giving in addition numerous information such as title of the work, related drawings, nearest stock size, material, name of draftsman checker, engineer, etc.

The functions generally assigned to the sheet metal worker varies according to plan of procedure, but although the schedule may differ the general procedure is roughly as follows:

To lay off work from drawings, and to make the various operation in a logical sequence. These various operations may require the use of numerous sheet metal machines of the type described in Chapter 5, in addition to a wide variety of hand tools such as: steel rules, calipers, scribers, knife edged blades, prick punches, dividers, micrometers, chisels, hack saws, drills, etc. each of which must be handled with skill and speed if a true and usable product is to be procured.

Procedure in Drilling of Holes.—Generally it can be said that holes larger than 1/8 or 3/16 in. should be drilled with a small drill and then enlarged successively until the desired size is obtained. For holes larger than 1/4 in. a taper reamer should be used if possible. When holes of 1 in. diameter or larger are to be drilled the procedure in the absence of special tools is to drill around the circumference of the marked circle as shown in fig. 1. After the drilling, the holes are connected with a chisel, after which a half-round suitable file should be used.

Sheet Metal Seams.—Seaming is one of the most important processes in sheet metal work, because a good seam not only joins two sheets of metal, but strengthens the product as well.

The choice of seam depends upon the equipment available and the strain to be resisted by the product. The seams shown in figs. 2 to 4 are most commonly used in light sheet metal work. Seams are frequently soldered along the edges to make them water-tight.

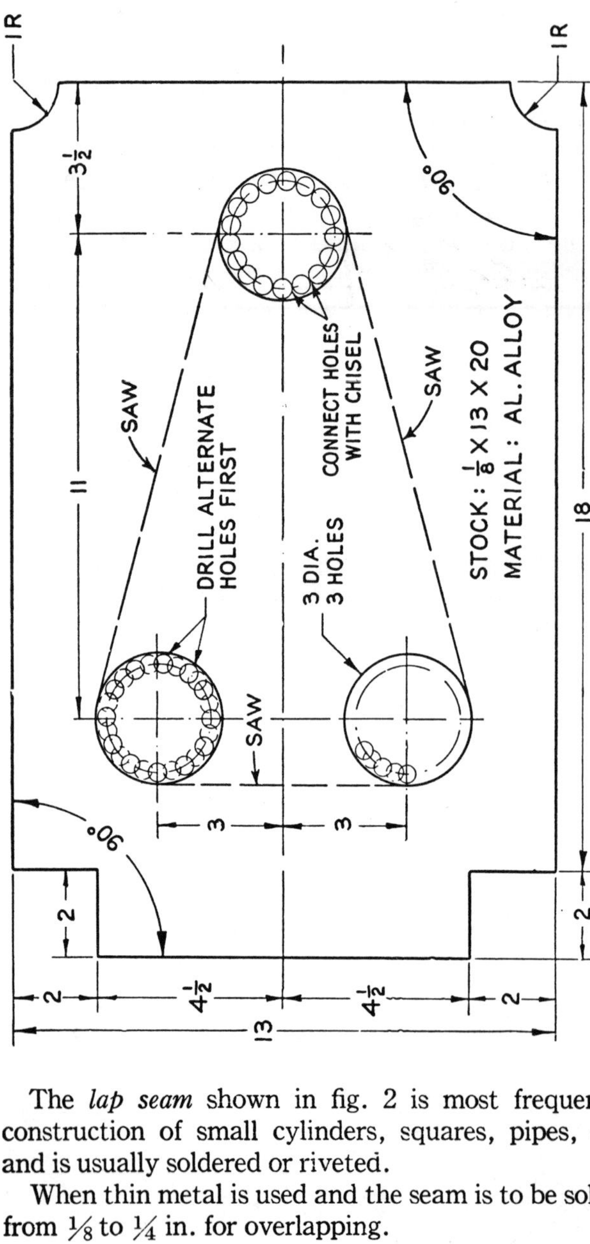

FIG. 1.—Proper procedure in drilling large holes. One important point to remember is to avoid drilling the small holes too close to circumference of large hole. After the small holes are being drilled, a sharp chisel should be used to connect the holes and the excess material should then be removed by a suitable half-round file. The sequence of operations are: *A*, squaring up stock; *B*, layout and work holes and cut-outs; *C*, cut and drill out corners according to specifications; *D*, drill out the 3 in. diameter holes if a 3 in. hole saw be not available. Drill a series of No. 30 holes which over-lap each other and remove the entire stock, then file to the scribed line; *E*, finally, saw connecting line between holes as marked. ***Note.***—Be careful not to scratch stock during process of operation.

The *lap seam* shown in fig. 2 is most frequently used in construction of small cylinders, squares, pipes, elbows, etc. and is usually soldered or riveted.

When thin metal is used and the seam is to be soldered, allow from ⅛ to ¼ in. for overlapping.

The *grooved lock seam,* fig. 3, is the most generally used for joining and locking the edges of longitudinal seams of sheet metal. Two single edges are hooked together and flattened with a small mallet to make them tight. At this point it is called a common *lock seam*. For positive assurance against unhooking, the seam is then *grooved* with a hammer and hand groover, fig. 13. In larger shops special grooving machines are usually available.

The *double lock seam,* fig. 4, is most commonly utilized in light gauge work to strengthen the bottom of articles. It is formed by hooking two single edges together and bending one edge to make a right angle.

Fuel Tanks.—Fuel tanks for aircraft are usually made of terne plate (a plate having a thin coating of lead on steel or iron to prevent corrosion) ranging in thickness from No. 24 (0.0156 in.) to No. 18 (0.050 in.) U. S. Standard gauge and joined by welded seams. Fittings, such as the filler neck, drain valve and fuel line connections are usually soldered to the tank. Fig. 16 illustrates a typical tank, although cylindrical tanks of the type shown in fig. 18 are often encountered.

The material used may in addition to the previously mentioned terne plate, be sheet metal, aluminum alloy, duralumin or stainless steel.

The tanks are in many instances supplied with inside baffle plates, as illustrated in fig. 17, to offset excessive splashing and foaming of the fuel when the aircraft is in motion.

Fuel Tank Cleaning.—With respect to repairs of fuel tanks, it is important that they be cleaned thoroughly as a safety precaution against the *explosion of gasoline or fumes* remaining in the tank. The usual procedure is to fill the tank with a solution containing an alkaline cleaner and then flush with steam. While flushing, keep all the fittings open to drain all sediment and to avoid pressure building up high enough to weaken or wreck the tank.

Capacity of Tanks.—Volume of tanks with rectangular or circular cross-sections are comparatively simple to calculate,

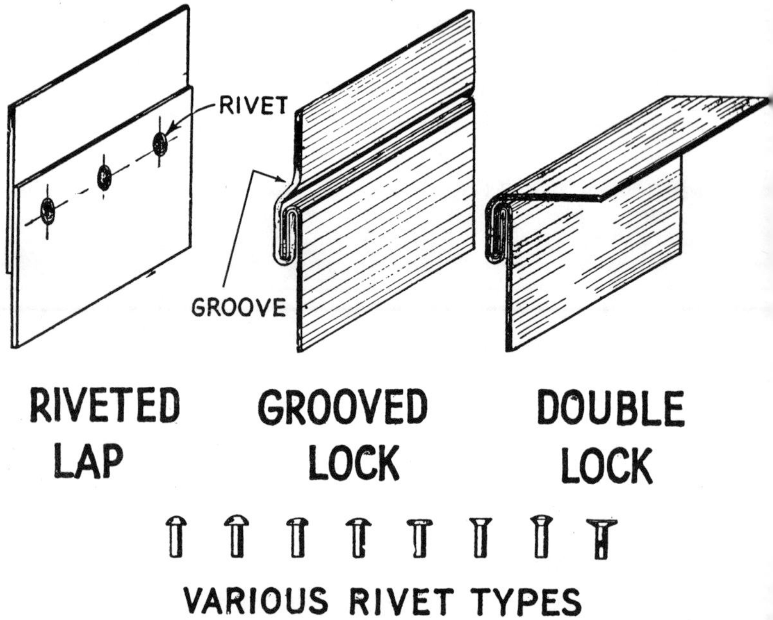

FIGS. 2 to 12.—Depicting various kinds of seams used in sheet metal work, and types of rivets met with. The rivets shown are termed respectively: *button head, round head, mushroom head, brazier* or *binder head, flat head, countersink head, oval head* and *tubular head.*

whereas various other shapes may require the use of mathematical formulae.

The volume of tanks with straight sides is equal to the cross-section times the length. If the three dimension (length, width, and depth) are in inches, the volume will be obtained in cubic inches. Since there are 231 cubic inches in one gallon, the volume (in cubic inches) should be divided by 231 to obtain the number of gallons.

Example.—If the average width, depth and length of tank illustrated in fig. 16 is 36, 24 and 30 inches respectively, what is its gallonage?

Solution.—The capacity of the tank in gallons =

$$\frac{36 \times 24 \times 30}{231} = 112.2 \text{ gal.}$$

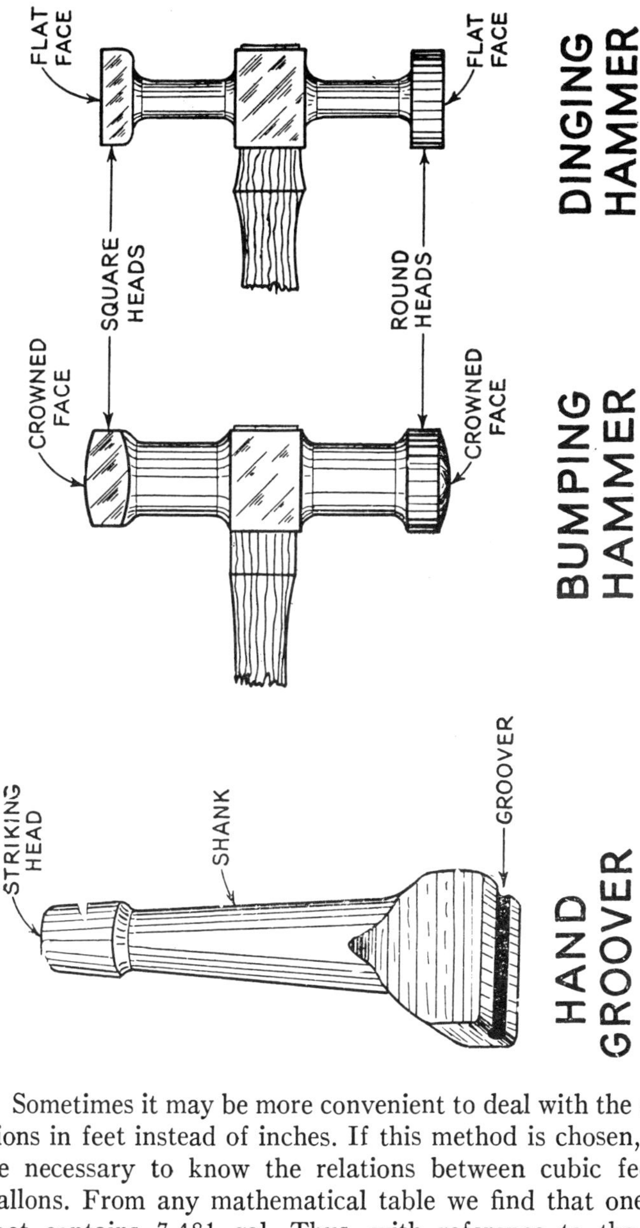

FIGS 13 to 15.—Tools used in sheet metal work. The hand groover is used in making certain kinds of seams. The bumping hammer shown in fig. 14 ranges from 14 ozs. to about 2 lbs. in weight and is used for heavy roughing-out work, whereas a dinging hammer, fig. 15, weighs about 10 ozs. and is used for lighter finishing work. Because the bumping hammer has a crowned face, it should be used on the inside of crowned surfaces, whereas the flat-faced dinging hammer is used on the outside.

Sometimes it may be more convenient to deal with the dimensions in feet instead of inches. If this method is chosen, it will be necessary to know the relations between cubic feet and gallons. From any mathematical table we find that one cubic foot contains 7.481 gal. Thus, with reference to the above

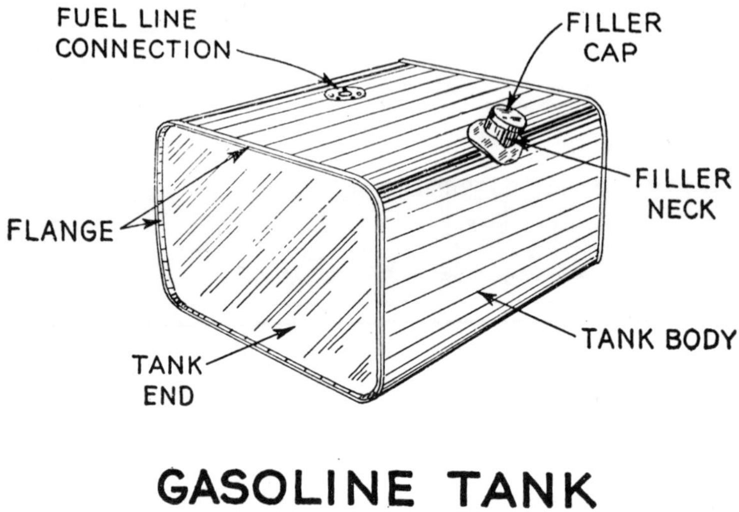

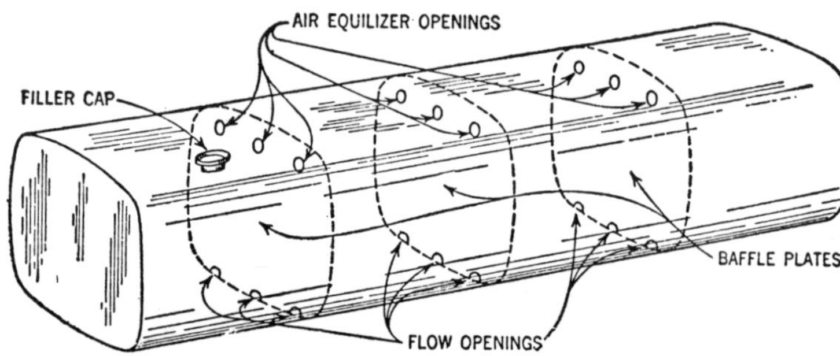

FIGS. 16 and 17.—View of fuel storage tank and tank showing filler opening and baffle plates.

example, we obtain the capacity of the tank in gallons as 3×2×2.5×7.481 = 112.215.

Example.—Calculate the material necessary to construct and show the layout of a circular 50-gallon gas tank.

Solution.—If the necessary attachments are omitted, and it is remembered that one gallon contains 231 cubic inches, then the volume of our tank=50×231=11,550 cu. ins. If it is

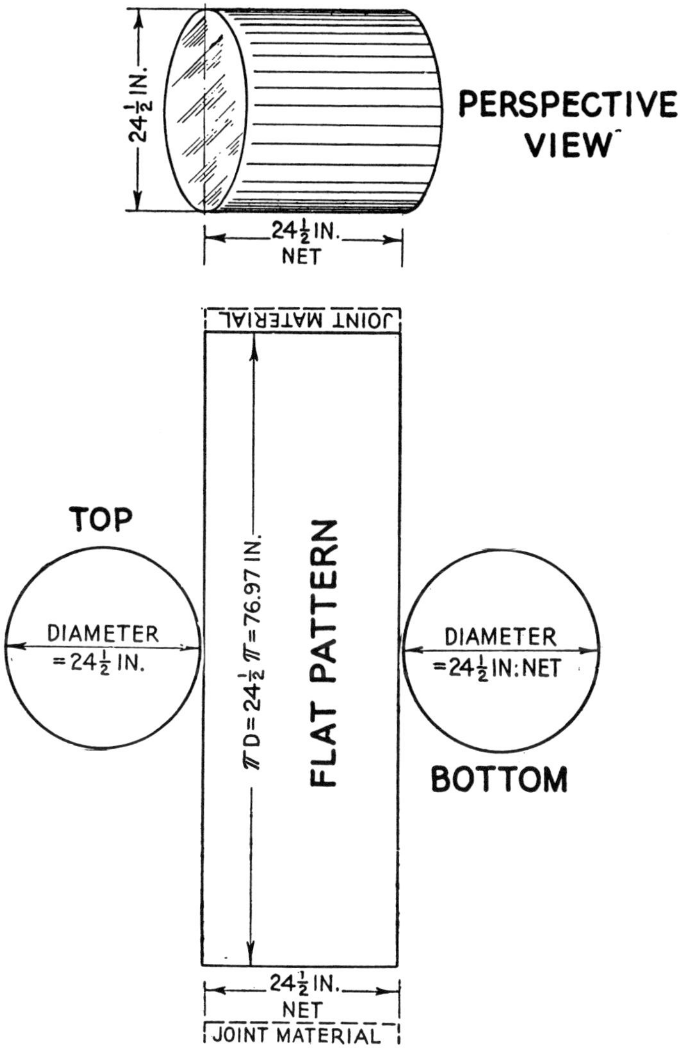

FIGS. 18 to 21.—Methods of development of cylindrical tank.

assumed that the diameter of the tank and its height be equal, then the dimensions will be obtained from the equation:

$$\text{Volume} = \frac{\pi D^2}{4} \times D, \text{ that is}$$

$$11{,}550 = 0.7854\ D^3$$

$$\text{and } D = \sqrt[3]{\frac{11{,}550}{0.7854}} = 24.5 \text{ inches}$$

With reference to fig. 21 the net material necessary is a rectangular sheet having a width of 24.5 $\times \pi$ or 76.969 say 77 inches and height of 24.5 inches. To this should be added the circular top and bottom each with a net radius of 24.5/2 or 12¼ inches, plus a sufficient lap for seaming on all three patterns.

Layout of Wing Tips and Wing Tip Bows.—Due to the fact that it is often deemed necessary for the aircraft sheet metal worker to know the fundamentals of wing structure layout, four principal wing tips are shown in figs. 22 to 25.

For the layout of the elliptical type the student is referred to pages 357 to 364 showing various methods for geometrical construction of the ellipse.

To make a layout of wing tips, fig. 25, begin to make a layout on a suitable sheet of paper of two parallel lines 60 in. apart, representing the leading and trailing edges respectively. From the leading edge draw quadrant NA, with radius as shown. Through the center of this circle-quadrant draw a line AC, parallel to the line representing the leading edge. Draw line NR perpendicular to AC. Now draw line HR and bisect as shown. Finally extend bisector until it intersects line AC at M. The length of line MA gives the required radius R.

Reading of Aircraft Blueprints.—Blueprints concerned with aircraft construction work are usually worked out by the engineering and drafting departments in closest cooperation and are usually classed according to their part in the system as a whole, such as: *Detail prints*, *assembly details*, *minor assembly prints*, *major assembly* and *installation prints*, *layout prints* and *special rework prints*.

The information conveyed on each should be complete without unnecessary repetition or duplication, to enable the workmen concerned to handle each phase of the work with speed

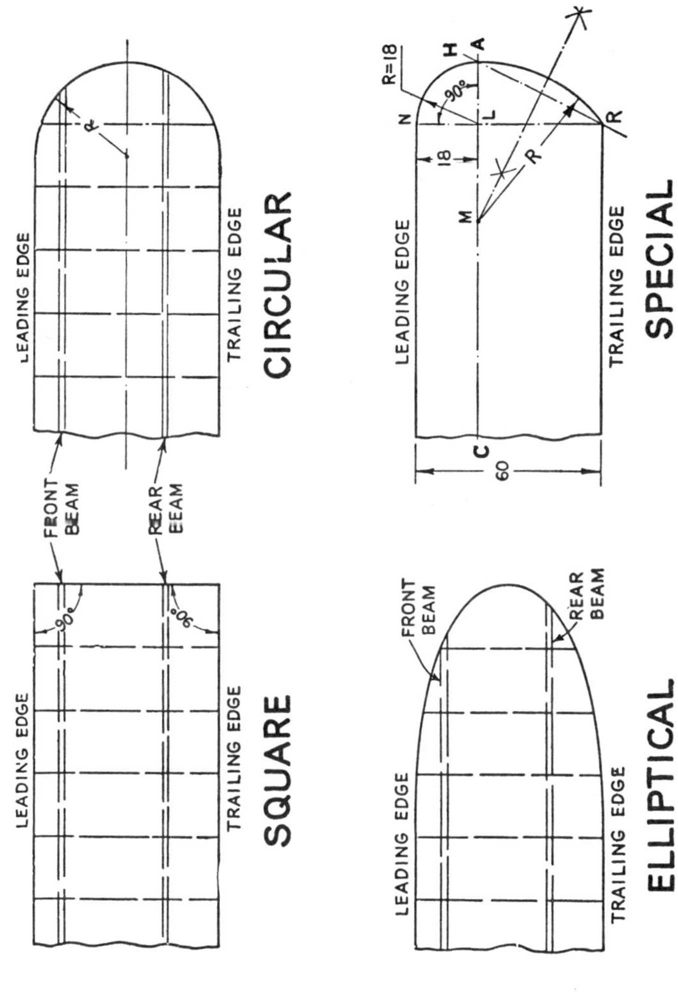

FIGS. 22 to 25.—Illustrating various forms of aircraft wing tips in common use. One of the most efficient and also one of the most difficult to build is the elliptical wing shown in fig. 24. The most inefficient is the square tip shown in fig. 22. A compromise between the elliptical and the round type is shown in fig. 25.

and dispatch, thus facilitating the fabrication of a true and faultless product.

Blueprint Face.— By definition the *face* of a blueprint is that space on the print enclosed by lines forming a rectangle in which are tabulated all the information data which are not directly placed on the drawing, and which are necessary for the mechanic to complete the part.

The arrangement of the face to accommodate the data placed therein depends upon the nature of the drawing and the practice of the manufacturer. In fact, it is so varied and given in so many different ways that no attempt is here made to show every arrangement.

The principal items which are arranged in sub-divisions of the face in "blocks" or columns are: 1. Title block. 2. Change block. 3. Material. 4. Scale. 5. Limits. 6. Model. 7. Dash number. 8. Part number. 9. Drafting checkers.

Size of Drawings.—In order to facilitate folding and filing in standard letter files, small drawing sheets are based on the 8½×11 or sometimes an even multiple of that size. Large drawings are frequently made in roll size form, that is, they are rolled to facilitate handling.

How to Represent Dimension Lines.—On all kinds of drawings dimension lines are of the utmost importance. In general

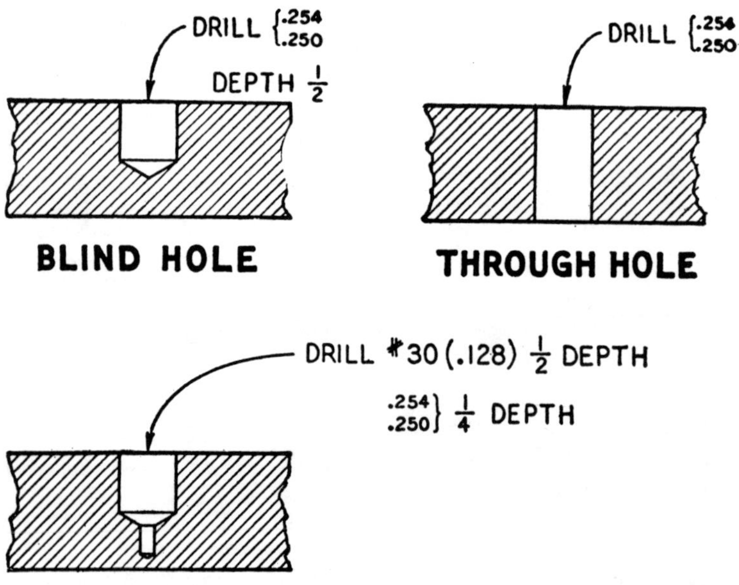

FIGS. 26 to 28.—Methods of indicating drilling.

the following points should be observed: ***A***. Avoid whenever possible the crossing of dimension lines. ***B***. Never repeat dimensions. ***C***. Never omit a necessary dimension. ***D***. Use fine solid dimension lines—never the dotted or dot-dash variety. ***E***. All dimensions should be expressed in inches, hence no inch marks are necessary and should by all means be omitted as they only serve to confuse. ***F***. Decimal points should be made sufficiently heavy to avoid mis-calculation. ***G***. Arrows should be made plain and their size should be in proportion to the length of the dimension lines themselves.

Drawing Scales.—The scales generally used for aircraft drawing are ¼, ½ full size, double size, four times size.

Finish Notations.—Information as to the finishing is usually given in the "Title Block" in the space with the listing of standard dimensional tolerances.

Drilling Notations.—Indications in regard to drilling are usually conveyed in the form shown in figs. 26 to 28 giving size, tolerances required for holes other than rivet or cotter holes. Figs. 26 and 27 represent a blind and through hole respectively, with drill limits of .254 and .250. Fig. 28 represents information necessary for drilling of a pilot hole.

Tabulated Drawings.—On certain drawings in which the parts vary only in size, such as bolts, nuts and similar items, much drafting time will be saved by tabulating the various dimensions. Thus, with respect to bolts, instead of making separate drawings for the different sizes, it will suffice to tabulate the various sizes using symbols such as A, B, C, D, etc. as headings on actual dimensions in their respective order, tabulated underneath.

Drawings (General).—A typical detail drawing giving the necessary information for the manufacture of a *wing, eye bolt*

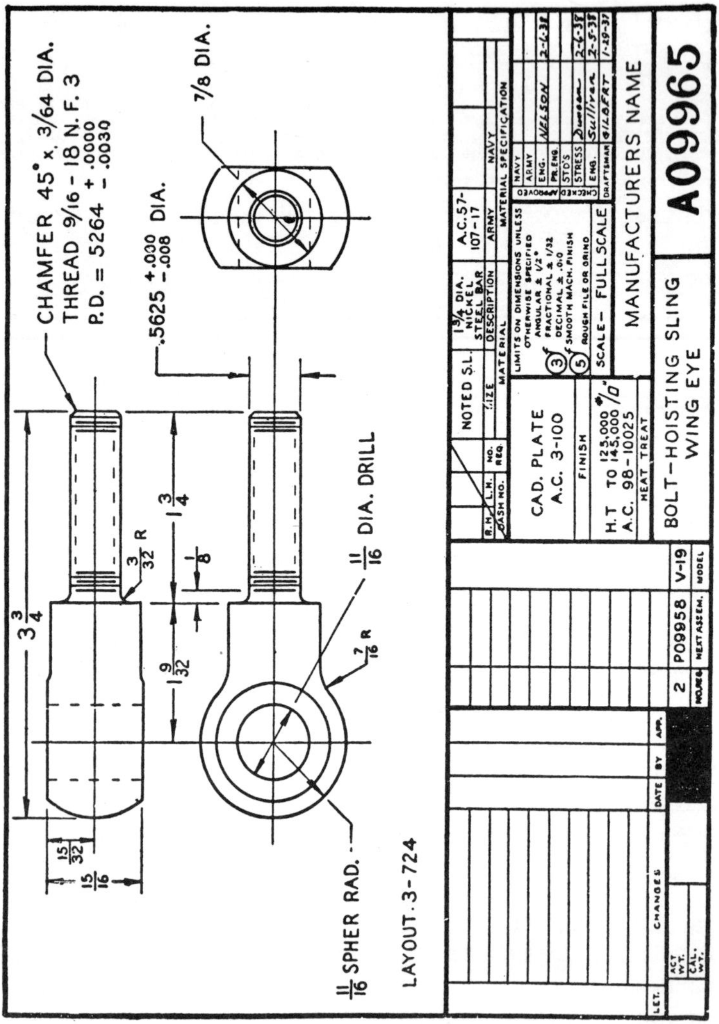

FIG. 29.—Drawing of a machine part showing title block notations.

together with the "Title block" is shown in fig. 29. Drawings (figs. 30 to 43) illustrate various sheet metal layouts.

In this type work it is customary to give a *perspective view* covering the article in question, together with the stock dimension and material used.

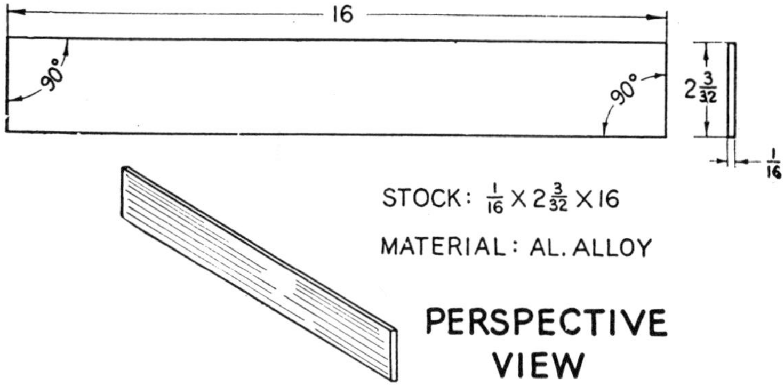

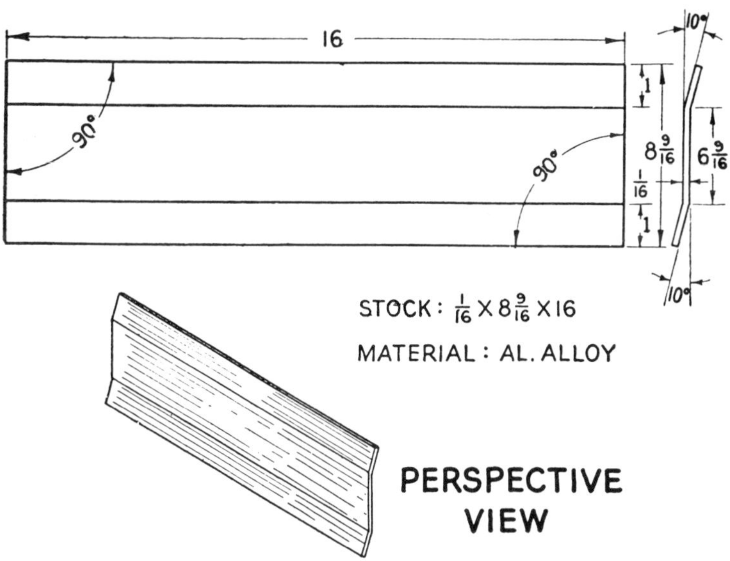

FIGS. 30 to 33.—Showing methods of layout instructions for manufacture of simple metal parts. With reference to fig. 32 the sequence of operation consists in squaring up stock as indicated, after which the bend lines are indicated by a suitable marking tool. The bending operation aiong the previously marked lines is usually performed in a leaf break machine.

Frustum Cone Development.—Since the frustum is quite frequently encountered in sheet metal work, it will often prove of value to be able to apply simple mathematical concepts in calculating sheet metal requirements of this kind.

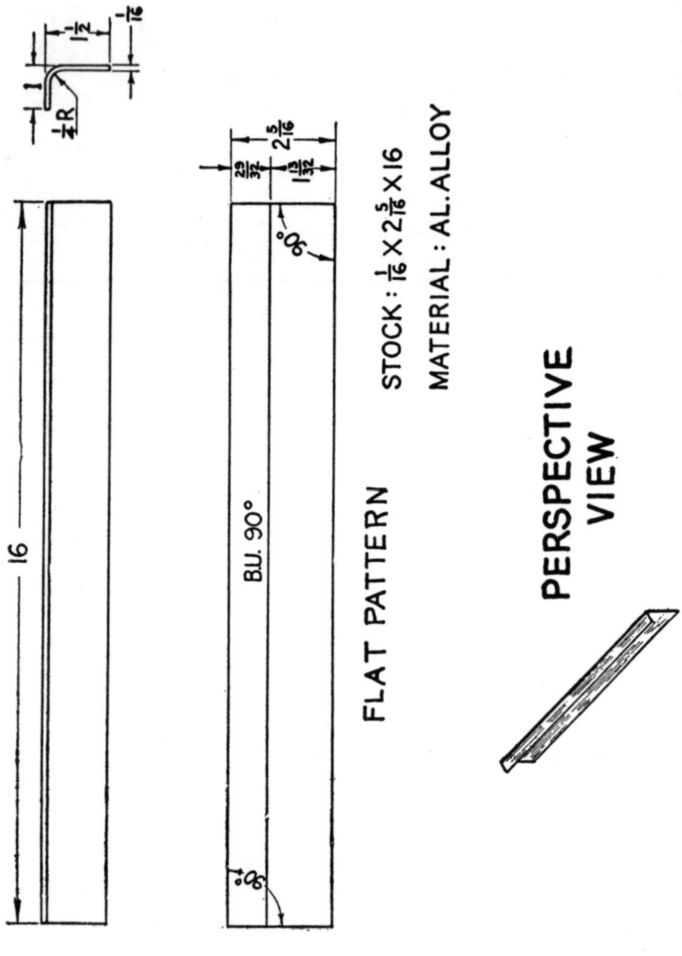

FIGS. 34 to 36.—Showing layout and necessary information for cutting and bending to size a 16 in. metal bar. If a perspective view be included with each piece to be fabricated it will greatly facilitate the instant understanding of the work to be performed. The sequence of operation follows those previously shown and is therefore not repeated for this layout.

Although there may be numerous methods by which a similar solution is arrived at, the specific problem given shows how a considerable saving in time and material will be obtained by the mathematical instead of by a geometrical or scale solution.

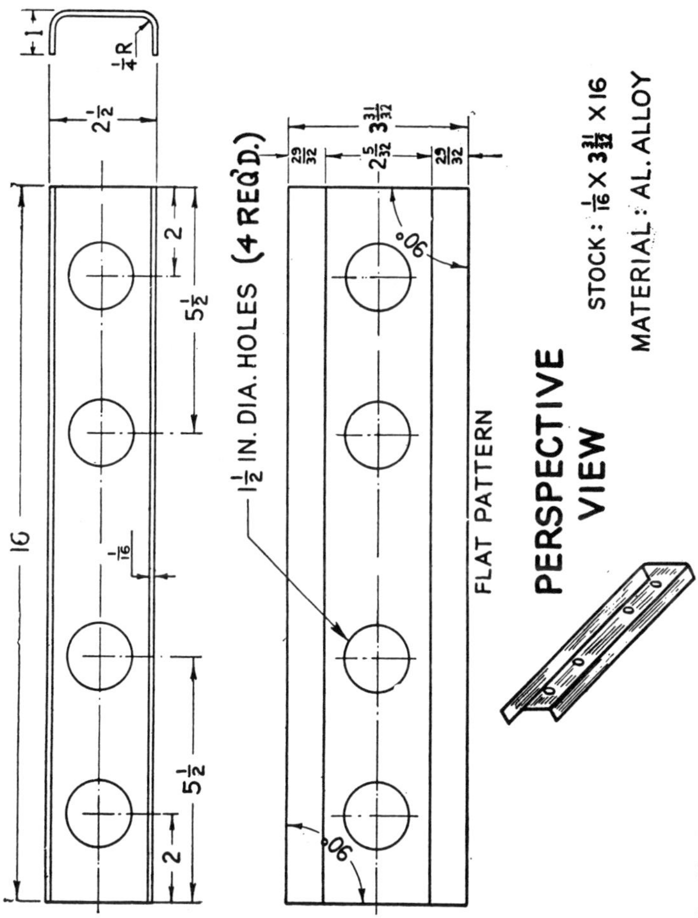

FIGS. 37 to 39.—Layout of simple metal bracket provided with four holes spaced as shown. If special tools for stamping of holes be not available the procedure shown in fig. 1 of this chapter should be followed. The working sequence should be: *A*, square up stock; *B*, layout and mark bend lines; *C*, layout and mark center line on which the 1½ in. diameter holes are located; *D*, punch the holes at the given distances; *E*, drill the holes and file to scribed line; *F*, form the two 1 in. angles on previously marked bend line.

Example.—A frustum fig. 44 with a height of 24 inches and a base and top diameter of 36 and 18 inches respectively is to be developed as illustrated. Find the sheet metal size required (a) without seams; (b) with seams permitted.

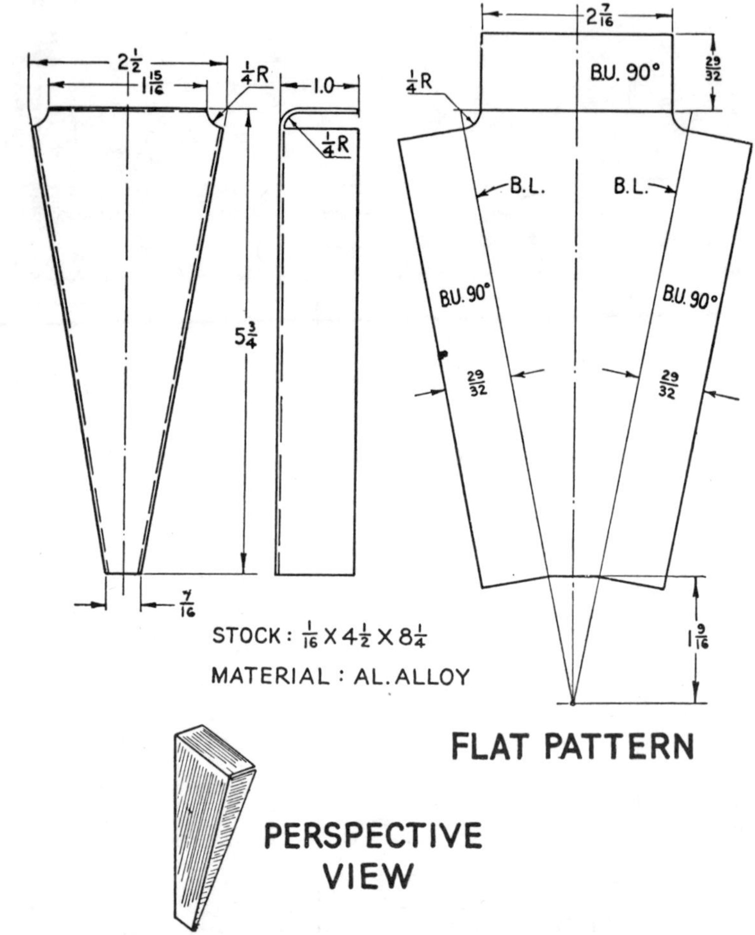

FIGS 40 to 43.—Showing cutting and bending operations of sheet metal piece along given dimensional lines.

Solution.—The slant height of the frustum is obtained by taking the square root of the sum of the squares of d and h, or,

(1) $s = \sqrt{h^2+d^2} = \sqrt{24^2+9^2} = 25.632$ in.

(2) The undeveloped angle Θ of the cone is:

$$\tan \Theta = \frac{9}{24} = 0.375 \text{ and } \Theta = 20.55°$$

(3) The subtended angle φ of the flat development is:

$$\varphi=\frac{d\times 360}{s}=\frac{9\times 360}{25.632}=126.40°$$

(4) The outside radius R, of the flat development is then,

$$R=\frac{s\times D}{d}=\frac{25.632\times 18}{9}=51.264 \text{ in.}$$

(5) The chord C, of the angle of the pattern,

$$C=\sin\frac{1}{2}\varphi\times R\times 2$$

$$=\sin 63.2\times 51.264\times 2$$

$$=0.8926\times 102.528=91.52 \text{ in.}$$

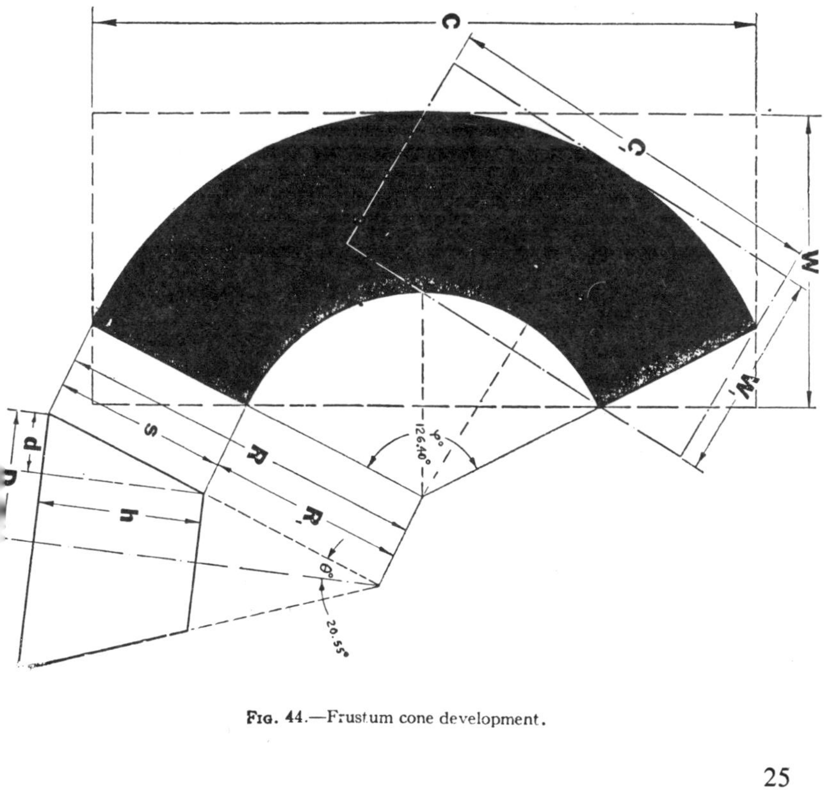

FIG. 44.—Frustum cone development.

A few standard abbreviations which appear on Air Craft Drawings are given below:

Aileron	Ail.	Exhaust	Exh.
Altitude	Alt.	Fillet	Fil.
Approximate	Approx.	Flat head	F.H.
Army & Navy	AN.	Front	Fr.
Assembly	Assem.	Horizontal	Hor.
Attachment	Att.	Horse power	H.P.
Bill of material	B/M	Inches	In.
Bracket	Brkt.	Instrument panel	Inst. panel
Bulkhead	Blkd.	Landing gear	Ldg. gr.
Cancelled	Can.	Longeron	Long.
Cantilever	Cantil.	Longitudinal	Long.
Casting	Cstg.	Material	Matl.
Center of buoyancy	C.B.	Miles per hour	M.P.H.
Center of gravity	C.G.	Naval Aircraft Factory	N. A. F.
Center of pressure	C.P.	Reinforcement	Reinf.
Change	Chng.	Revolutions per minute	R.P.M.
Countersink	Csk.	Segment	Seg.
Design	Des.	Shock absorber	Sh. abs.
Designation	Desig.	Square inches	Sq. in.
Developed length	D.L.	Stabilizer	Stab.
Developed width	D.W.	Stiffener	Stif.
Dimension	Dim.	Superseded	Sup.
Direction finder	D/F	Trailing edge	T.E.
Elevator	Elev.	Transverse	Transv.
Engine	Eng.	Vertical	Vert.

(6) The width W, of the rectangular pattern is obviously,

$$W = R - \cos\frac{1}{2}\varphi \times \frac{R}{2}$$
$$= 51.264 - \cos 63.2^\circ \times 25.632$$
$$= 51.264 - 0.4509 \times 25.632 = 39.707 \text{ in.}$$

Thus we have found, that the required size of the sheet metal for making a frustum of the specified type is (a) 91.5×40 inches (approximately).

In many sheet metal processes a considerable saving of material will be obtained if the sheet metal pattern be divided into two or more separate parts and joined in the manufacturing process.

If, in the present example it be desired to make the frustum in two pieces, the width of such a plate is the perpendicular distance between the outer and the inner chord, the formula for our calculation will be as follows:

$$W_1 = R - R_1 \cos \frac{1}{4}\varphi$$

$$= 51.264 - 25.632 \times 0.8517$$

(b) $$= 51.264 - 21.831 = 29.433 \text{ in.}$$

Thus the width of the plate if seams be permitted will only be 29½ in. approximately against 40 inches without seaming of the material.

The chord C_1 of this smaller pattern is:

$$C_1 = 2R \times \sin \frac{1}{4}\varphi$$

$$= 2 \times 51.264 \times 0.524 = 53.73 \text{ in.}$$

It is interesting to compare the area required in both cases.

Thus, in case (a) the area is:

$$A = C \times W = 91.52 \times 39.707 = 3{,}634 \text{ sq. in.}$$

In case (b) the area is,

$$A_1 = C_1 \times W_1 \times 2 = 53.73 \times 29.433 \times 2 = 3{,}163 \text{ sq. in.}$$

The net saving in metal if two equal plates of 53.73×29.433 in. be used instead of one plate 91.52×39.707 in. would therefore be approximately 15%.

CHAPTER 2

Principles of Pattern Cutting or Development of Surfaces

By definition, a development is *a full view drawing of an object in which all the surfaces of the object are shown in the plane of the drawing board:* that is, the development is a drawing which represents the surfaces of the object unfolded or unrolled or spread out in the plane of the drawing board so that the entire surface is seen in one plane true size without any foreshortening.

In sheet metal work solids are divided into two general classes, according as the surfaces of the solid are

1. Elementary, or
2. Warped.

Solids having elementary surfaces may be developed accurately, and those having warped surfaces, only approximately. Objects having elementary surfaces may be formed by simply folding or rolling the metal pattern, whereas if the object have warped surfaces, the metal pattern must undergo the operation of *raising* or *bumping* to bring the pattern to the true shape of the object when folded or rolled.

By definition an elementary surface is one in which *a straight edge may be placed in continuous contact in one direction,* as in figs. 9,410 and 9,411.

The line of contact is an element of the surface. Elementary surfaces may be either

1. Plane, or
2. Curved

as shown in figs. 9,412 and 9,413. By definition a plane surface is *one in which elements may be drawn in any direction,* as *ab, cd, ef* in fig. 9,412. A curved surface is *one in which no three consecutive* elements lie in the same plane,* as in fig. 9,413.

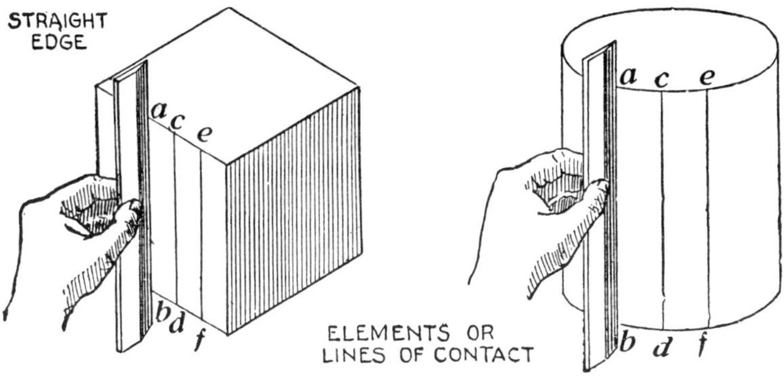

FIGS. 9,410 and 9,411.—Elementary surfaces or surfaces in which a straight edge may be placed in *continuous contact* with the surface, as for instance in positions *ab, cd, ef,* etc. The imaginary line of contact of the straight edge with the surface is called an *element* of the surface.

In the figure let *ab*, *cd* and *ef* be three consecutive elements, then if a plane LARF pass through *ab* and *ef*, the intervening element *cd* will not lie in this plane. The curved surface shown in fig. 9,413 is a *cylindrical* surface or curved surface having parallel elements, as distinguished from another class of curved surfaces which does not have parellel elements, as for instance a conical surface. The distinction is shown in figs. 9,414 and 9,415.

By definition, surfaces of the second general class (known as *warped surfaces*) are those in which *a straight edge may be placed in contact at only one point,* as, for instance, the sphere shown in fig. 9,416.

Patterns for such surface can be cut only to the approximate shape, as for instance in fig. 9,417 the pattern *abdfec*, for a section of the surface of a sphere. This pattern, as can be seen, must be warped or hammered to the shape *abd'fec'*, so that its surface will coincide with the surface of the sphere. The figure clearly shows the shape of the pattern before and after warping.

From the explanations thus far given it must be evident that patterns may be cut accurately for objects having elementary surfaces and only approximately for objects having warped surfaces.

*NOTE—***Consecutive elements*** are those which lie infinitely close to each other. Thus in fig. 9,410 the elements *ab, cd* and *ef* are so close to each other that no other elements could be drawn or imagined to lie between them. It should be noted, however, that in fig. 9,413, these elements are drawn quite far apart to clearly illustrate the principle which defines a curved surface

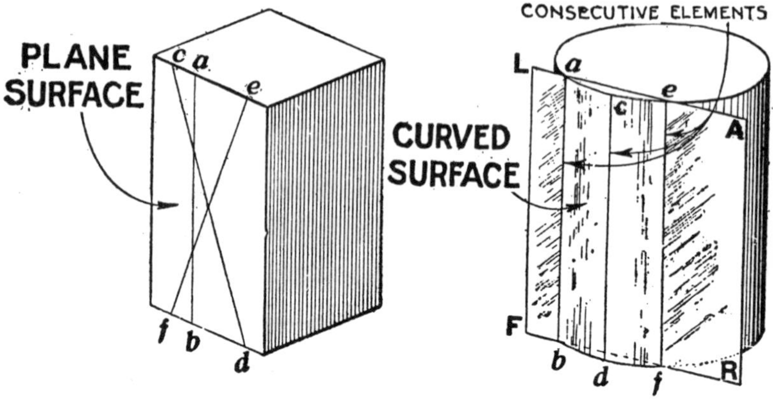

FIGS. 9,412 and 9,413.—Plane and curved surfaces. Elements of a plane surface may be drawn in the surface in any direction as *ab*, *cd*, *ef*, as in fig. 9,412. In a curved surface no three *consecutive* elements lie in the same plane as in fig. 9,413.

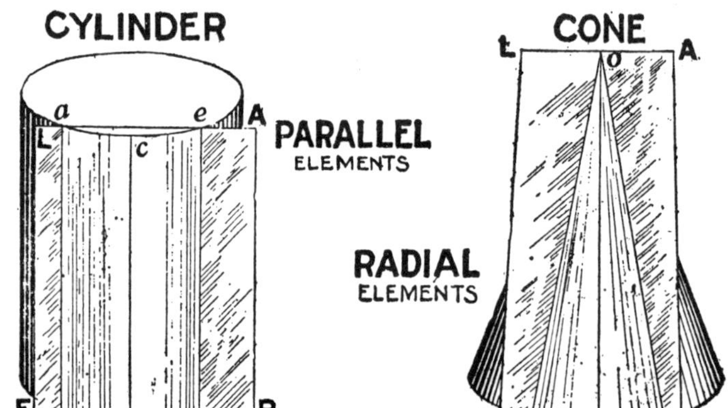

FIGS. 9,414 and 9,415.—Distinction between ***cylindrical*** and ***conical*** surfaces. Fig. 9,414, elements parallel; fig. 9,415, elements radial. Both surfaces being curved surfaces, no three consecutive elements lie in the same plane as indicated by plane **LARF**, passing through the first and third of the three consecutive elements *ab*, *cd*, and *ef*.

Considering these two classes of surfaces, various methods are employed in developing the surfaces in laying out patterns as

1. Methods applied to objects having elementary surfaces.
   - *a*. Parallel line
   - *b*. Radial line
   - *c*. Triangulation
2. Methods applied to objects having warped surfaces.

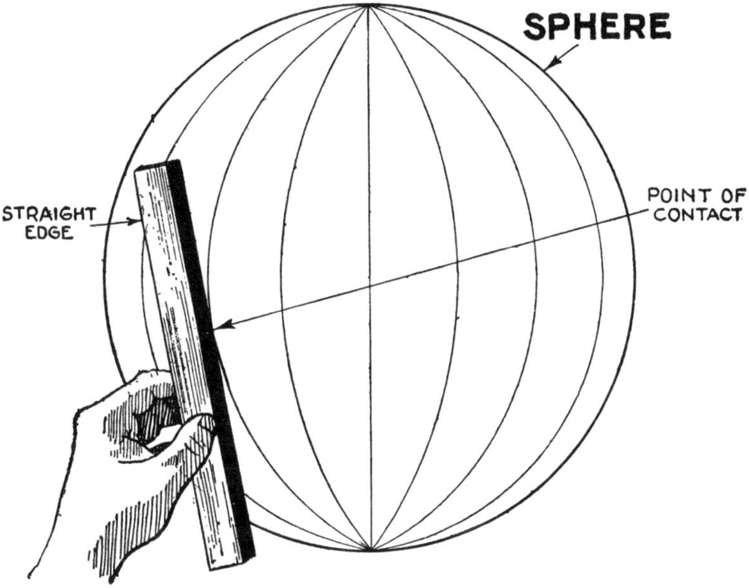

FIG. 9,416.—***Warped surface*** or *surface in which a straight edge can be placed in contact only at a point.*

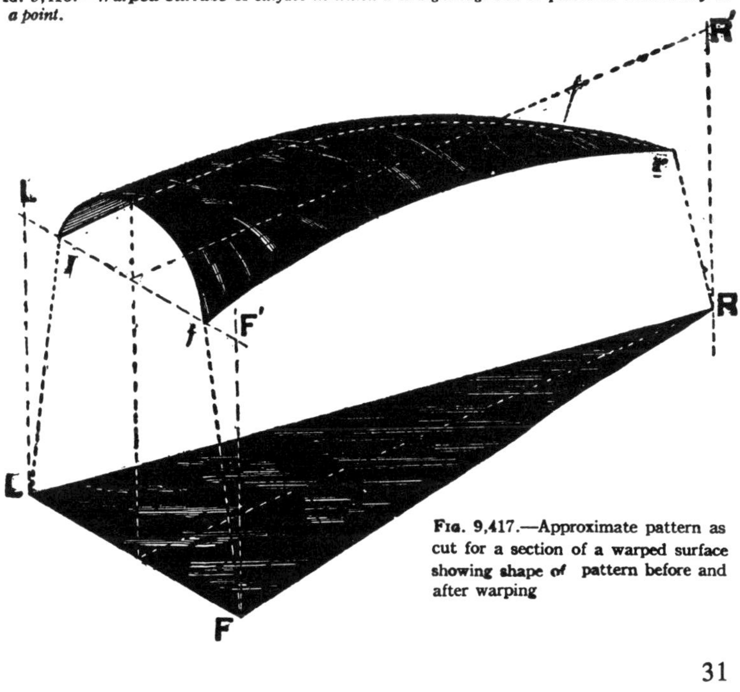

FIG. 9,417.—Approximate pattern as cut for a section of a warped surface showing shape of pattern before and after warping

1. Elementary Surfaces

Development of Patterns by Parallel Lines.—In any development a plan and elevation of the object is first drawn, and from these views the development or shape of the pattern is obtained by laying off what is called a "*stretch out*" which is simply the outline of the object unfolded and laid out flat.

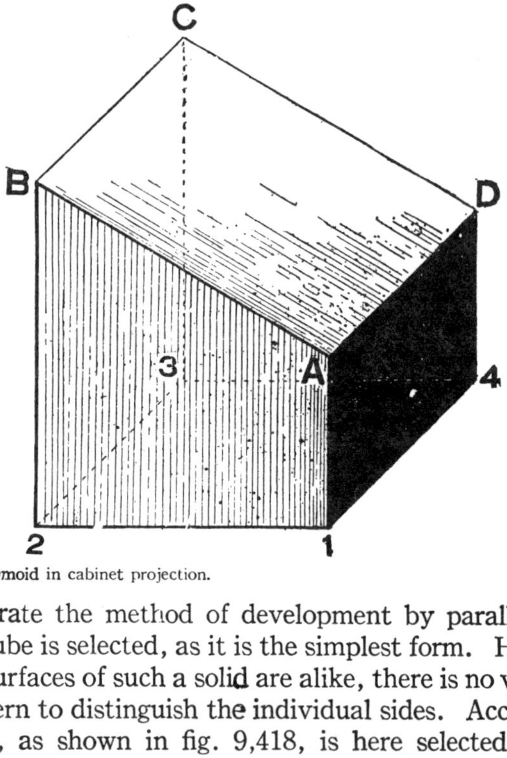

Fig. 9,418.—Prismoid in cabinet projection.

To illustrate the method of development by parallel lines, usually a cube is selected, as it is the simplest form. However, as all the surfaces of such a solid are alike, there is no variation in the pattern to distinguish the individual sides. Accordingly a prismoid, as shown in fig. 9,418, is here selected for the first example. Note carefully the general shape of the solid, especially the shape of the sides.

To develop, draw in fig. 9,420 a base line at the elevation of the base and on this base line lay off points 1, 2, 3, 4 and 1. The distances between these points are obtained from the plan. The distance between points 1 and 2, in stretchout equal distance between points 1 and 2, in plan; between 2 and 3, in stretchout equal distance between 2 and 3, in plan, and so on all the way around to the starting point 1. Erect perpendiculars at the points thus obtained and project over points A,D,B and C, from the elevation by the dotted lines parallel to the base line. The intersections of these dotted lines with the perpendiculars give the heights of the perpendiculars corresponding to the heights of the edges of the prismoid,

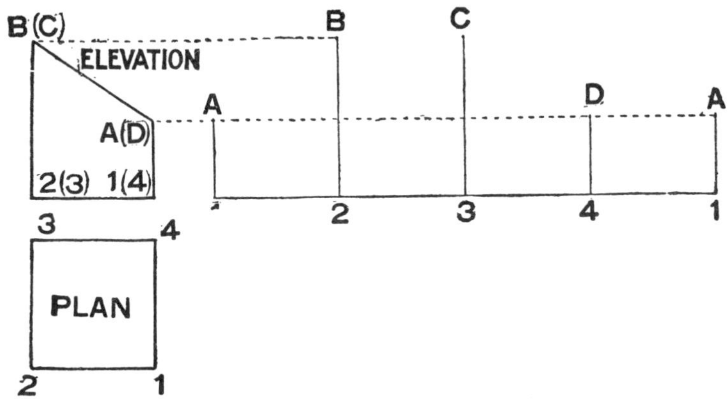

FIG. 9,419 to 9,421.—Development of pattern *by parallel lines* for the prismoid shown in fig. 9,418; base line and perpendiculars

that is, A1, in stretchout equal A1, in elevation; B2, in stretchout equal B2, in elevation, etc.

The stretchout is now completed by joining the points A,B,C,D and A. The completed stretchout is shown in fig. 9,428. The moving picture (figs. 9,422 to 9,426) shows the progressive development of the pattern.

Example.—Develop a pattern for the sides of the oblique parallelopipedon shown in fig. 9,430. ***Case 1.*** Development on line A1; ***Case 2.*** Development referred to base line.

Case 1.—First draw an elevation and plan of the parallelopipedon in orthographic projection as shown in figs. 9,431 and 9,433, lettering these views to correspond with the numbers and letters of fig. 9,418.

Draw A1, equal to and parallel to A1, in elevation. Draw 12, equal to 12, in elevation making angle A12, equal 60°.

Through points 2 and A, draw lines respectively parallel to A1, and 12, obtaining point B, and completing the first section of the pattern.

Similarly draw in lines 3C, 4D and 1A. Through points 3,4 and 1, draw lines parallel to 1A. Through B, draw line parallel to 23, obtaining point C; through C, draw line parallel to 34, obtaining line CD; and through D, line parallel to 41, obtaining line DA, thus completing the stretchout.

Case 2.—Draw in elevation and plan as in Case 1, and continue base 21 of elevation, giving a base line of reference LF, to which the angles are referred, as shown in figs. 9,434 to 9,436.

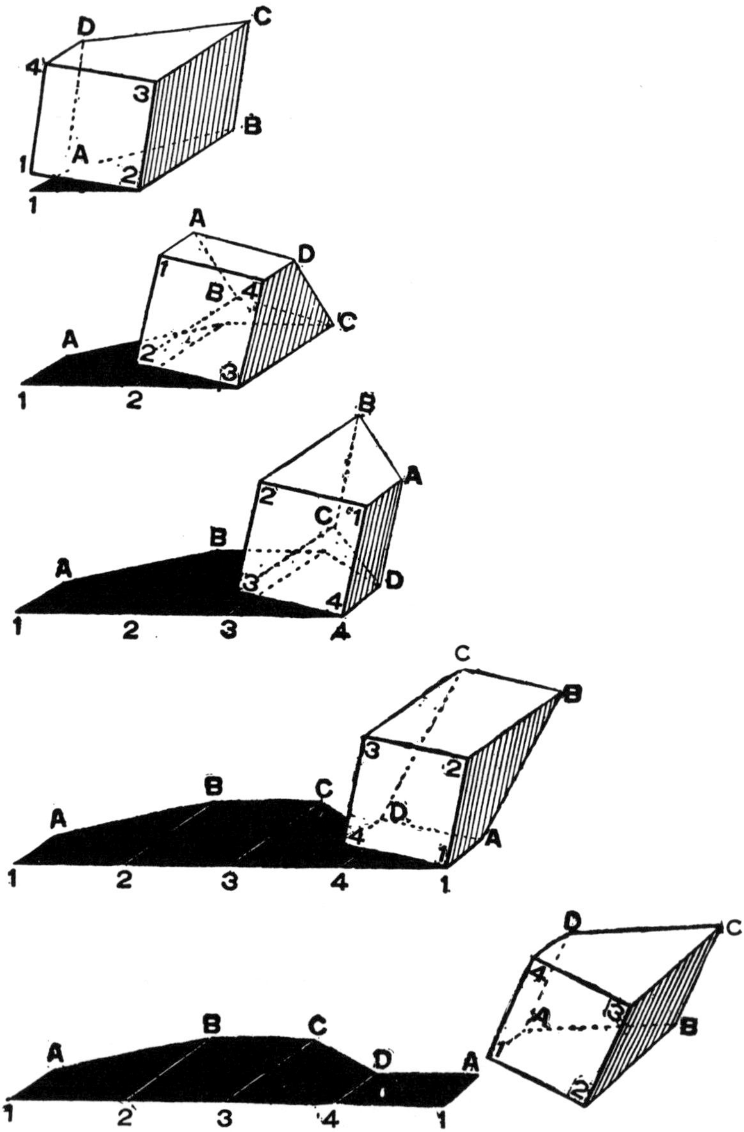

FIGS. 9,422 to 9,426.—Moving picture of development of pattern by parallel lines for the prismoid of fig. 9,418, showing prismoid rolling over on its sides and pattern progressively developed.

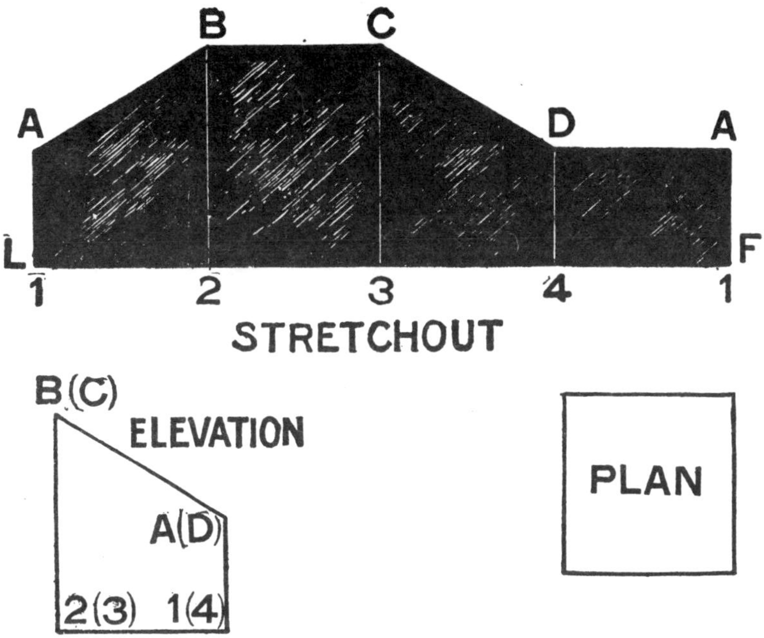

FIGS. 9,427 to 9,429—Development of pattern *by parallel lines* for the prismoid shown in fig. 9,418; stretch out or pattern completed.

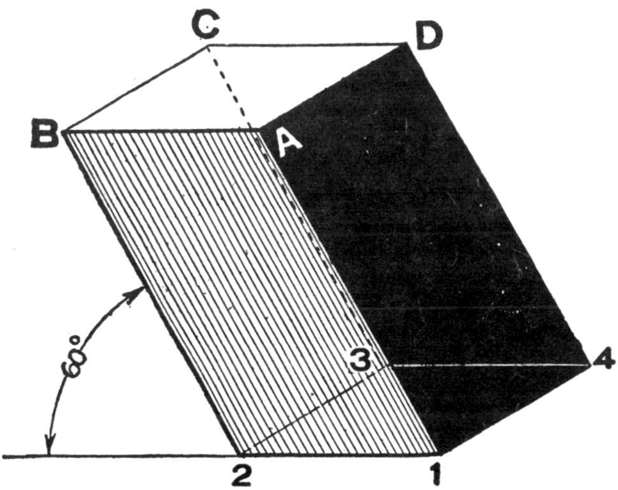

FIG. 9,430.—Oblique parallelopipedon in cabinet projection.

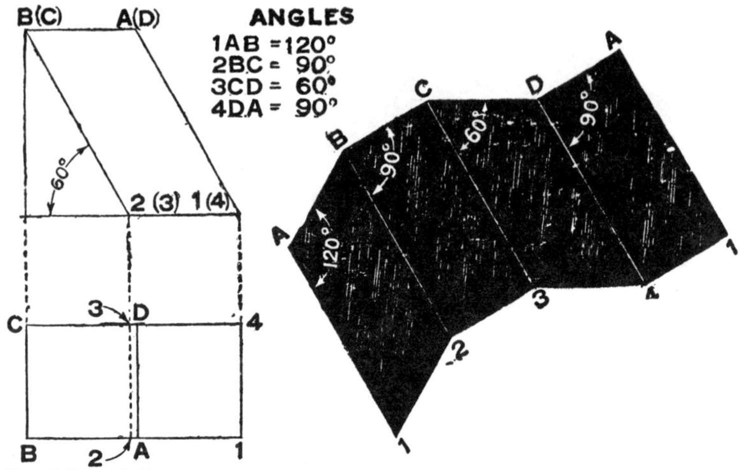

FIGS. 9,931 to 9,933.—Development of pattern ***by parallel lines*** for the oblique parallelopipedon shown in fig. 9,430 ***case 1.*** Development on line A1.

On LF, lay off points 1 and 2, spaced as in the elevation. Through 1 draw a line at an angle of 60° with LF, and project over point A in elevation, giving length 1A, on stretchout equal to length 1A in elevation. Through points 2 and A draw lines parallel to 1A and 12, respectively, completing the first section of the pattern 1AB2, in stretchout which is the development of one side of the solid.

Construct in a similar manner the remaining sections thus completing the stretchout 1ABCDA14321.

Figs. 9,442 to 9,445 show the various angles and the required angles referred to the "laying off line" LF, and fig. 9,446 the method of laying off these angles by use of T square and 30-60° triangle.

Example.—Develop a pattern for cylindrical surface of the cut cylinder shown in fig. 9,447.

In fig. 9,448, take any number of points on the circle as 1,2,3, 4,5,6, 7 and 8, and project up in elevation obtaining points A(I), B (H), C(G), D(F) and E. Here some of the points coincide, that is some lie directly back of the others. Thus point 2, or its projection (B), lies back of 8, or its projection H, the parentheses indicating the fact.

Draw base line for the stretchout and lay off on this line points 1,2,3,4,5, 6,7,8,1, spaced equal to the lengths of the arcs between points 1,2;2,3,etc., in plan.

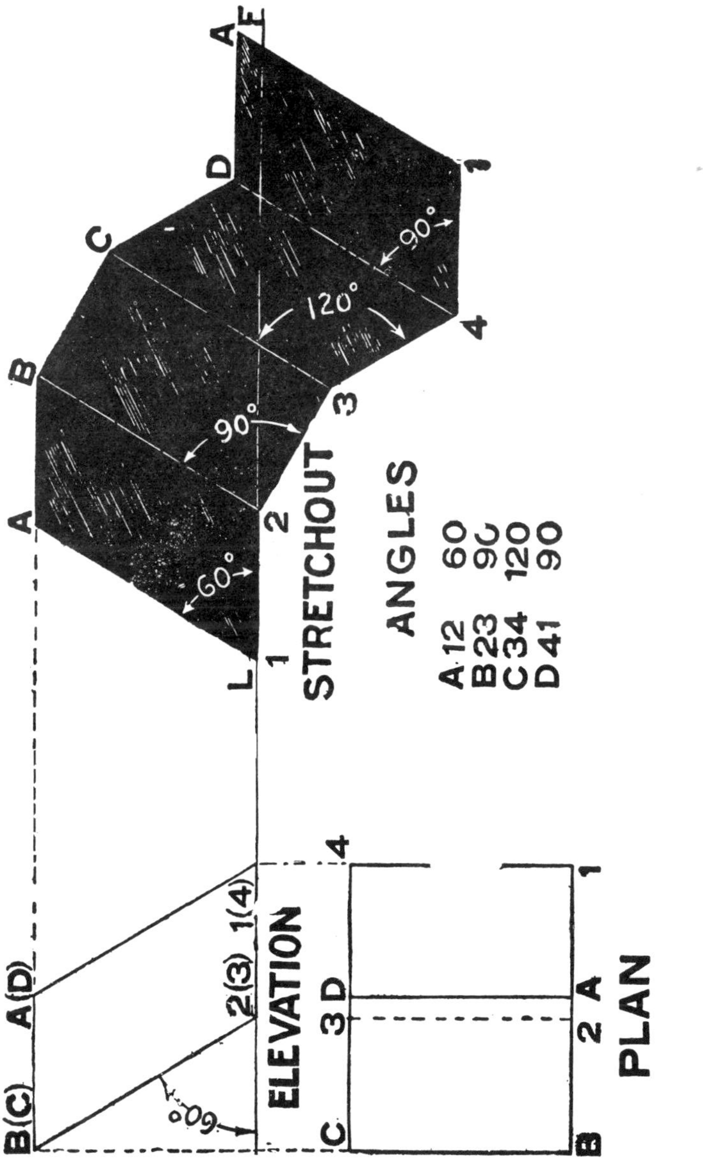

FIGS. 9,434 to 9,436.—Development of pattern *by parallel lines* for the oblique parallelopipedon shown in fig. 9,430. *Case 2.* **Development referred to base line.**

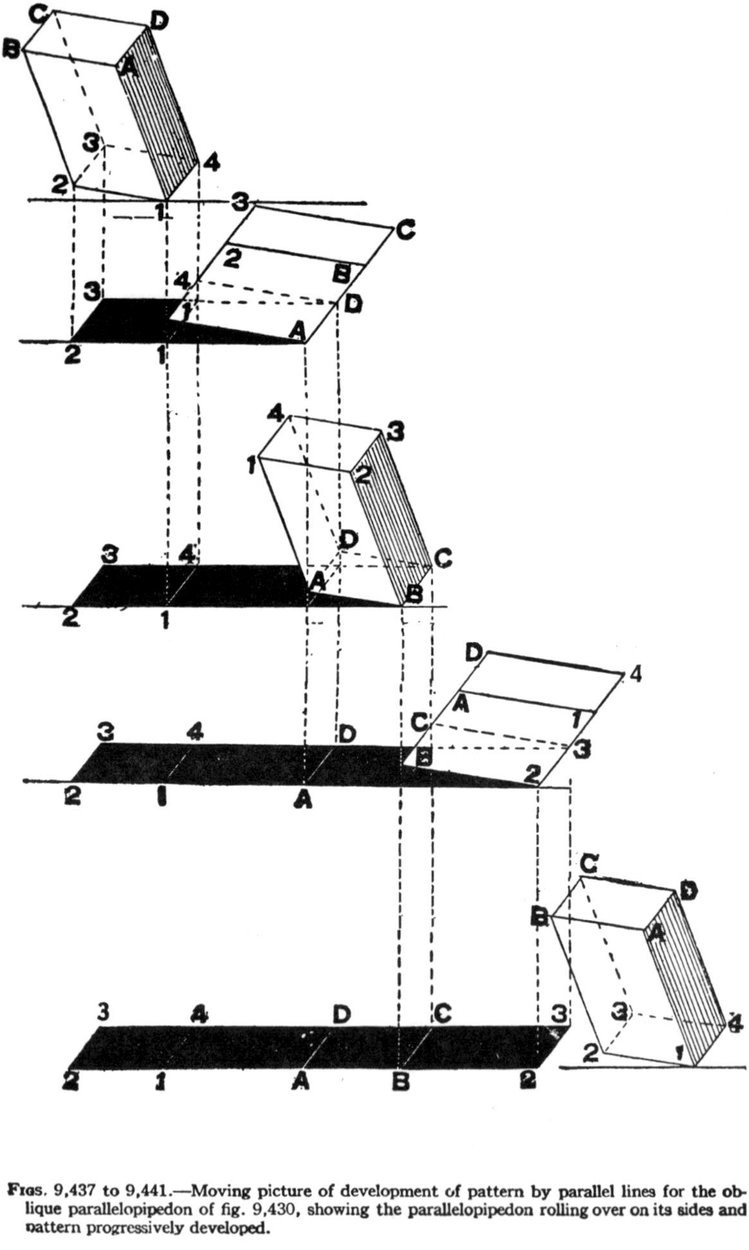

FIGS. 9,437 to 9,441.—Moving picture of development of pattern by parallel lines for the oblique parallelopipedon of fig. 9,430, showing the parallelopipedon rolling over on its sides and pattern progressively developed.

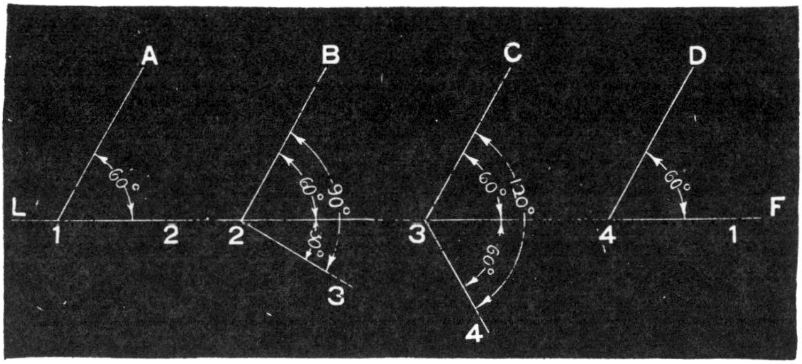

FIG. 9,442 to 9,445.—Angles of the oblique parallelopipedon shown in fig. 9,430 and the required laying off angles referred to laying off line LF. The reason for obtaining values referred to LF, is so that the angles may be laid off by use of T square and 30-60° triangle as shown in fig. 9,446.

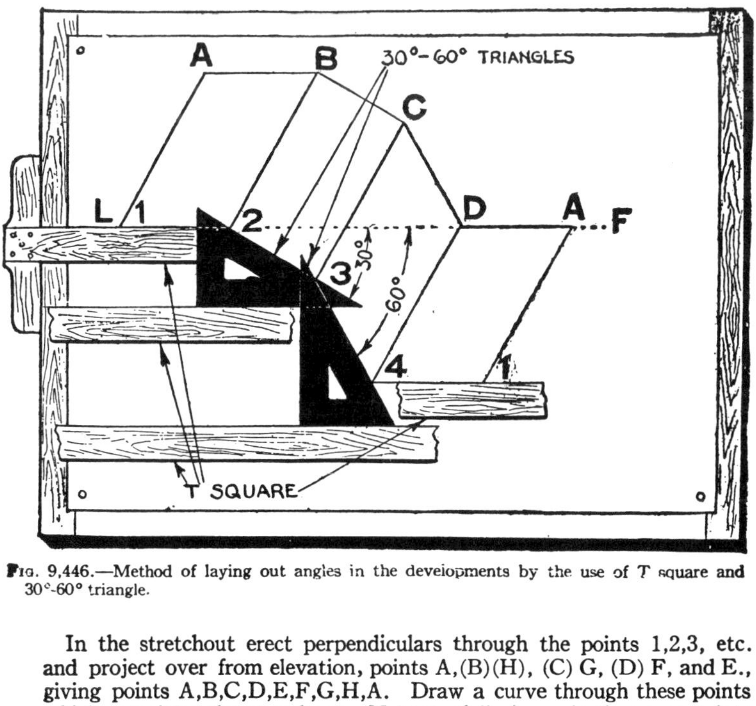

FIG. 9,446.—Method of laying out angles in the developments by the use of T square and 30°-60° triangle.

In the stretchout erect perpendiculars through the points 1,2,3, etc. and project over from elevation, points A,(B)(H), (C) G, (D) F, and E., giving points A,B,C,D,E,F,G,H,A. Draw a curve through these points which completes the stretchout. Note carefully in projecting over points

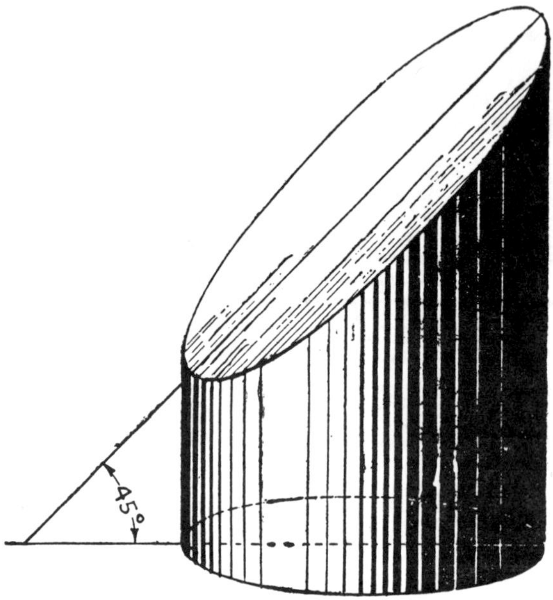

FIG. 9,447 —Cabinet projection of cylinder of revolution cut by plane inclined 45° to the base.

from elevation, their location on stretchout. Thus point (B), is projected to perpendicular through 2, and point H, to perpendicular through 8; this is evident from the plan.

Where a large number of points are taken the length of the spacing of the point in stretchout may be obtained by setting the dividers by the chord method as in fig. 9,451, but it should be understood that this is only an approximation and the stretchout will never be the full length no matter how many points are taken.

For precision, especially where only a few points are taken (as in. fig. 9,448) the dividers should be set by calculation, called line setting (figs. 9,453 and 9,454). A comparison of the two methods is shown in figs. 9,455 and 9,456. To illustrate the method of line setting suppose the diameter of the cylinder be 6 ins. and the 8 points be taken as in fig, 9,454. Then

$$\text{circumference} = 6 \times 3.1416 = 18.85 \text{ ins.}$$

that is, the length of the stretchout = 18.85 and the distance between points = 18.85 ÷ 8 = 2.36 ins. Draw a line and measure off accurately a distance LF = 18.85 ins.

Now set dividers to 2.36 ins. and make a trial spacing. The result obtained will probably be as on line A. Note magnitude of the error. Make additional trials as indicated on lines B,C,D, etc., until the true setting is obtained as on line LF.

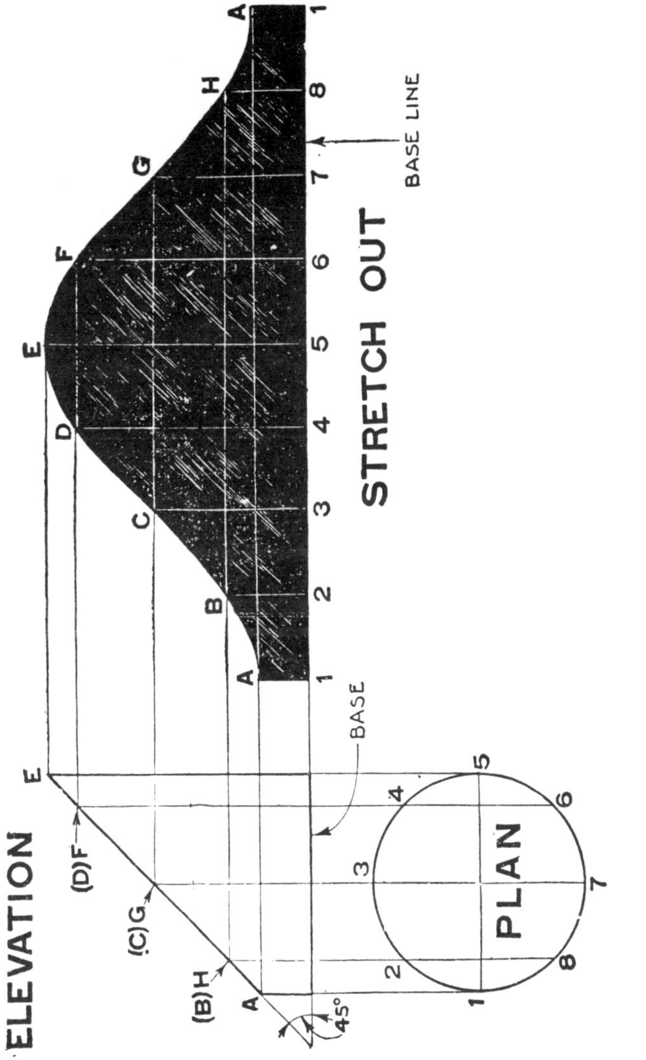

FIGS. 9,448 to 9,450.—Development of pattern by *parallel lines* for the cylinder cut by 45° plane as shown in fig. 9,447.

Development of Patterns by Radial Lines.—For objects such as pyramids, cones, etc., whose elements converge toward

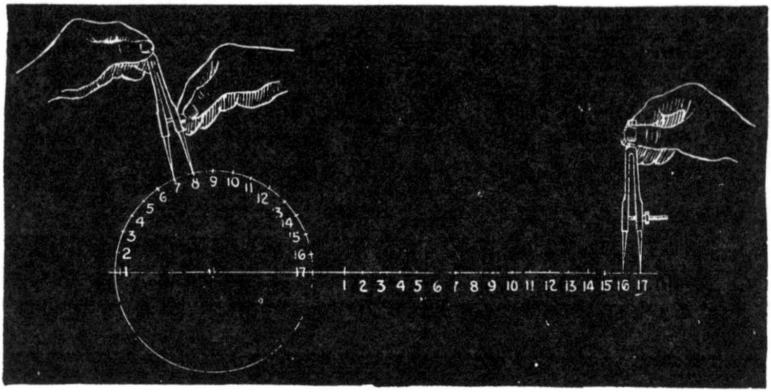

FIGS. 9,451 and 9,452.—Chord method of spacing.

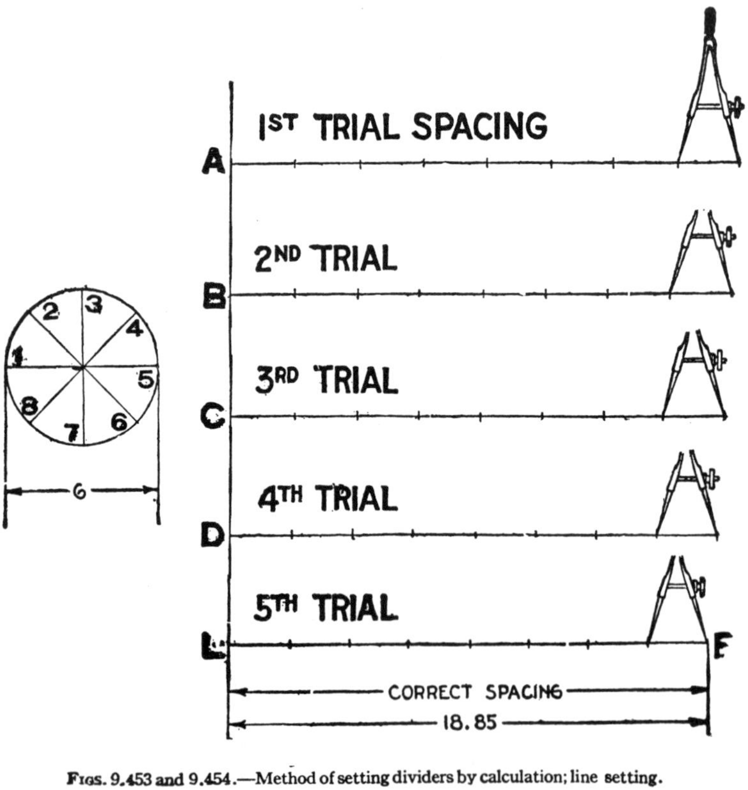

FIGS. 9.453 and 9.454.—Method of setting dividers by calculation; line setting.

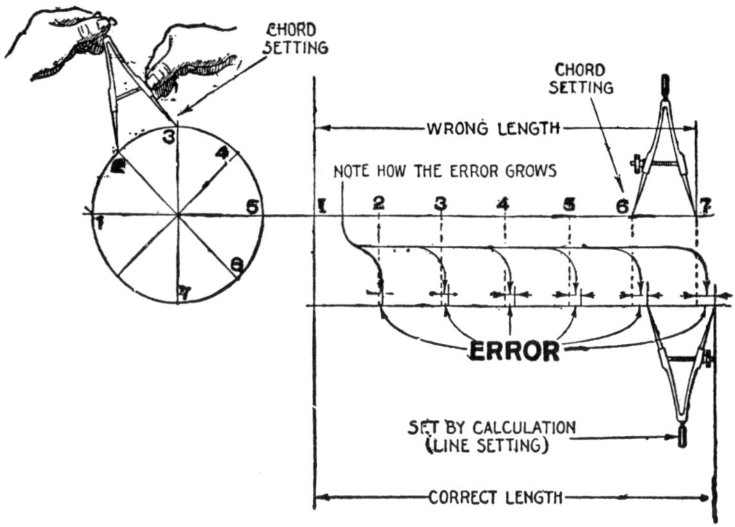

FIGS. 9,455 and 9,456.—Comparison of chord and line setting of dividers showing magnitude of error due to its multiplication in spacing.

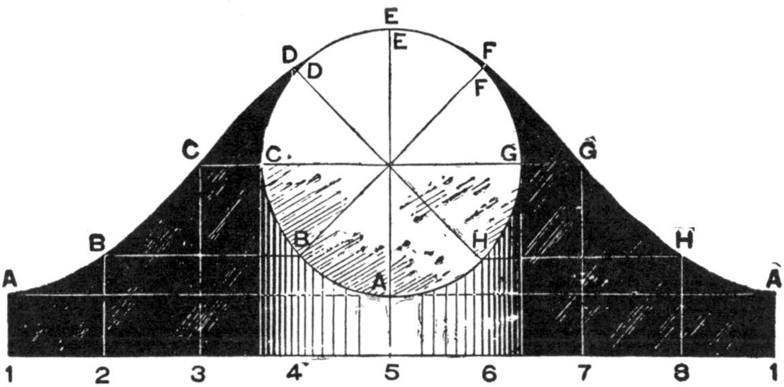

FIG. 9,457.—Pattern for cylinder (as obtained in figs. 9,448 to 9,450) in flat and rolled position showing its appearance in these positions.

a common point, the development of a pattern is made by the method of radial lines as illustrated in the examples following.

Example.—Develop a pattern for the sides of the pyramid shown in fig. 9,458.

First draw an elevation and plan as shown in figs. 9,459 and 9,460. By examination of the pyramid (fig. 9,458) it will be seen that the edges

(H1, H2, H3, H4) are of equal length, and since these edges (which are elements of the surface) converge to a common point, their extremities in the development will lie in an arc of which the common point or apex is the center. Since the elevation does not show the true length of the elements, revolve one element as H4, into the plane of the elevation. To do this, describe an arc through point 4, in plan with D4, as radius and where this cuts axis MS, at *s*, project up to H′, and draw HH′, which gives true length of the elements. Develop the pattern, take H′, as center and H′H, as radius describe arc LF.

Set dividers to distance of one side obtained from plan and space off points 1, 2,3,4,1 on LF. Connect these points with H′, and draw lines connecting points 1,2; 2,3; 3,4; 4,1. The figure thus obtained is the pattern required.

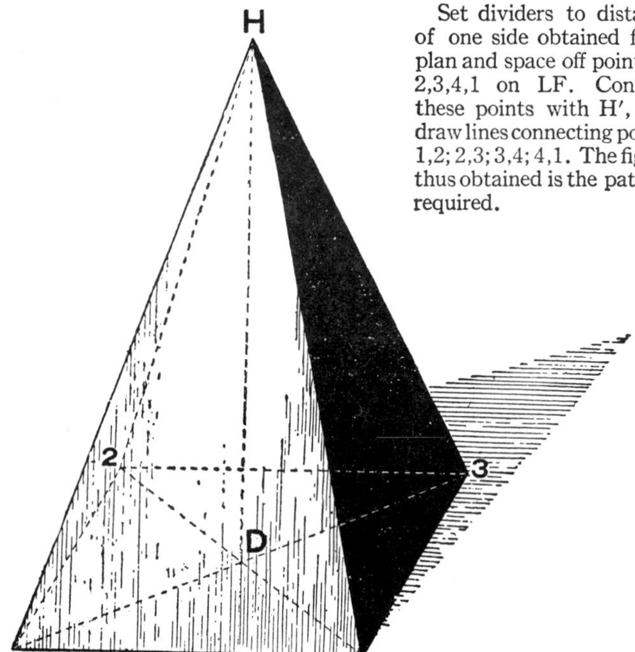

FIG. 9,458.—Pyramid in cabinet projection.

Example.—Develop a pattern for the slant surface of the oblique cone shown in fig. 9,461.

Draw elevation and plan as in figs. 9,462 and 9,463. The element H1, appears in true length in elevation and may be taken as the beginning of the pattern.

The other elements with exception of 5, do not appear in true length. Hence, revolve H′2, to MS, in plan and project up to 2′, in elevation, giving H2′, as true length of H2. Similarly obtain H3′ and H4′, true lengths of H3 and H4. With H, as center and radius H1, describe arc L; with radius H2′, arc A; with radius H3′, arc R; with raduis H4′, arc F; with radius H5, arc G. Set compasses to common distance between elements in plan as

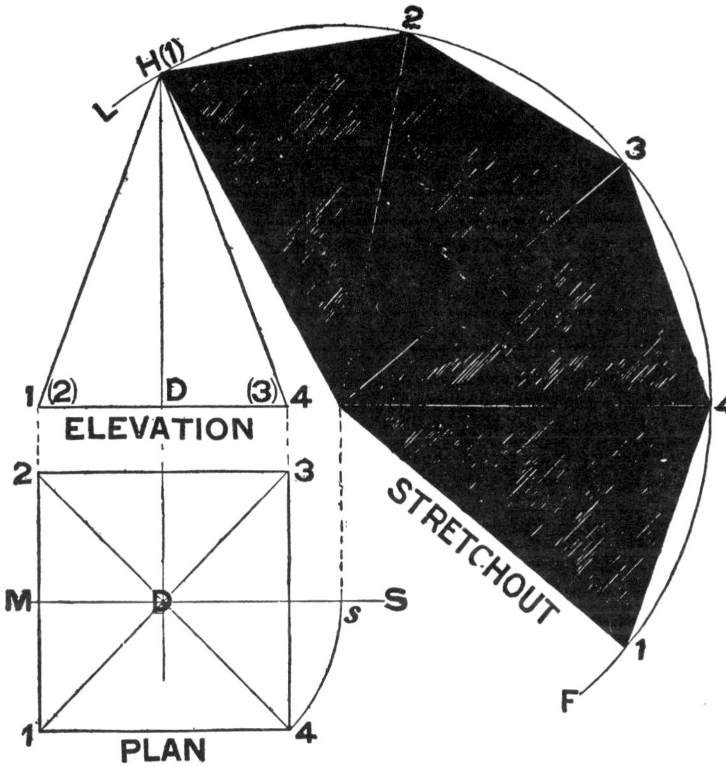

FIGS. 9,459 and 9,460.—Development of pattern ***by radial lines*** for the pyramid shown in fig. 9,458.

distance between points 1 and 2. With dividers set to this distance and with 1 (in stretchout) as center describe arc *a*, cutting A at 2; with 2, as center, arc *r*, cutting R, as center; with 3, as center, arc *f*, cutting F at 4; with 4, as center, arc *g*, cutting G at 5. This gives points for half of the pattern of which the other half is similar, the points 6, 7 and 8. being obtained by the intersection of arcs *f′*,*r′*,*a′*, with F′,R,A. Join all the points by a curve and draw 1′H, thus completing the pattern.

Development of Patterns by Triangulation.—There are some forms of elementary surfaces so shaped that although straight lines can be drawn on them such straight lines when drawn would neither be parallel nor be inclined toward each other with any degree of regularity.

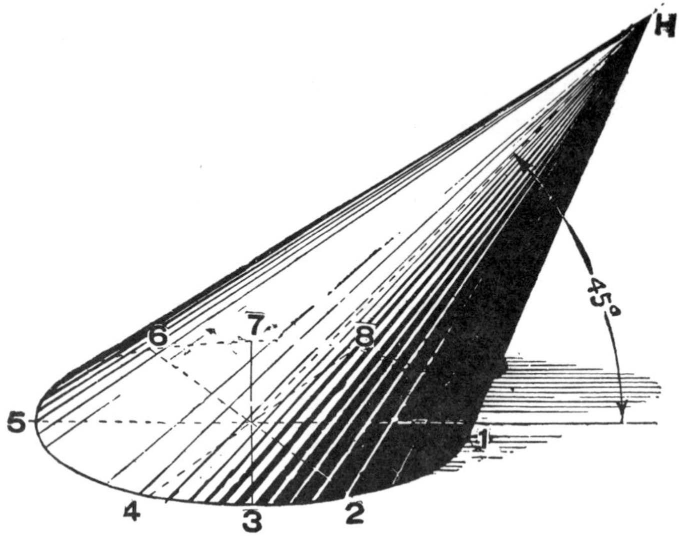

FIG. 9,461.—Oblique cone in cabinet projection.

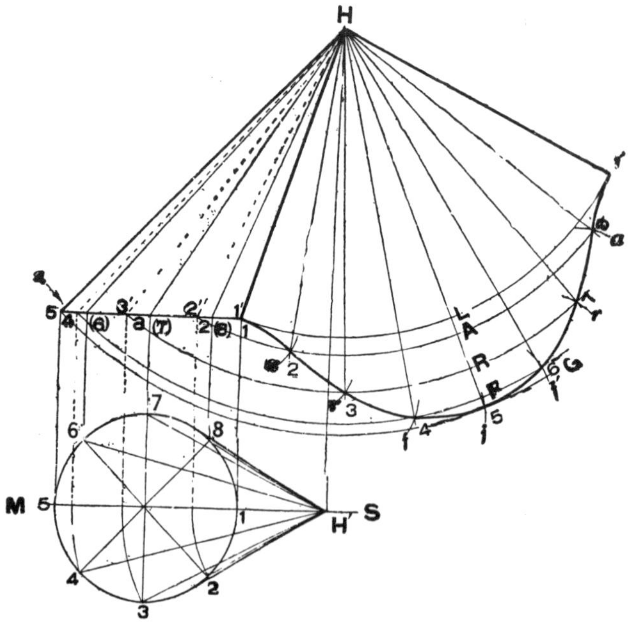

FIGS. 9,462 and 9,463.—Development of pattern *by radial lines* for the oblique cone shown in fig. 9,461.

In developing a pattern for such surfaces the surface is divided up into a number of elementary triangles (hence the name *triangulation*). Next the true lengths of the sides of the triangles are found, and the triangles reproduced in the pattern.

Example.—Develop a pattern for the slant surface of the irregular elementary solid shown in fig. 9,464.

Draw elevation and plan as in figs. 9,465 and 9,467. Select any number of points on the bottom edge and the same number of similarly located points on the top edge. For simplicity only 8 points are taken on each edge (though in practice a greater number are taken).

If points be taken for instance at the intersection of similar axes with the edges, they will be similarly located. These points are 1,3,5,7 for the bottom and A, C, E, and G, for the top. Determine the true lengths of the elements by constructing for each element a right angle triangle whose base is equal to its projection on the base or length in plan, and its altitude to the vertical height of the element in elevation. The hypotenuse of such triangle will then equal the true length of the element. To construct these triangles continue over to right the top and base in elevation and the distance between these lines will equal common altitude of the triangles. Beginning with element A1, its true length appears in elevation, hence it is not necessary to construct a triangle to find its true length.

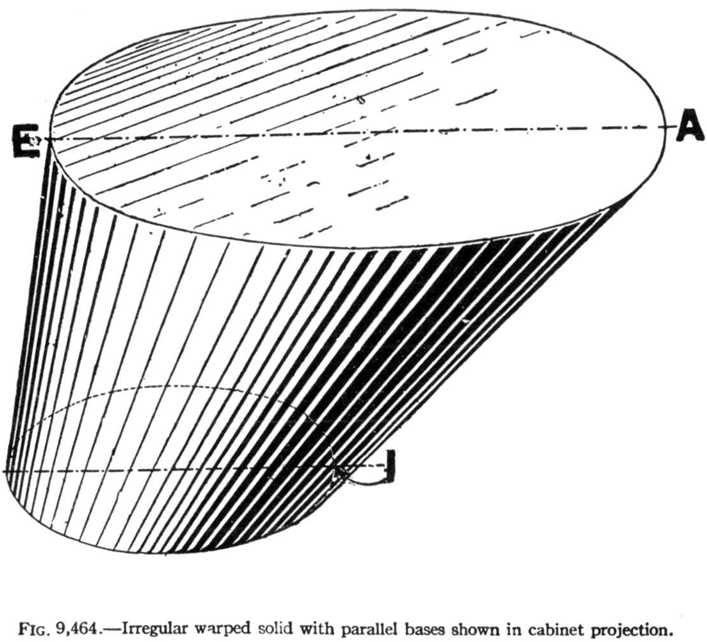

FIG. 9,464.—Irregular warped solid with parallel bases shown in cabinet projection.

For the next element B1, set dividers to distance B1, in plan, and mark this distance on base line of triangle layout as *b*1. Draw perpendicular *b*B, and join B1, thus completing the triangle. Its hypotenuse B1, then, is the true length of the element B1, which appears foreshortened in both plan and elevation.

In similar manner the true length of all the other elements are found. Next lay out the pattern, using the true lengths of the elements just found.

Begin laying out pattern by drawing A1 in true length.

With A, as center and radius equal to chord distance AB, in plan describe arc *l*, and with 1, as center and with radius equal to true length of of element B1, as found in the triangle layout, describe arc *f*, intersecting arc *l*, at B. Join A and 1, to B, thus completing the first triangle A1B.

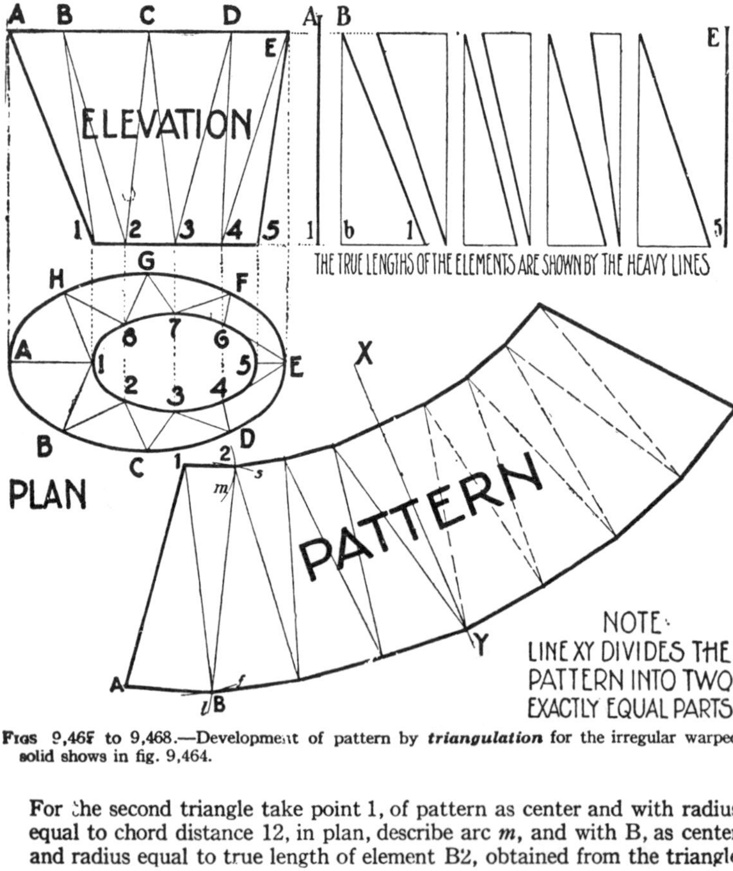

FIGS 9,465 to 9,468.—Development of pattern by ***triangulation*** for the irregular warped solid shows in fig. 9,464.

For the second triangle take point 1, of pattern as center and with radius equal to chord distance 12, in plan, describe arc *m*, and with B, as center and radius equal to true length of element B2, obtained from the triangle layout, describe arc *s*, intersecting *m*, at 2. Join 1 and B to 2, thus completing the second triangle 1B2. Continue in same manner until the pattern is completed.

Example.—Develop a pattern for the slant surface of the irregular warped solid shown in fig. 9,469.

This is virtually the same problem as worked out in figs. 9,465 to 9,468 except that the top is inclined to the base. The same method is used and the drawings of each problem are similarly lettered. The only difference

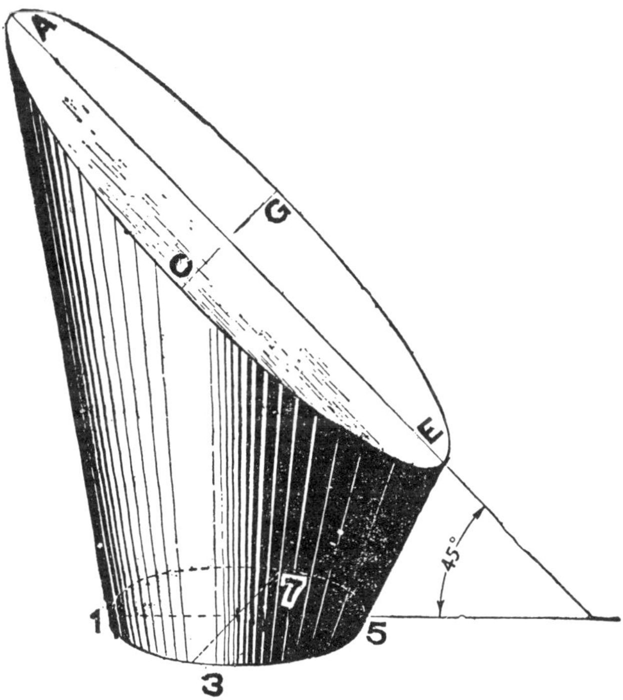

FIG. 9,469.—Irregular warped solid with inclined top shown in cabinet projection.

in laying out the lines is in the triangle layout. Here as will be seen in figs. 9,470 to 9,473 the triangles have different altitudes depending on the location of the points on the top edge.

Elements A1, and E5, appear in their true length, hence no triangles are necessary for these.

2. Warped Surfaces

A warped surface has been defined as one in which *a straight edge may be placed in contact only at a point*, as for instance the

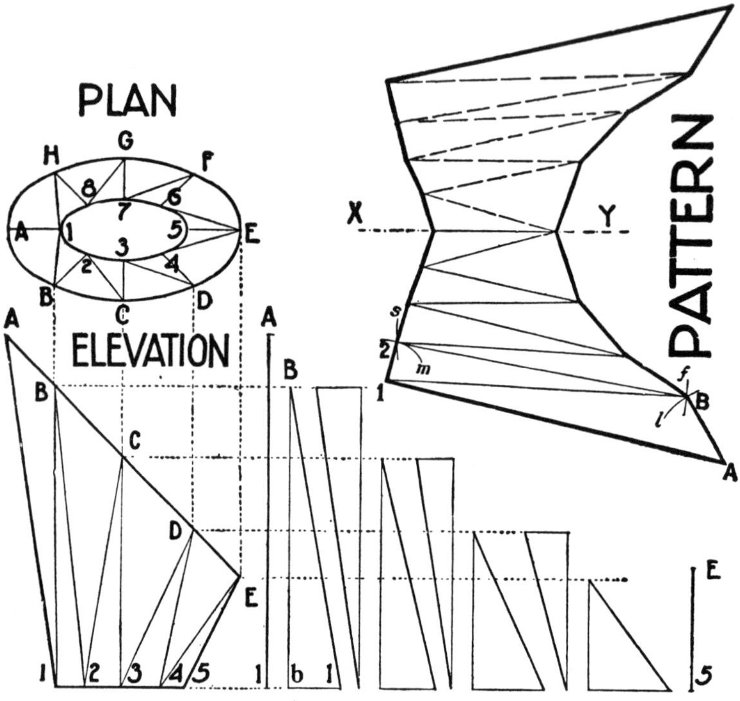

FIGS. 9,470 to 9,473.—Development of pattern *by triangulation* for irregular warped solid shown in fig. 9,469.

sphere shown in fig. 9,466. It is not possible to develop a pattern that will fit such surface because the surface does not contain elements. Accordingly after developing a pattern approximating the surface, it is necessary to *raise* the surface of the metal pattern by hammering to shape so that when the pattern is in position it will coincide with the warped surface of the solid. A sphere is a typical example of a warped surface and to avoid repetition, only part of it will be considered.

Example.—Develop a pattern for the warped surface of a hemisphere or half-sphere.

The warped surface may be divided into as many *sections* as desired, the greater the number, the nearer will the pattern of each section approach to the shape of the warped surface when placed in position; that is, the less will be the amount of hammering necessary to *raise* the surface of the pattern so it will coincide with the warped surface. The warped surface may be divided into

1. Zones, or
2. Segments.

***Case 1.**—Zone method.* In fig. 9,475, draw elevation of sphere and divide it into zones A,B,C. Zones A and B, will be frustums of cones and

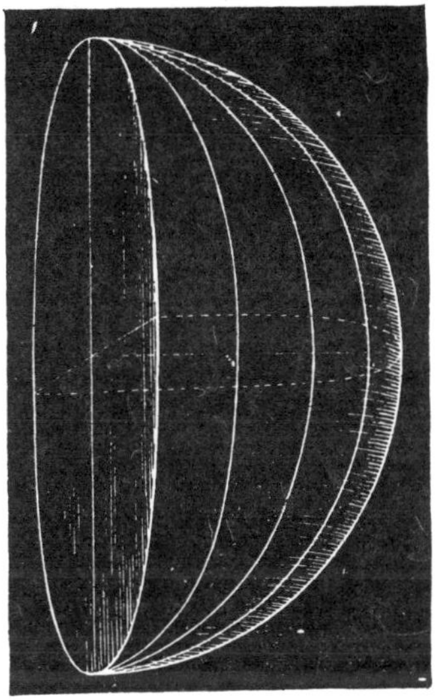

FIG. 9,474.—Hemisphere shown in cabinet projection.

C, a cone. Continue slant surface of frustum A, till it intersects the axis at H′, giving radius center for frustum A. With H′, as center describe the two arcs M and S.

Draw end view which gives the boundaries of the zones. In end view rectify arc 12, and space off on arc S, the points 1, 2, 3, 4 and 1. At 1, draw radial line connecting M and S, thus completing pattern for frustum A.

The patterns for frustum B, and cone C, are obtained in a similar way, no further explanation being necessary.

***Case 2.**—Segment method.* Fig. 9,480 shows appearance of ¼ of the hemisphere which is divided into two segments. The shaded surface 1H2 being one of these sections. It is only necessary to develop a pattern for one section as all the others are of the same shape.

In fig. 9,481 which shows the same portion of the hemisphere, partly in cabinet projection and partly in elevation, draw zone circles F and L, spaced as shown.

In plan draw 1H2, projection of the segment for which a pattern is to be developed. Project into plan points *l* and *f*, and describe zone L and F′, which is the projection in plan of L and F, in elevation.

To develop the pattern, draw an axis HD. Rectify arc H*h*, and mark off the rectified length as H*h*, on pattern. With H as center, describe arc

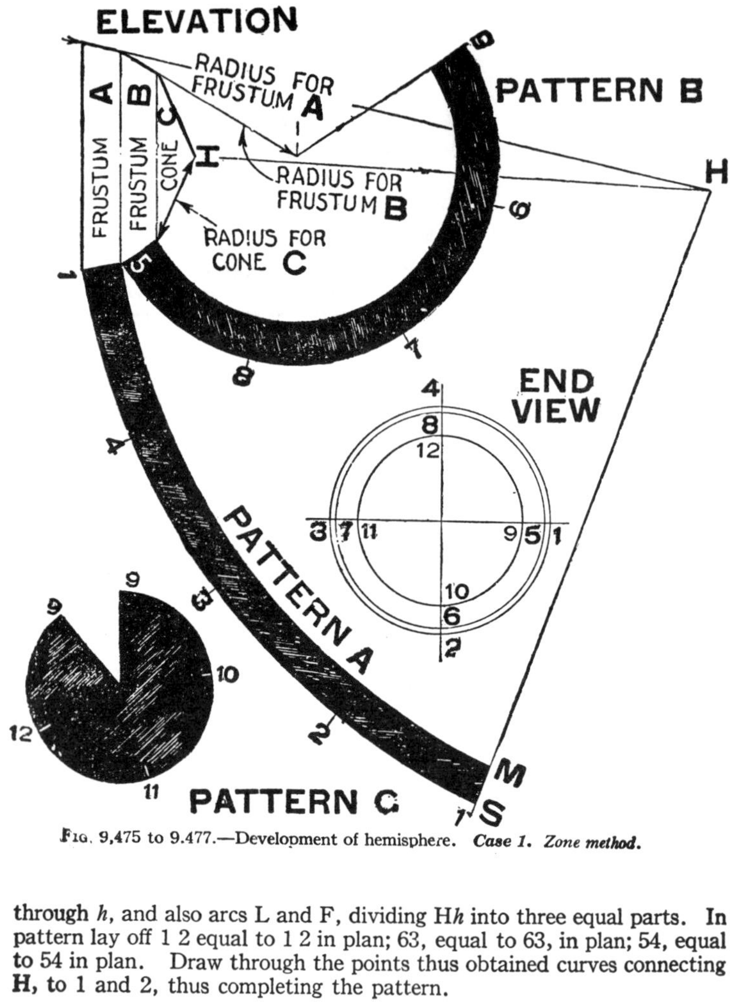

FIG. 9,475 to 9.477.—Development of hemisphere. *Case 1. Zone method.*

through *h*, and also arcs L and F, dividing H*h* into three equal parts. In pattern lay off 1 2 equal to 1 2 in plan; 63, equal to 63, in plan; 54, equal to 54 in plan. Draw through the points thus obtained curves connecting **H,** to 1 and 2, thus completing the pattern.

NOTE.—*The Sphere.*—A semi-circle revolving about its diameter generates a solid called a *sphere*. Every section of a sphere made by a plane is a circle. The circle is called a *great circle* if the plane passes through the center, a *small circle* in all other cases. The part of a sphere contained between two parallel planes is called a *spherical segment*, and the part of the surface of the sphere contained between the planes is called a *zone*. The circular sections made by the planes are called the *bases* of the segment, and the distance between them is called the *height* of the segment or zone. A single plane divides a sphere into two parts called *segments of one base*, and the surface into two parts called *zones of one base*. A great circle divides a sphere into two equal segments called *hemispheres*. The portion of a sphere generated by the revolution of a circular sector about one of its radii is called a *spherical sector*.

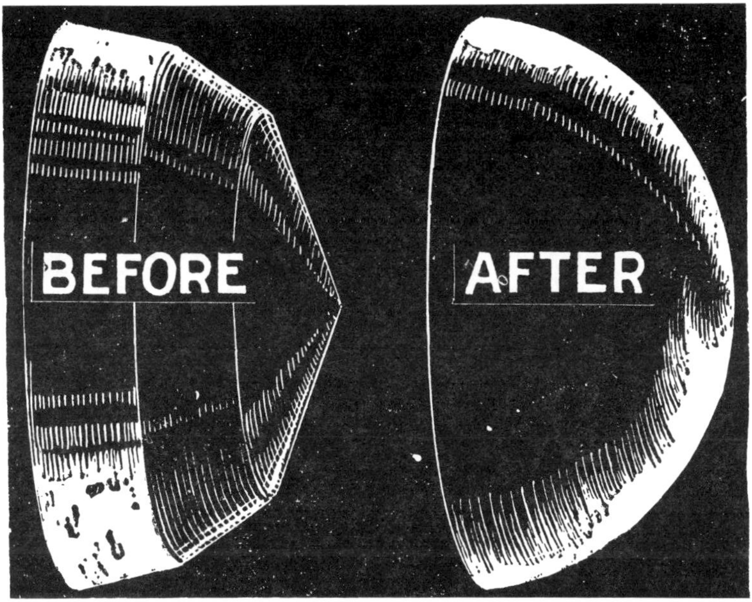

FIGS. 9,478 and 9,479.—Appearance of pattern for hemisphere before and after hammering or raising to the warped shape of the hemisphere. ***Case 1. Zone method.***

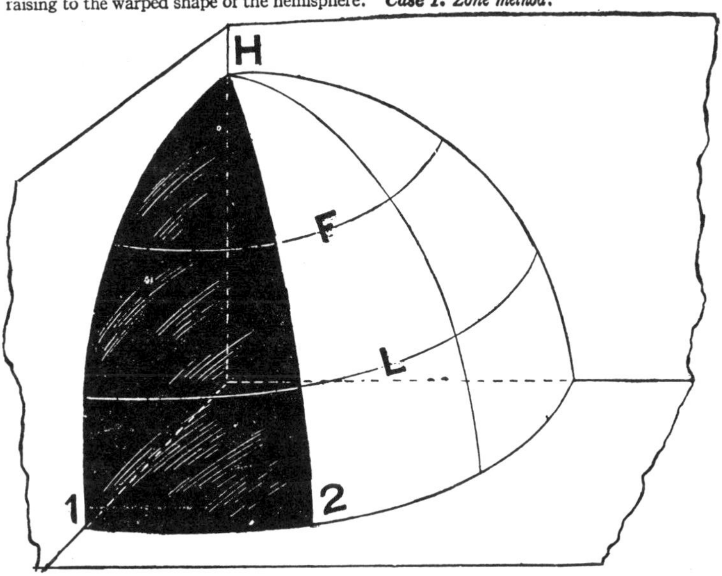

FIG. 9,480.—Quadrant of a hemisphere showing surface divided into segments.

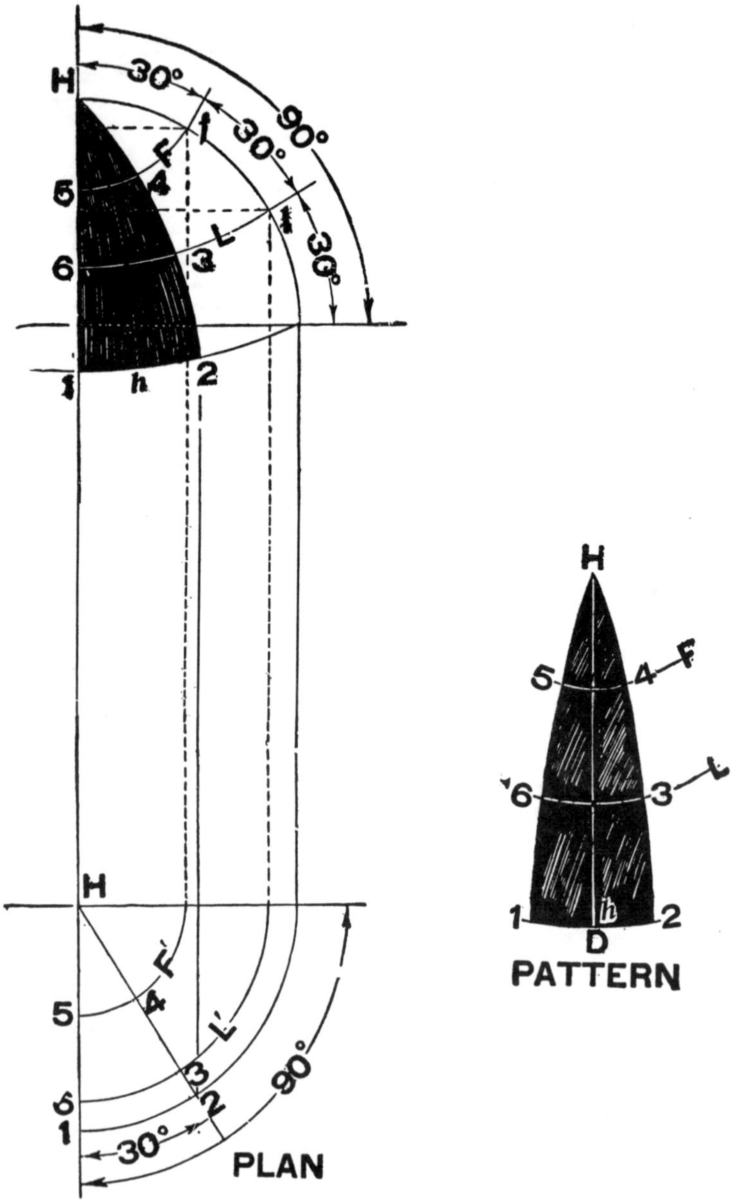

FIGS. 9,481 to 9,483.—Development of patterns ***by approximation*** for a hemisphere or warped surface. ***Case 2.*** *Segment method.*

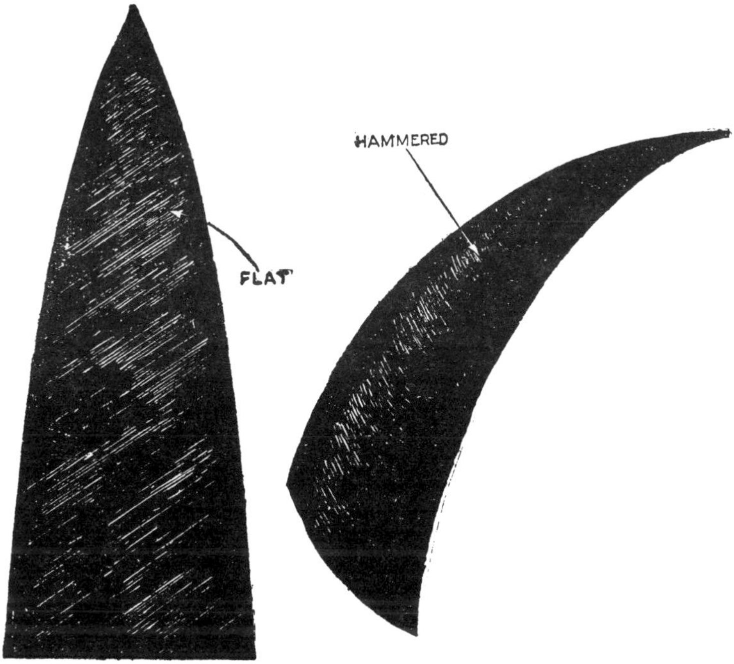

FIGS. 9,484 and 9,485.—Appearance of pattern for hemisphere before and after hammering *Case 2. Segment method.*

CHAPTER 3

Sheet Metal Work Layout

In this chapter numerous examples are given, further illustrating each of the methods explained in the preceding chapter so that the student may obtain proficiency in developing patterns for objects having *elementary*, or *warped* surfaces.

The author is indebted to Dr. John Weichsel, professor of drawing and sheet metal work, for the problems in this chapter. They are arranged in groups under each method of development, although in some cases more than one method is employed.

1. Elementary Surfaces

Development by Parallel Lines

Problem 1.—Pattern for flushing Tank. Figs. 9,486 to 9,490.

Draw the plan, front and end views of the required tank, full size, taking care to make all angles exactly right angles (90°).

The end view is not necessary for developing the pattern but is given just for practice in orthographic projection.

In developing the pattern, first lay out the rectangle 1 2 3 4, exactly equal to the plan IJKL. This rectangle will provide the bottom part of the tank. Extend the four lines of this rectangle, indefinitely, upward and downward, also to right and to left. Then set the dividers to a distance AB, equal to the height of the tank, and with this distance as radius, describe arcs at the four corners of the rectangle, using the points 1,2,3,4 as centers. These arcs cut off the extended lines of the bottom part, thus giving the rectangles 6 1 4 5, for the front part; 2 7 8 3, for the rear part; 1 10, 9 2, for the left and 4 3 11, 12, for the right end of the tank.

Additional strips of stock should be added along the edges 6 1, 2 7, 3 8, and 4 5, for laps of proper width.

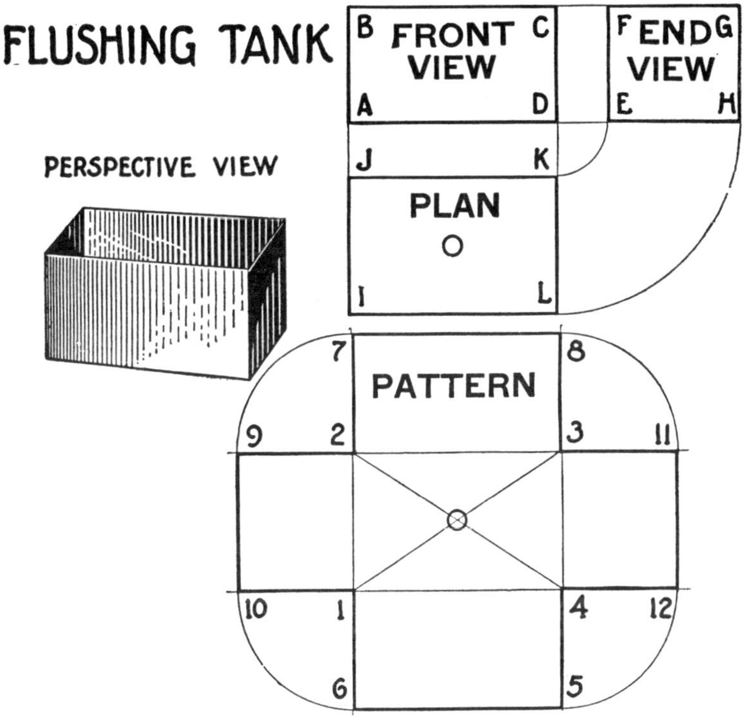

FIGS. 9,486 to 9,490.—*Problem 1.* Flushing tank and development of its pattern.

The point of intersection of the two diagonals in the bottom part furnishes the center of the hole for the flushing pipe.

There being no cover to this tank, its pattern is made up of only five rectangles. When a pattern is to be made for a closed, square container, six rectangles are required, as in problem 2.

This is virtually the same as problem 1 except the additional rectangle 1 2 3 4, forming the top or cover, considering the cube as a container with a cover.

Problem 3.—Pattern for a wedge shaped drainer. Figs. 9,493 to 9,496.

Draw front and end views as in figs. 9,493 and 9,494.

All the angles in the front view should be right angles or 90°. In the end view, the angle at the corner F, is a right angle. To lay out the required pattern, draw the rectangle 1 2 3 4, exactly equal to the front

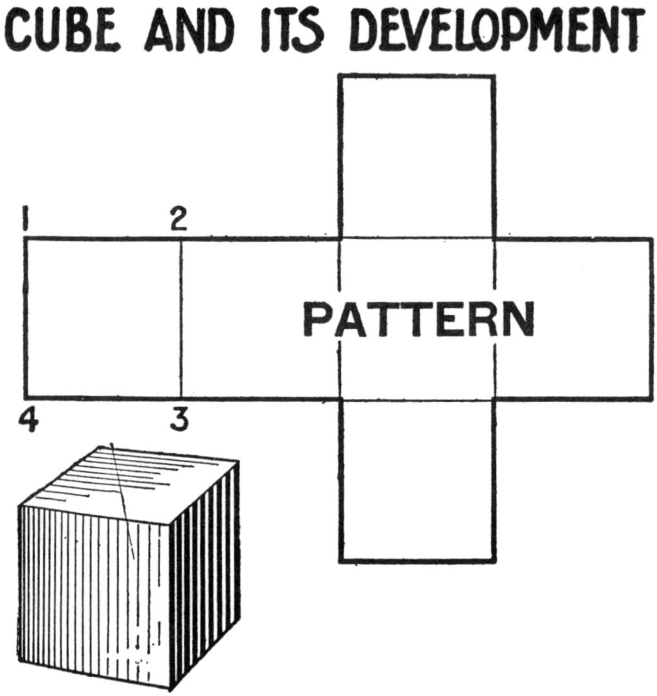

FIGS. 9,491 and 9,492.—*Problem 2.* Cube and development of its pattern.

Problem 2.—Pattern for a cube. Figs. 9,491 and 9,492.

view A B C D. Extend the line 2 3, to right and to left, also extend downward the lines 1 2 and 4 3.

Set dividers to a distance equal to the oblique line G E. With this distance as radius, describe the arc 7 5, with point 1 as center. Then describe the arc 8 6, with point 4 as center. These arcs determine the rectangle 7 1 4 8, for the front part of the drain as well as the triangular parts 1 5 2 and 4 3 6, for the two ends of the drain.

Additional strips of stock should be added along the edges 1 5 and 4 6, for laps of a suitable width.

Problem 4.—Pattern for a rear fender for an automobile. Figs. 9,497 to 9,500.

As now manufactured, a full crowned fender is made with the aid of dies and rolls and when thus formed the shape presents a warped surface.

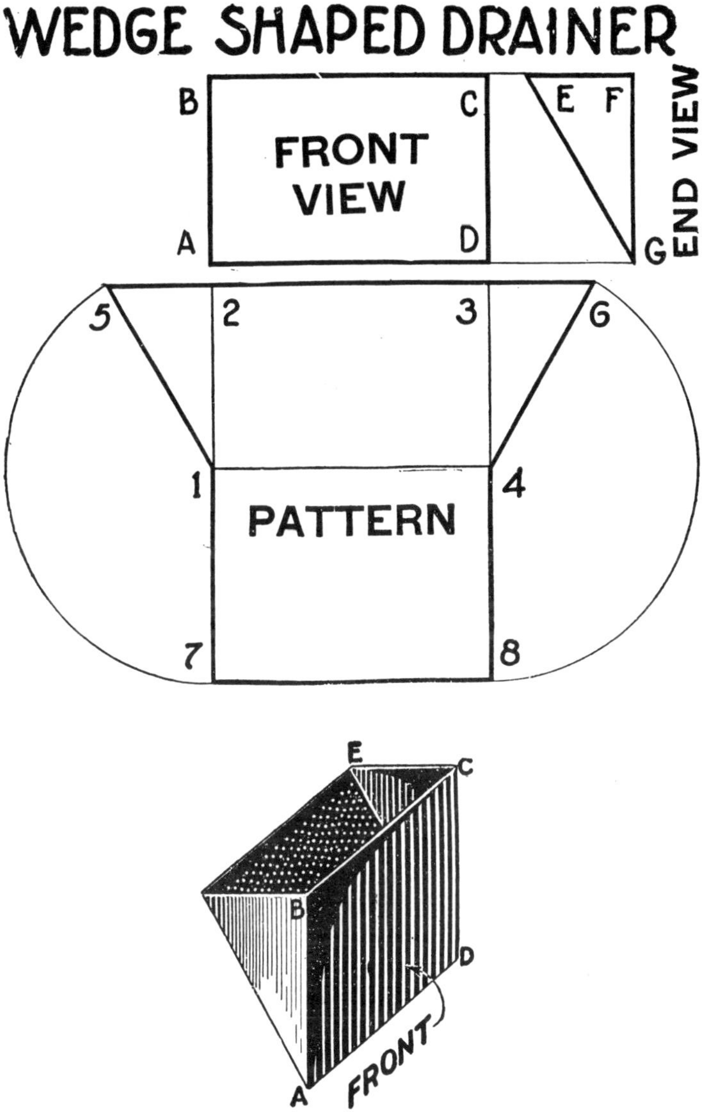

FIGS. 9,493 to 9,496.—***Problem 3.*** Wedge shaped drainer and development of its pattern.

The fender here considered has an elementary surface. The body of the fender is provided with a flange for wire and beads, a rib being wired into the fender.

DEVELOPMENT OF A REAR FENDER FOR AN AUTOMOBILE

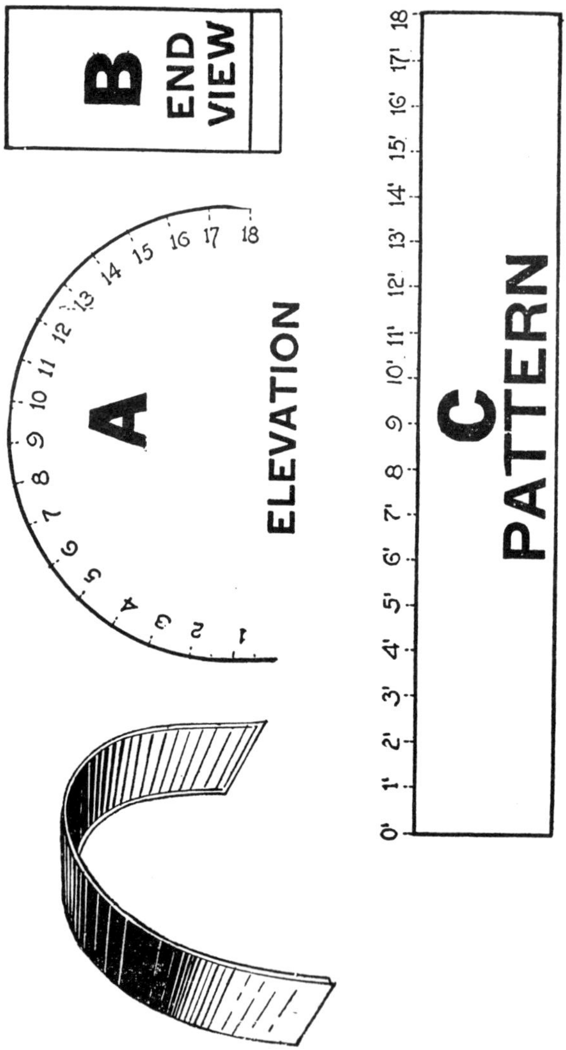

FIGS. 9,497 to 9,500.—*Problem 4.* Rear fender for automobile and development of its pattern.

The development here described refers to the body of the fender. At A, the profile of the fender is divided into a number of equal parts, marked 1,2,3, etc. The width of the fender body being given at B, the pattern is made up of a rectangular strip, shown at C, whose width is equal to that of B, and the length, to the combined length of all the divisions into which the profile at A, is divided.

In rectifying the profile to obtain the length of the pattern, chord distances should not be taken but the line method used as explained in figs. 9,453 and 9,454.

Problem 5.—Patterns for a Tee pipe. Figs. 9,501 to 9,506.

The problem involves the development of two intersecting cylinders, of equal diameters. The solution here given is the more important because it applies to a large class of elbows and other objects composed of intersecting cylinders. The method here used, may also be used where the intersecting cylinders are not of the same diameter.

A hole must be cut in the veritcal or *run* pipe to provide an opening of suitable shape for the *branch* pipe and the end of the branch pipe must be shaped to fit the opening in the run.

NIJM, shows the development of the run pipe. The length IJ, is that of the circumference of the pipe, half of which is shown in the plan. The equally measured divisions on this circumference are put together to make up the line IJ. At right angles to this line, the lengthwise edges of

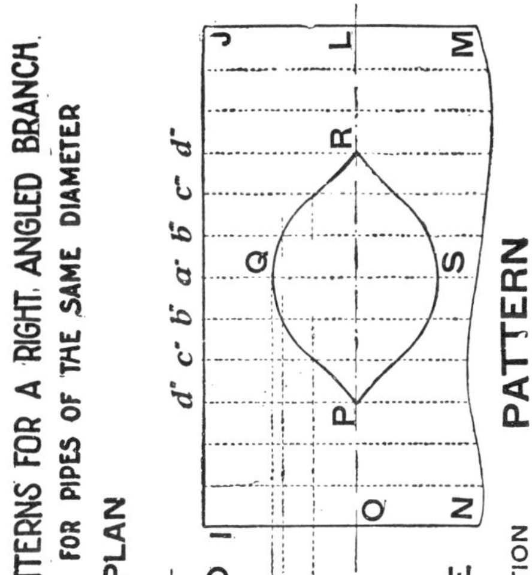

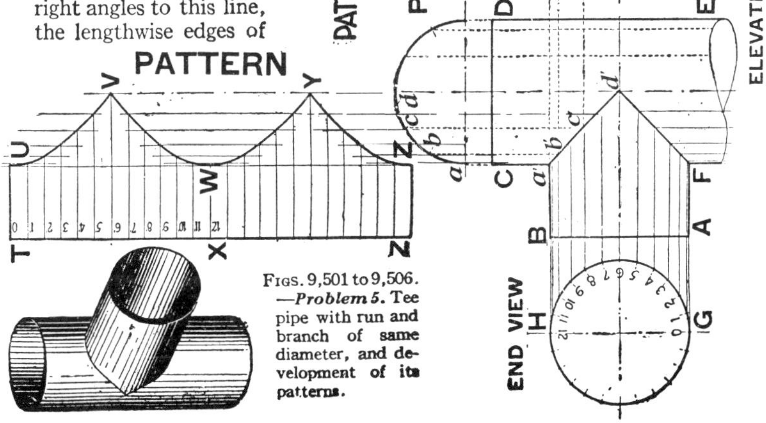

FIGS. 9,501 to 9,506. —*Problem 5.* Tee pipe with run and branch of same diameter, and development of its patterns.

the development of the run pipe are shown by the lines IN, and JM. Only a part of the length of this pipe is developed, the part around the required opening PQRS.

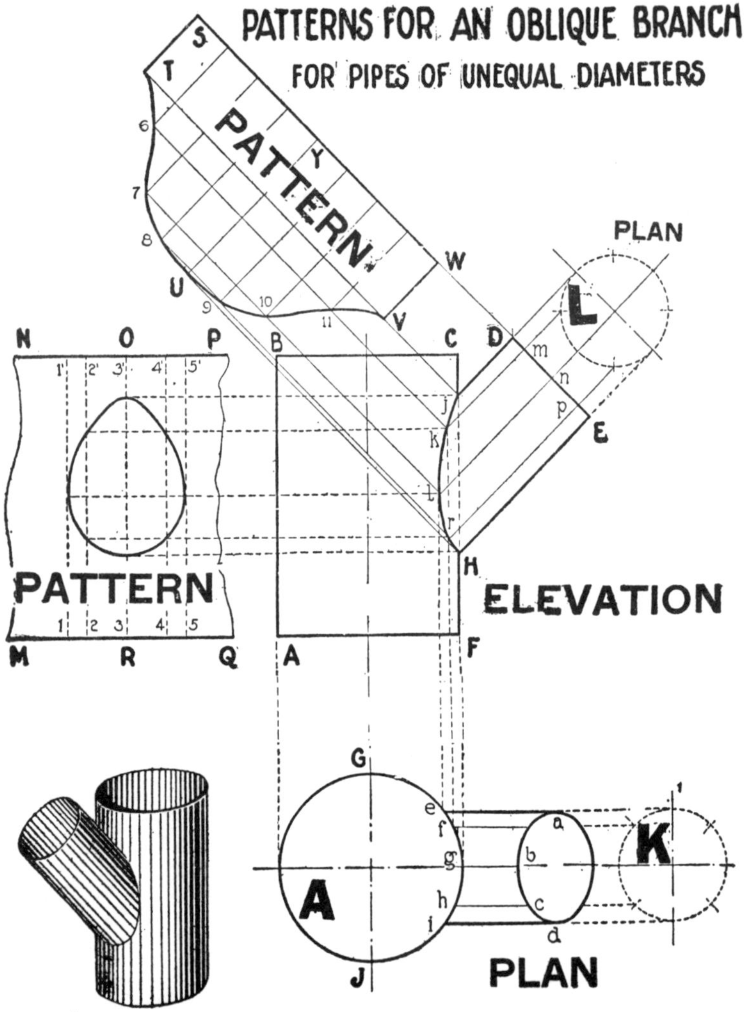

FIGS. 9,507 to 9,513.—*Problem 6.* Y branch with run and branch pipes of unequal diameters and development of its patterns

From the points of division in the plan, the longitudinally drawn projecting lines furnish on the joint line a′d′, the points a′, b′, c′, d′. From these, another series of projecting lines is drawn to the development, intersecting the longitudinal lines on the development which start from the points d″, c″, b″, a″, b″, c″, d″, thus giving points defining the outline of the opening PQRS as clearly shown.

In a like manner the other pattern, Z,T,U,Z′ is obtained from A,B,*a′*,*b′*,*c′*, *d′*. The circumference of the branch pipe GH, is divided at the points

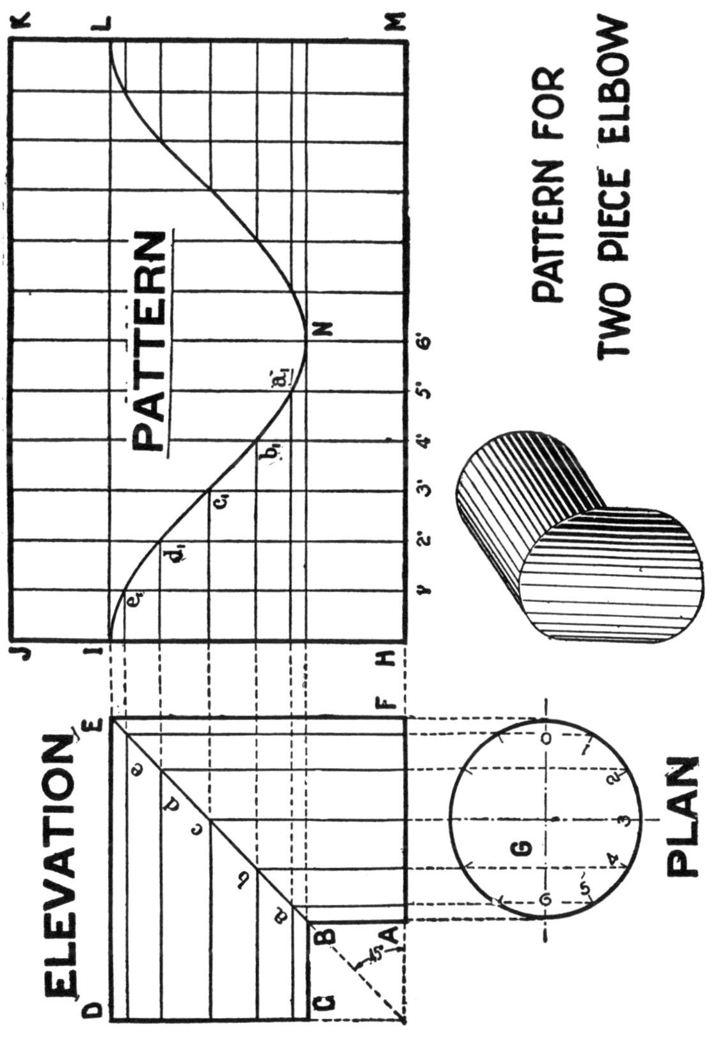

FIGS. 9,514 to 9,517.—*Problem 7.* Two piece elbow and development of its patterns.

0,1,2,3, etc., into equal parts. All these divisions are laid down, together, on the line ZT, so that the length ZT, is equal to the entire circumference of the branch pipe.

The procedure for the tracing of the outline Z′YWVU is the same as for the opening PQRS, it is clearly shown by the lines.

The patterns do not show any laps for joints.

Problem 6.—Patterns for a Y branch with run and branch of unequal diameters. Figs. 9,507 to 9,513.

Draw plan A, then draw the outlines of the elevation ABC*j*DEHF. To these outlines the curved joint edge between the main pipe and the branch may be added as follows: Divide the circumference of the branch pipe given at K, and also at L, into a number of equal parts and through the division points pass parallels along the branch pipe in the plan and elevation. In the plan A, these parallels cut the circumference of the run in the points *i,h,g,f,e*.

On the obliquely situated branch in the elevation B, the parallels should be drawn so as to pass for some distance into the elevation of the run. Then draw vertical projection lines from the points *i,h,g,f,e*, into the elevation of the run intersecting in the points H,*r,l,k,j*. These points when joined, give the view of the intersection between the branch and run pipes.

Now, proceed to lay out the pattern VUTSW, for the branch. Extend the end line DE, of the branch, indefinitely, toward the desired pattern and upon this extended line, make WS, equal in length to the circumference of the branch. At every division point, erect a perpendicular to WS, thus dividing the space for the pattern into a number of elementary parts. Now, from the points H,*r,l,k,j* (on the elevation) project a number of lines upon the pattern, parallel to the stretched out line SW. These projection lines cut the elementary lines on the pattern in the points T,6,7,8,U,9,10,11, V and the points in the curved outline of the pattern of the branch.

The run pipe has to be cut out on an oval line to receive the branch. For the plotting of this oval on the pattern, draw the short part MNPQ, of the development of the run. In the middle of this, erect the perpendicular center line RO. Starting from this line, set off to left, the distances 32 and 21 respectively equal to *gf* and *fe* (in plan) and do the same in opposite order, to the right of the center line, so that 34, equals 23, and 45, equals 21.

Through the points 1,2,3,4,5 draw the vertical elements 11′, 22′, 33′, etc.

In the elevation, project points *j,k,l,r*, H, horizontally over to pattern of run. These horizontal projection lines cut the vertical elements 11′, 22′, etc., in a series of points, which, when joined, give the oval shape of the hole that is to be cut in the main pipe. Add to the patterns thus obtained proper lap for joints.

Problem 7.—Patterns for a two-piece elbow. Figs. 9,514 to 9,517.

TWO PIECE ELBOWS

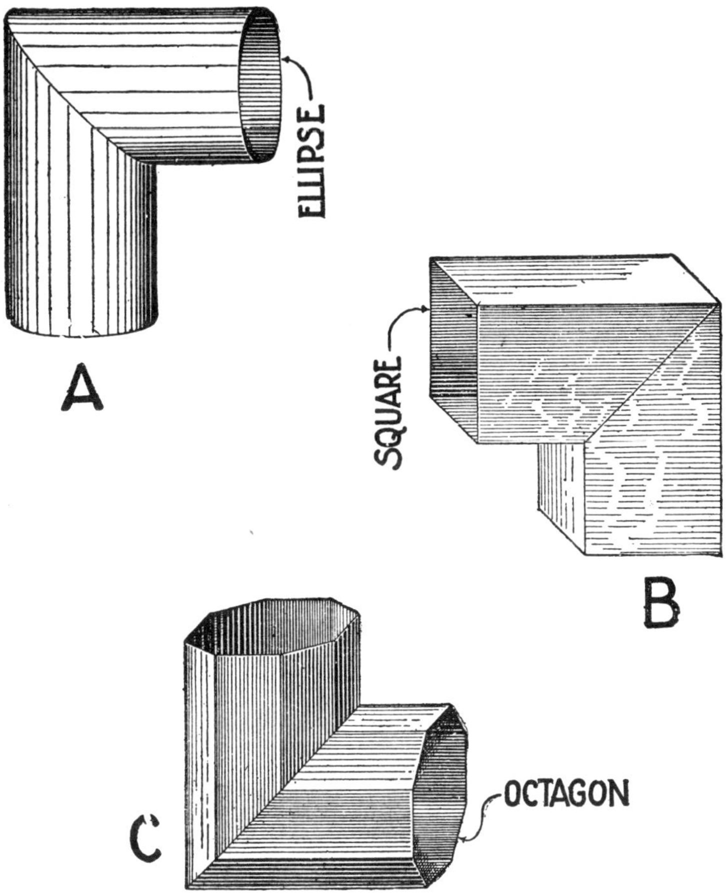

FIGS. 9,518 to 9,520.—*Problems 8 to 10, for practice.* Two piece elbows. **A**, elliptical; **B**, square; **C**, octagonal.

A two piece elbow for round pipes may be imagined to have been made up of two adjoining parts of one pipe that was cut at a miter of 45°. Hence, the patterns for the two halves of the elbow may be obtained first by developing the whole pipe of which the elbow parts are to be derived, and secondly, by dividing this development into two portions, in a suitable manner.

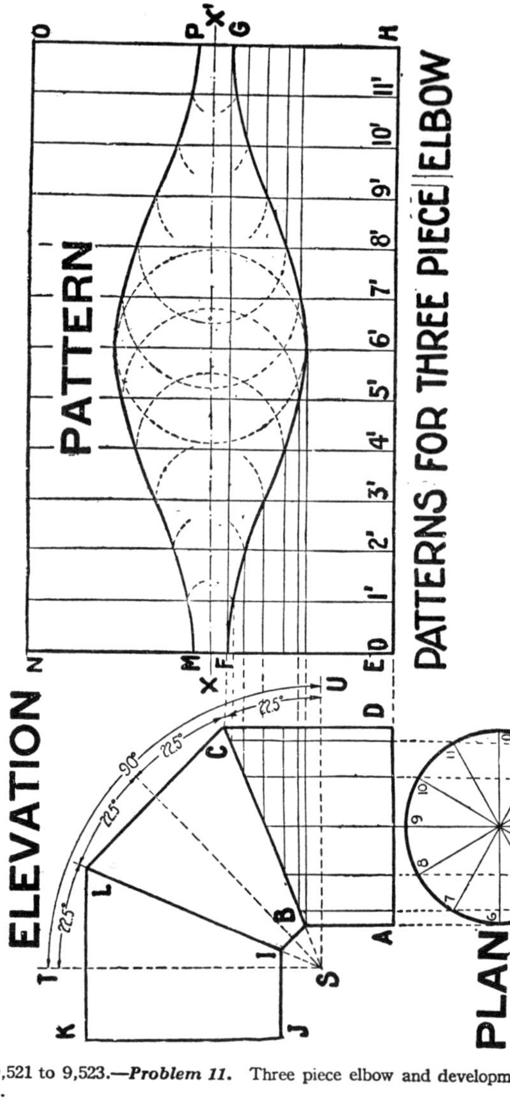

FIGS. 9,521 to 9,523.—***Problem 11.*** Three piece elbow and development of its patterns.

In the elevation, ABEF, is one part of the elbow and the other part BCDE, is identical to the first part.

In the plan is shown the circumference of the elbow with a convenient number of equal divisions.

In the pattern development, HJ, is taken equal to AB + FE (in elevation). HJ, then represents the length of the single pipe that is to furnish the two component parts of the elbow. Hence develop the cylinder whose length is HJ, in the rectangle HJKM, wherein the divisions upon HM, are reproductions of the divisions on the circumference shown in plan, so that HM, is equal to the stretched out circumference.

Vertical or elementary lines (1′2′3′ etc.) are drawn from the points just obtained. Project point 1,2,3, etc., in plan to cut the miter line EB (in elevation) in points B,*a*,*b*,*c*, etc., whence, in turn, horizontal projecting lines are drawn to intersect the elementary lines upon the development in the points N, *a*,*b*, etc., thus giving the curve IN, as well as its counterpart NL. This curve divides the development into the two halves of which the elbow is to be made up.

The pattern does not show which laps should be provided for the joint. For method of making this provision see Problem 8.

Problem 11.—Patterns for a three-piece elbow. Figs. 9,521 to 9,523.

The general shape of the elbow is shown in the elevation. The imaginary cylinder from which the three parts of the desired elbow are to be had, will be equal to the combined lengths CD, IB, and KL, in elevation. This combined length is shown as EN, in the development wherein EF, is equal to DC; FM, equal to BI; MN, equal to KL. The development of the imaginary cylinder is shown as the rectangle ENOH, EH being equal to the length of the circumference of the pipe shown in plan.

The development of the pattern is by elementary lines through points 0. 1′, 2′, 3′, etc., parallel to EN. and spaced equal to the arc distance, between similar points 0,1,2,3, etc., in plan.

By projecting points 1,2,3, in plan up to miter line BC, in elevation, thence, over to pattern, the intersections with corresponding elementary lines 0,1′, 2′,3′, etc., will give points defining the curve FG.

To obtain the curve MP, draw the center line XX′ so as to bisect the distances MF and PG. Then, set off upon each of the elementary lines 0,1′, 2′,3′, etc., above XX′, the amount which the curve FG, deviates from XX′, thus obtaining similar points which define curve MP.

Problem 12.—Patterns for a four piece elbow. Figs. 9,524 to 9,526.

In the elevation, the four pieces forming the elbow are AKSI, KXTS, XYZT, and Y*fd*Z. Of these four parts, the two larger parts, AKSI and Y*fd*Z, are equal. The same is true of the two remaining smaller parts KXTS and XYZT.

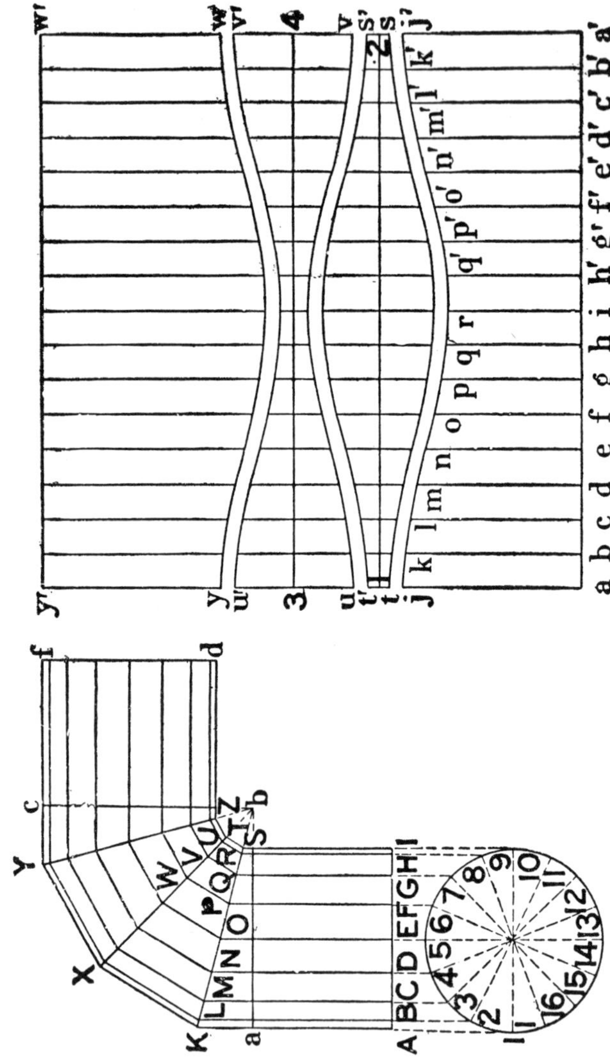

FIGS. 9,524 to 9,526.—*Problem 12.* Four piece elbow and development of its patterns.

To lay out these parts in the elevation a right angle *abc*, is drawn, the sides of which intersect at right angles, the two largest branches of the joint. It is evident that the point *b*, must be equidistant from both pipes.

The right angle *abc*, is divided first into three equal parts and then each one of these parts is divided in turn into two equal parts; the right angle is

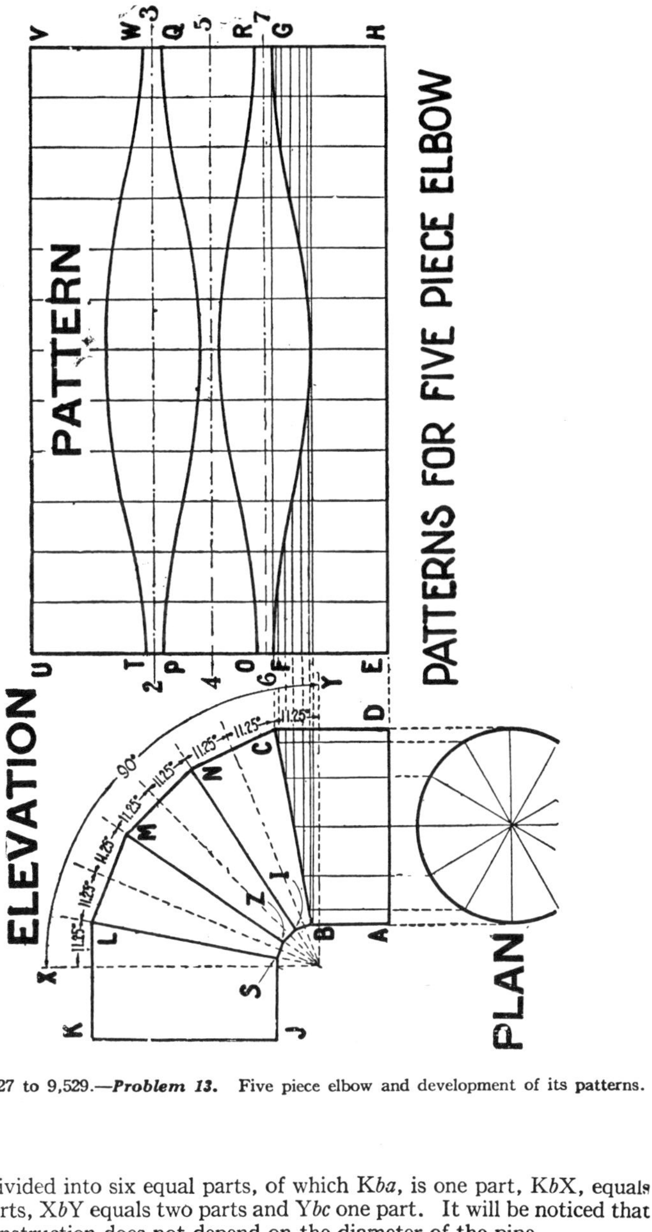

FIGS. 9,527 to 9,529.—*Problem 13.* Five piece elbow and development of its patterns.

thus divided into six equal parts, of which K*ba,* is one part, K*b*X, equals two parts, X*b*Y equals two parts and Y*bc* one part. It will be noticed that this construction does not depend on the diameter of the pipe.

The problem of developing the four part elbow resolves itself into developing two only of its parts, one large branch and one smaller part of the elbow, the remaining parts being correspondingly equal to these.

The circumference of the pipe, as seen in plan, is divided into sixteen equal parts by the points 1,2,3,4,5, etc.

Through these points are drawn lines parallel to the center line of the pipe which is to be developed.

In the development, the vertical branch of the elbow, (AKSI, of the elevation), will be taken up for the purpose. The parallels upon the surface of this branch are AK, BL, CM, DN, EO, FP, GQ, HR, and IS. Through the points K,L,M,N,O,P,Q,R, and S, draw parallels for the part KXTS, which will be next developed; some of these parallels are ST, RU, QV, PW.

To develop the vertical branch of the four piece elbow set off, upon a straight line *aa'*, sixteen equal parts, which altogether are equal to the circumference of the cylinder, which is to be developed.

Let the division points *a,b,c,d,e,f*, etc. correspond to the division points, 1,2,3,4, etc., upon the circle in plan. Through the points, *a,b,c,d,e*, etc., draw vertical lines equal to the parallel lines drawn upon the surface of the vertical branch of the joint; thus *aj*, is made equal to AK, *bk*, equal to BL; *cl*, equal to CM and so on until *ri* is made equal to SI.

The part laid out so far is *ajklmnopqri*. This is one-half of the development; the other half, *irj'a'* being exactly the same as the first one, may be laid out in the same way.

The part *tt'ss'* is the development of the small part of the elbow. It is evident that its length, *ts*, must be equal to the circumference of the pipe in the elbow. The lines in the pattern, *tt'ss'* drawn at right angles to the center line of it, and bisected by it, are made equal to the parallel lines, ST, RU, QV, PW, etc., drawn upon the surface of the part, KXTS, in the development.

It is plain that the part, *uu'vv'* is equal to the part *tt'ss'*, with the difference that the small parallels in it are laid out above the large parallels in the other part; in the same manner, the part *yy'ww'* is equal to the part *aja'j'*.

Laying out the pattern in this manner makes it possible to cut out the complete elbow from the square piece of metal, *ay'w'a'*. The spaces between the patterns are left for laps, which are necessary for joining all parts.

Problem 13.—Patterns for a five-piece elbow. Figs 9,527 to 9,529.

The five parts of the elbow may be thought of as so many parts of one long pipe, cut to the miter angle at proper distances. The length of that cylinder, this time, will be made up of the sum of the alternately consecutive outlines of the five parts of the elevation of the required elbow. Thus, in the development of the whole cylinder shown at EUVH, the vertical edge EU, is equal to the combined lengths of DC,BI,MN,ZS and KL, laid off on the development as the lengths EF,FO,OP,PT and TU.

Along the horizontal edge of the development, on EH, lay off all the equal parts into which the circumference of the elbow pipe is divided and, from the points of division on the edge EH, draw vertical elementary lines across the development.

From the division points on the circumference, projecting lines are drawn upward to the miter line BC, cutting it in a number of points from which, in turn, horizontal projecting lines are drawn meeting the vertical elementary lines on the development in points forming the curve FG. The miter line, as is seen in the elevation, has an angle of 11¼°.

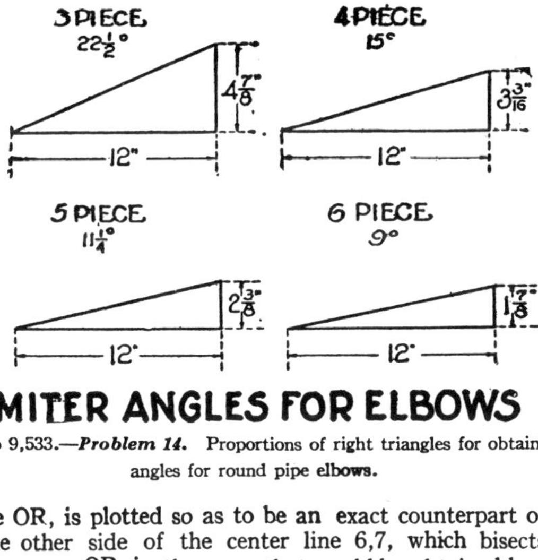

MITER ANGLES FOR ELBOWS

FIGS. 9,530 to 9,533.—*Problem 14.* Proportions of right triangles for obtaining miter angles for round pipe elbows.

The curve OR, is plotted so as to be an exact counterpart of the first curve, at the other side of the center line 6,7, which bisects OF, and RG; that is, curve OR, is the curve that would be obtained by revolving curve FG, 180° on 67, as an axis.

Having obtained the second curve OR, a second center line 45, is drawn bisecting OP and RQ, and, above this center line, the curve PQ, is plotted so as to deviate from the center line, along each vertical elementary line exactly as much as the curve OR deviates from the center line.

In a like manner, with the aid of the third center line, 23, the curve TW, is laid out opposite and equal to the curve PQ , the center line 23, bisects PT and QW. For simplicity in explaining the development no laps are provided for joints. The method of making this provision is explained in Problem 8.

Problem 14.—Miter angles for round pipe elbows. Figs. 9,530 to 9,533.

In all elbows, the miter upon the end piece depends upon the number of pieces of which the elbow is to be made up.

For a 2 piece elbow, the miter angle is 45°.
" " 3 " " " " " " 22½°.
" " 4 " " " " " " 15°.
" " 5 " " " " " " 11¼°.
" " 6 " " " " " " 9°.

The angles can be traced with the aid of a protractor. However, sufficiently accurate angles may be laid out for the different miters by the handy method of right triangles shown in figs. 9,530 to 9,533. The miter angle for a 3 piece elbow may be obtained by constructing a triangle, one leg of which is 12 ins. long, the other leg 4⅞ ins. long. The hypotenuse of this triangle gives the desired miter.

The proportions for the various other miters are given in figs. 9,530 to 9,533. The miter for any elbow not given in the illustrations may be found as follows:

***Rule.**—Divide by 90 the number of pieces in the elbow less one, multiplied by two.*

Thus for a three piece elbow, 90 ÷ (3 — 1) 2 = 22½°.

Problem 15.—Patterns for an offset on a rectangular pipe. Figs. 9,535 to 9,538.

The general appearance of this offset is shown at L, in fig. 9,534.

In making an offset on any pipe it is necessary to keep in mind that all parts of it should offer the same area for the even flow of the water within the pipe. Hence, in the elevation, the cross section of the oblique part of the pipe, FGJK, has the same width as the vertical part. To obtain this result, the joint FK, must be drawn so as to bisect the angle EFG, and likewise, the joint GJ, must bisect the angle KJI.

For the patterns of the offset, the four sides of the plan ABCD, are laid off, consecutively, upon the line MNOPQ, and in the points of measurement, elementary lines are drawn at right angles to the line MQ. Upon these lines are projected the vertical measurements of the part to be developed. Thus ML, is equal to GH, and SN, is equal to JI. The drawing of the pattern LMQR, plainly shows its derivation from the upright part GHIJ, by means of the projection lines.

For the development of the oblique part of the square pipe FGJK, project lines from points F,G,J,K, at right angles to the length of the oblique section. These lines cutting the line UV, in the points V,W,T,U, give starting points for horizontal projectors that determine the lengths of

FIG. 9,534.—Leader construction on house showing general appearance of parts for problems 11 and 12. L, offset on rectangular leader (problem 11); F, branch from leader (problem 12).

of the different portions of the development WUYZ. Joint laps should be added to the pattern, thus obtained.

Problem 16.—Patterns for round branch from rectangular leader. Figs. 9,539 to 9,543.

The general appearance of this branch is shown at F, in fig. 9,534.

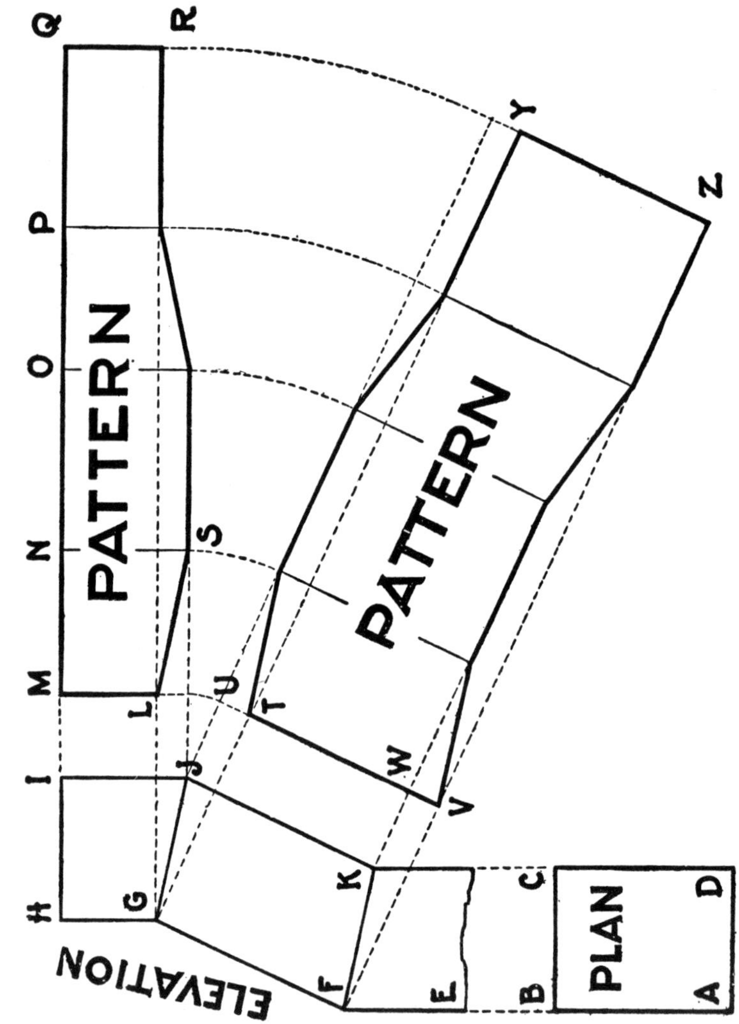

OFFSET FOR A RECTANGULAR PIPE

FIGS. 9,535 to 9,538.—*Problem 15.* Offset on a rectangular leader and development of its patterns.

PATTERN FOR Y BRANCH ON LEADER

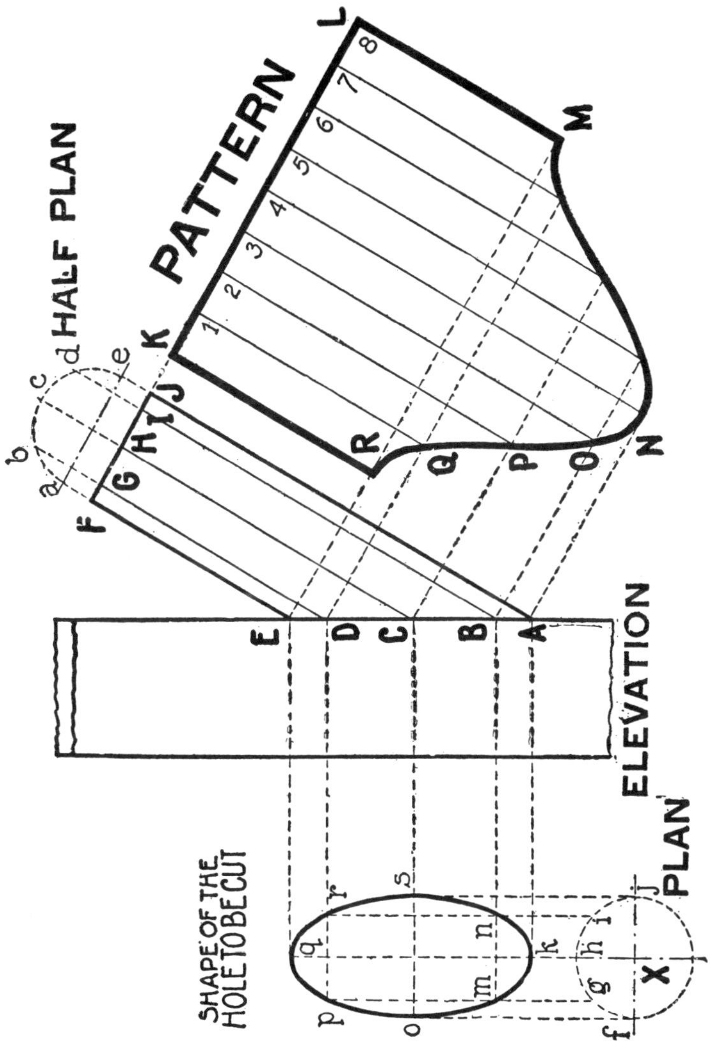

FIGS. 9,539 to 9,543.—***Problem 16.*** Round branch from rectangular leader and development of its pattern

In the elevation the round branch AEFJ, cuts the rectangular leader along a curved line that is symmetrical lengthwise and crosswise; the curve is an ellipse.

In the half plan of branch the circumference of the branch is divided into a number of equal arcs (half of the circumference being shown with the dividing points *a,b,c*, etc.).

Through *b,c*, and *d*, project the elementary lines GD, HC, and IB. The elements through the point of division intersect the square pipe in the points A,B,C, etc.

Now, for the pattern of the round pipe, make the line KL, a continuation of the line FJ, and equal to the combined length of the divisions on the circumference of the oblique pipe.

Lay off points 1,2,3, etc., on KL, spaced equal to arc distances *ab, bc, cd*, etc., in half plan, through points 1,2,3, etc., draw elements perpendicular to KL, and project over points A,B,C, etc., parallel to KL, and intersecting the element at R,Q,P, etc. Through these points describe the curve RNM, and join points K,R and L M, thus completing the pattern.

Laps for joints must be added.

The opening to be cut in the rectangular leader is an ellipse. It is obtained by projecting points A,B,C, etc., from elevation and points *f,g,h*, etc., from plan X. The intersections give points *k,m,o*, etc., which define the ellipse.

Problem 17.—Pattern for a box shaped leader trough. Figs. 9,544 to 9,548.

Draw, plan, elevation and end view of the trough full size. Note that the plan and elevation is divided into two halves by a center line VU.

In the development, draw first center line 21, 22. Since the trough consists of an upright part BCDE, surmounting the tapering part ABEF, the pattern will have two corresponding portions. The rectangle 8145, provides the pattern for the two end parts 1827, and 6345, as well as for the rear part 7236, of the right band. The front part of the latter is attached at the bottom of the pattern. It is marked 20,17,18,19 and is exactly equal

NOTE.—***Leaders.*** For quickly computing the size to be allowed for a leader, 1,800 sq. ft. of roof area for a 3 in. round leader or its equivalent in area for a square shaped leader. From 1,800 to 2,250 sq. ft. for a 3½ in. round leader or its equivalent in square leader. From 2,250 to 3,000 sq. ft. for a 4 in. round leader or its equivalent in square leader. From 3,000 to 5,000 sq. ft. for a 5 in. round leader or its equivalent in square leader. From 5,000 to 7,000 sq. ft. for a 6 in. round leader or its equivalent in square leader. Horizontal leaders should be larger and should be set with as much inclination as possible from the horizontal.

NOTE.—Judgment should tell what size leader to use when the roof area passes from one size or factor to another. For it is more economical to use, say, a 4 in. leader for 3,000 sq. ft. of roof area; but, a 5 in. leader would give a greater factor of safety in case of an unusual rainfall.

NOTE.—It is not considered good practice to use leaders less than 3 ins. in diameter because of the danger of stoppage or freezing. 2 in. leaders, however, are often used for small porch roofs or the gutters on turret skylights. In corrugated leader, the corrugations are not figured but the smallest diameter of the pipe is called the size of the leader

PATTERN FOR A BOX SHAPED LEADER TROUGH

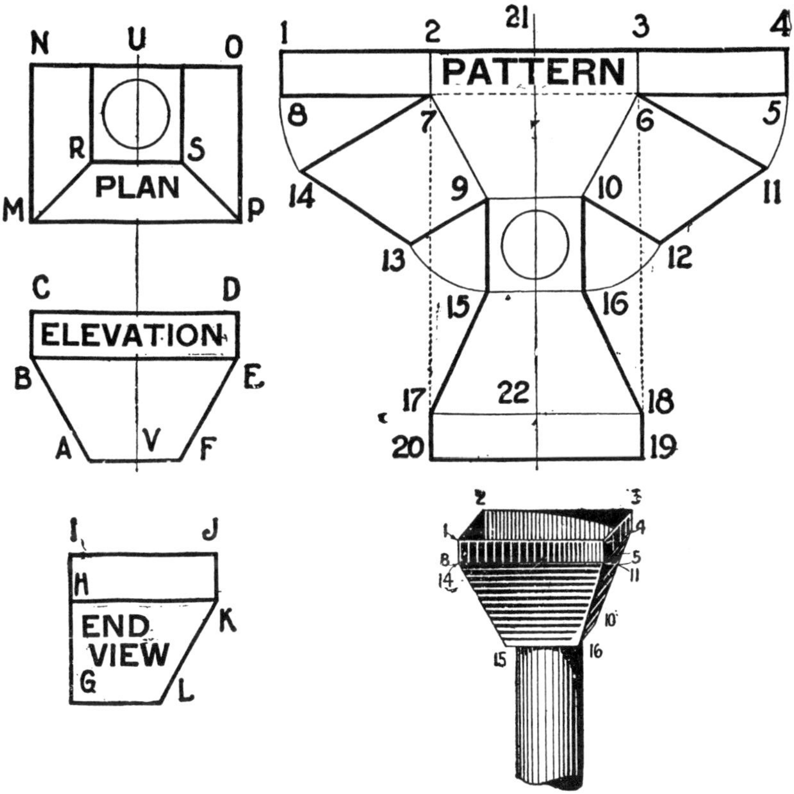

FIGS. 9,544 to 9,548.—*Problem 17.* Box shaped leader trough and development of its patterns.

to 7236. Note that the center line 22,21, passes through the middle of the upper band 8145, as well as through the middle of the part 20,17,18,19.

To lay out the pattern for the tapering trough, make 976,10, exactly equal to the projection in a vertical plane of ABEF, the tapering part of the front view.

Since the elevation does not show the true length of this tapering part but its projection in a vertical plane, 976,10, is made equal to ABEF.

To the lower edge, 9,10, then, attach the rectangle 15,9,10,16, which is equal to the bottom RUS, of the trough. At the points 7 and 9, erect the lines 7,14, and 9,13, at right angles to the line 9,7. Similarly, at the points

10 and 6, erect the lines 10,12 and 6,11, at right angles to the line 10,6. Then the irregular figure 13,14, 79, forms the left side of the tapering part, while the figure 12,10,6,11, forms the right side of the tapering trough.

Draw lines 2,20 and 3,19, parallel to the center line 21,22, and with a radius equal to 12, 11, from center 16, describe an arc cutting the line 3,19, at the point 18, and, with the same radius, from center 15, describe an arc cutting the line 2,20, at the point 17. Connect the points 17 and 18, by a straight line and obtain the front face 17,15,16,18, of the trough.

Finally, to the edge 17,18, attach the band 20,17,18,19, equal to 7236.

At the center of the square 976,10, describe a circle for the opening, whose diameter is equal to the diameter of the round leader.

Add to the pattern proper laps for the seams.

Problem 18.—Pattern for a pan. Figs. 9,549 to 9,552.

Draw elevation and end view. The flare on all sides being equal in this pan, the oblique edges in both views make the same angle with the bottom lines. Also, these oblique lines J,HG,DA and CB, are equal in length.

The pattern is to be made of one piece. Furthermore, its joints are not to be soldered but made water tight by turning some of the stock of the

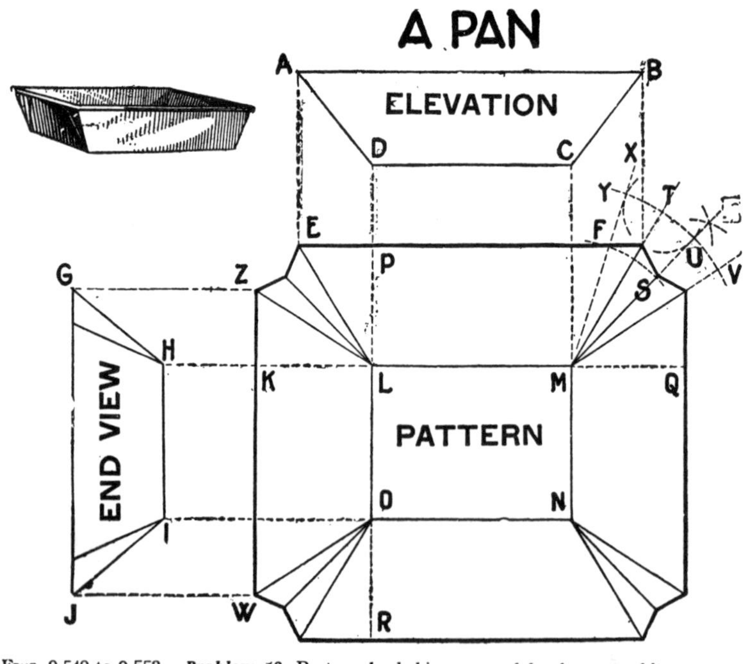

FIGS. 9,549 to 9,552.—*Problem 18.* Rectangular baking pan and development of its pattern.

sides flatly upon the end walls of the pan, and this in such a manner that the folded metal should reach exactly to the wired edge.

To draw the pattern, project the corner points I,H,D and C, of the end view and elevation so as to form, by intersection, the rectangle OLMN; this is the bottom of the pan.

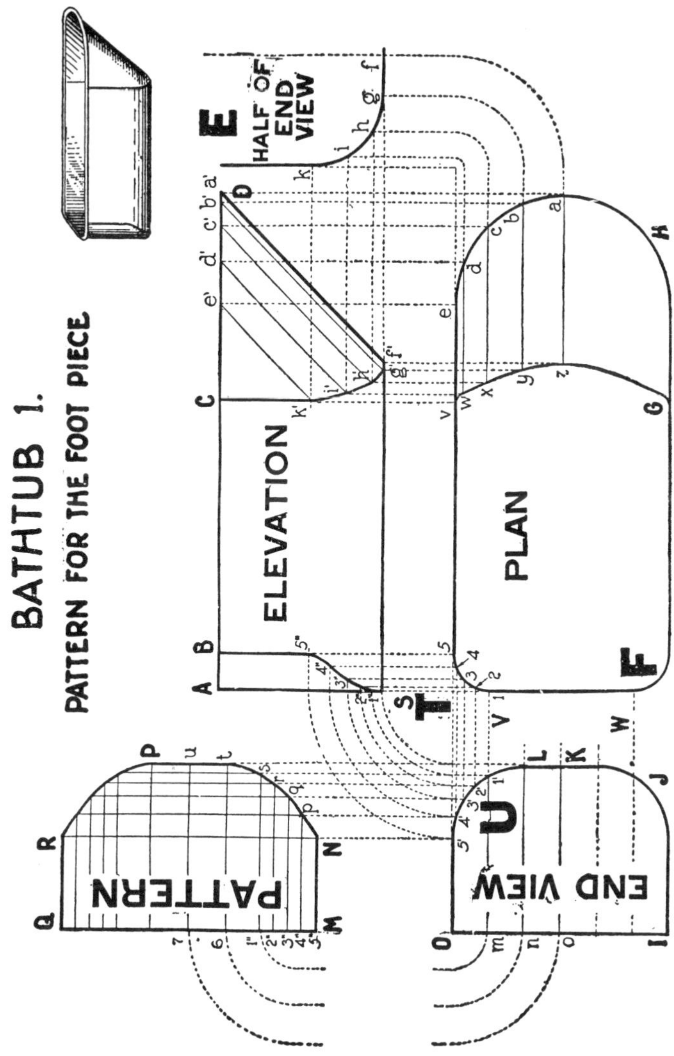

FIGS. 9,553 to 9,558.—*Problem 19.* Foot piece of bath tub and development of its pattern.

Space off distance LK, LP, MQ and OR, each equal to the oblique line HG. Draw lines through these points, and project from points J,G,A,B, lines intersecting at W,Z,E, etc., giving outer edges of the pan. Connect points W,Z,E, etc., to the rectangle OLMN, which gives the four sides of the pattern. This KL, LP, MQ and OR, is each equal to the oblique line GH.

Upon these outer edges now project the distances JG and AB, to the points W,Z,E and so on, which points, when connected to the corners of the rectangle ONML, furnish the shapes for the four side parts of the pattern.

The metal between the edges LZ, LE, etc., has to be so shaped that, upon upon being folded up on the side wall, it should reach exactly to the wired edge. The procedure for this purpose is explained in the upper right corner of the pattern. The acute angle between the oblique edges of the sides, that is the angle TMV, is bisected by the line MU. Then one half of the bisected angle is laid off alongside of it. Thus the angle UMT, is laid off at TMX.

This is done by making the arc TY, equal to the arc TU. Now, from the point F, where the edge of the added angle cuts the edge EF, an arc is drawn with the radius MF, from the point M, as center, cutting the bisecting line MU, in the point S, which determines the angular shape to which this corner, as well as all the other corners, should be cut.

For purposes of wiring it is necessary to add along the outer edges a strip sufficient partly to enclose the wire. The amount required for this purpose is equal to approximately three fourths of the circumference of the wire.

Problem 19.—Pattern for foot piece of bath tub. Figs. 9,553 to 9,558.

The bath tub here considered is of the type intended to be encased in wood. The foot piece is at an angle of 45° with the bottom.

The laying out of the patterns for the tub can not be accomplished without first drawing a plan, elevation and end view, including the joint edge between the body piece and the foot and head pieces.

After laying out the outside lines of the three views, with standard dimensions (or, for practice, to any convenient small scale) proceed to plot the joint edges 1″,2″,3″,4,″5″, and *g′,h′,i′,k′*, on the elevation, and *v,w,x,y,z*, on the plan.

Divide the corner T, in the plan, into any number of equal parts, say four parts, marked 1,2,3, etc. Project these points into the end view, to the corner U, at the points 1′,2′,3, etc., and which points in turn, project by means of arcs, to the edge SA, of the elevation and further on by a series of parallels of indefinite length toward the desired joint edge S5″. The intersection of these parallels with the projection lines 11″, 22″, 33″, etc., give points on the desired curved joint 1″,2″,3″, etc.

Now at the opposite end of the bath tub, divide the semi-circular outline into equal parts, by the points *a,b,c*, etc., from which, by projection, obtain the points a',b',c', etc. From these points draw the elementary lines $b'g'$, $c'h'$, etc., on the line AC, of the elevation. In line with the elevation, repeat a half of the end view, at E. By means of arcs, as shown, carry the division a,b,c, etc., to the points f,g,h, etc., and project these points by means of horizontal parallels ff',gg',hh', etc., into the elevation toward the line Ck'. These parallels intersecting with the oblique lines $b'g',c'h'$, $d'i'$, etc., give points on the joint edge f',g',h', etc.

Again, the vertical projection lines drawn downward from the points f',g',h', etc. to the plan intersect the horizontal projection lines az,by,cx, etc. (in plan) in the points z,y,x, etc., which gives one half of the joint line Gv for the bottom of the bath tub.

With the three views completed, the pattern for the foot piece is plotted as follows: Extend the line O1, indefinitely, and upon its extended part from M, to Q, set off in succession, the stretched out length of the corner T (in plan), then, the distance WV, that is, the flat portion of the foot piece, and the length of the arc around the corner F. The stretched-out length of the arc of the corner T (which is the same as that of corner F), is composed of the four equal divisions marked 1,2,3, etc. Thus, starting at M, measure these divisions from 5″ to 1″, thence, from 1″ to 7″, lay off one half of VW, in this manner gaining one half of the edge length of the desired pattern. The other half of that edge 7Q, is an exact counterpart of the first half.

From all divisions upon the edge MQ, draw, at right angles to it, a series of parallels toward the opposite edge of the pattern tuP, which is located upon the extension of the line KL (of the side view). Where this series of parallels is cut by the projection lines that start from 1′,2′,3′, etc., at the points s,r,q, etc., will be the curved outline of the desired pattern. It is obvious that this curve is repeated from P to R.

Problem 20.—Pattern for body of bath tub. Figs. 9,559 to 9,562.

The outline of the end view may be regarded as a cross section of the bath tub and therefore its stretched out length will furnish the length of the pattern for the body. In the end view, the cross length of the body as seen is composed of two times the length marked by A, plus two times that of the corner B, plus once the length C; that is 2A + 2B + C. On the pattern, starting at the edge N′ O′, make N′Q′ equal to the length A; next to it, starting at Q′, lay off the distance E′, equal to E; then, from the point P′, to F′, is one half of the distance D′, that is a half of the distance D. The straight edge P′J′, on the pattern, marked by D′, is equal to D, the flat portion of the foot piece. The curve P′Q′, within the distance E′, contains the sum of all the parts into which the corner E (on the side view) had been divided, that is, the distances 1′2′,2′3′,3′4′, and 4′5′. The drawing of the pattern plainly shows how the projection lines from the points 1,2,3,4,5 (plan view), in intersecting the short parallels drawn from the divisions upon E′, define the curve P′Q′.

The opposite longitudinal edge of the pattern is obtained by means of projection lines from the points O,M,K,I, and G, and with the aid of parallels drawn across the pattern from the points N′,Q′,P′,F′,H′,J′,L′. The

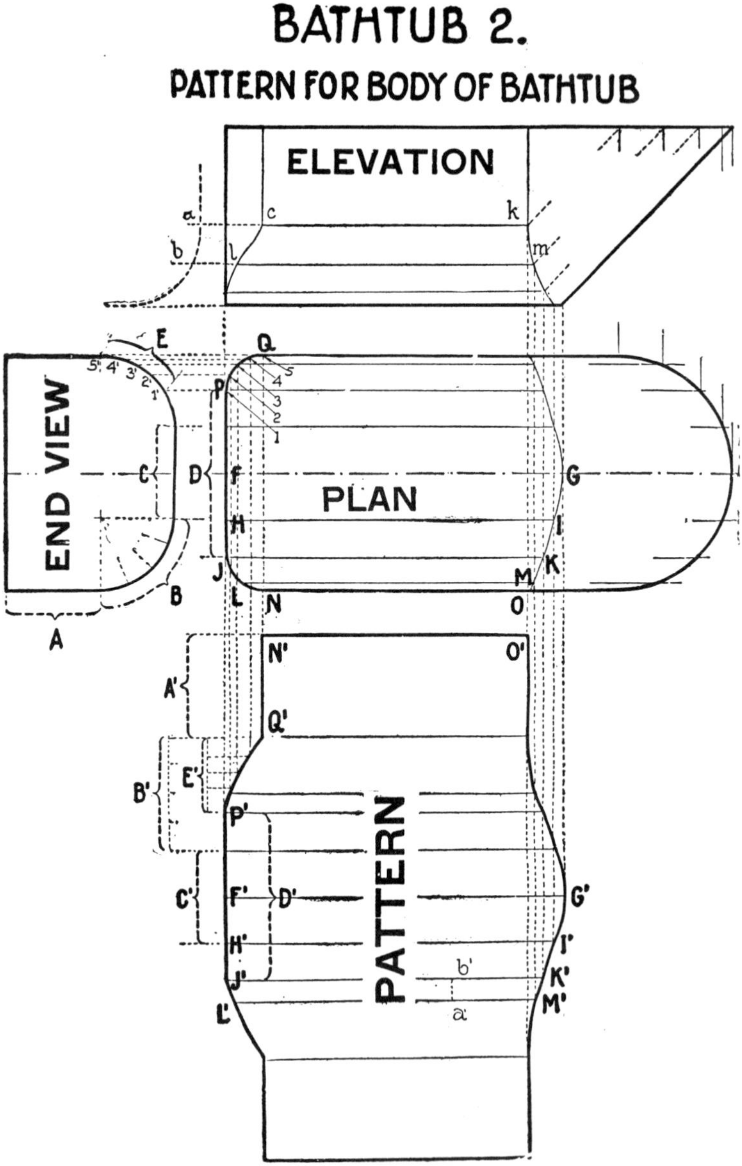

FIGS. 9,559 to 9,562.—*Problem 20.* Body of bath tub and development of its pattern.

distance *a′b′*, separating the lines L′M′ and J′K′, is equal to the distance *ab* obtained from the front view, whence the lines *lm* and *ck* were projected outward, upon the arc *ab*, which arc is a reproduction of a part of the side view. Referring to the pattern, the line F′G′, is equal in length to FG (of the plan view); the line H′I′ is equal to HI; J′K′ is equal to JK and L′M′ is equal to the length of the line LM. The center line F′G′ divides the pattern into two exactly symmetrical parts.

It should be understood, that in this case, as in all problems where curve shaped patterns are to be plotted, best results are achieved when each curve is determined by a large number of points. However, in the illustrations, for the sake of simplicity, we are constrained to use only a limited number of such defining points.

Problem 21.—Pattern for head piece of bath tub. Figs. 9,563 to 9,567.

The head piece is developed by the method used for the plotting of the pattern for the slope sheet of a boiler. First, obtain a cross section through the head piece on the line S*a*, which is perpendicular to the oblique outline of the head piece. One half of this cross section is shown in the drawing. Its different points are located on the parallels that are continuations of the oblique lines 1′*b*,2′*c*,3′*d*, etc. The distance 67, is equal to 89 (in plan). The vertical distance from the point 4, to the line 85, is measured from the line 5″7, up to 4″, upon the extension of the oblique parallel *e*4′. The distance from 3, to the line 85 is laid off from the line 5″7, to the point 3″, along the next oblique parallel. The distance from 2, to the line 85, is set off from the line 5″7, to the point 2″, on the next oblique parallel,and finally, the distance from the point 1, to the line 85, is measured from the line 5″7, to the point 1″, on the next parallel. This procedure is illustrated in the drawing by means of the series of projection lines and arcs which start at the dividing points on the semi-circular edge in the plan and which lead toward the half cross section, that is to right of the front view,where they intersect the series of oblique parallels drawn from the points 1′,′2,3′, etc., thus setting off the required distances.

The cross section through the head piece gives the curvature of the direction of the line S*a*. When stretched out, this curvature furnishes the longest measurement on the development of the head piece, namely, the line *a′a″*, on the pattern. This distance *a′a″*, is made up by putting together all the parts of the outline of the cross section, so that, starting at *a′*, along the line *a′a″*, are laid off, one after another, the lengths 61″, 1″2″, 2″3″, 3″4″ and 4″5″, thus reaching the middle line of the pattern, the line E*f′*. On the other side of E*f′* the same divisions are laid off in reversed order, up to the point *a″*. Through these divisions on the line *a′a″*, are passed the elementary lines A*b′*,B*c′*,C*d′*,D*e′* and E*f′*, etc.

The terminations of the elementary lines are obtained by means of the projection lines *bb′*,*cc′*,*dd′*,*ee′* and *ff′*, on one side of the pattern, and by the projection lines 1′A,2′B,3′C,4′D and 5′E, on the other side of it.

All proper edges of the patterns for the bath tub should be provided with additional stock for necessary joint locks.

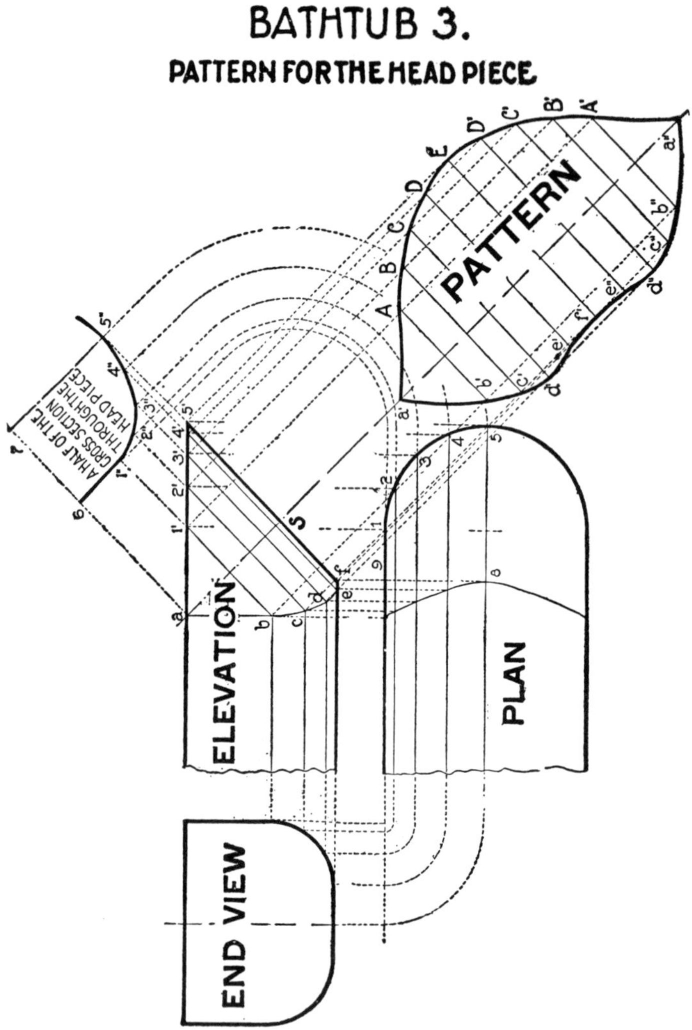

FIGS. 9.563 to 9,567.—*Problem 21.* Head piece for bath tub and development of its pattern.

Development by Radial Lines

The second class of elementary surfaces are those whose elements are not parallel to each other but *radiate from a com-*

mon point. This class may be sub-divided into two groups, as

1. Surfaces having three or more consecutive elements in the same plane

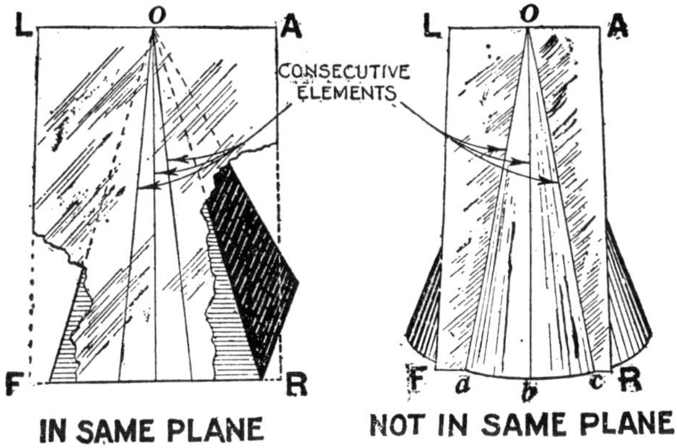

FIGS. 9,568 and 9,569.—Examples illustrating the two groups of elementary surfaces having radial elements. In the figures *oa, ob, oc,* are the radical elements and LARF, the plane.

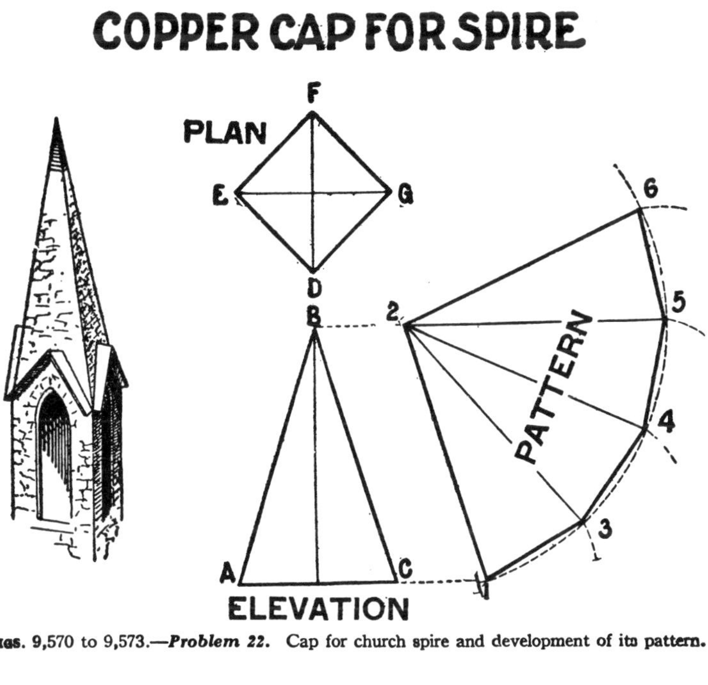

FIGS. 9,570 to 9,573.—*Problem 22.* Cap for church spire and development of its pattern.

2. Surfaces having no three consecutive elements in the same plane.

Surfaces such as pyramids belong to the first mentioned group, and cones to the second group, as shown in figs. 9,568 and 9,569.

The following examples are given to illustrate the method of development of such surfaces.

Problem 22.—Pattern for cap of church spire. Figs 9,570 to 9,573.

The required cap will have the shape of a four sided pyramid and the pattern for it is obtained by drawing the development of this pyramid.

With the oblique height of the cap CB, as radius, from the point 2, as center, describe the arc 16. Upon this arc mark off four distances, each equal to DG. The required pattern is 2 1 3 4 5 6.

All pyramids, no matter how many faces they might have, as long as all faces are equal one to another, can be developed into patterns in the manner employed in the above problem. In all cases, an arc is drawn with a radius equal to the oblique height of the pyramid, and, on this arc is laid

DEVELOPMENT OF A TRUNCATED OCTAGONAL PYRAMID

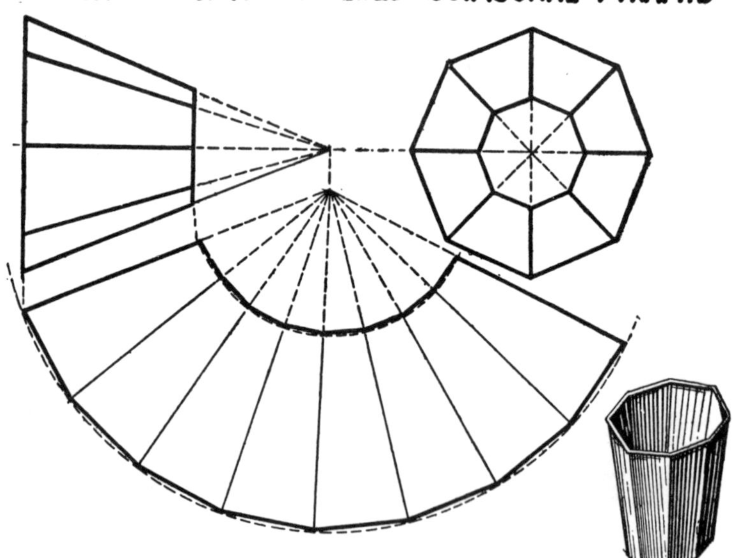

FIGS. 9,574 to 9,577.—***Problem 23.*** Truncated octagonal pyramid and development of its pattern.

off the bottom width of one of the faces as many times as there are faces in the pyramid.

Problem 23.—Pattern for truncated octagonal pyramid. Figs. 9,574 to 9,577.

CONICAL HOOD FOR A PIPE

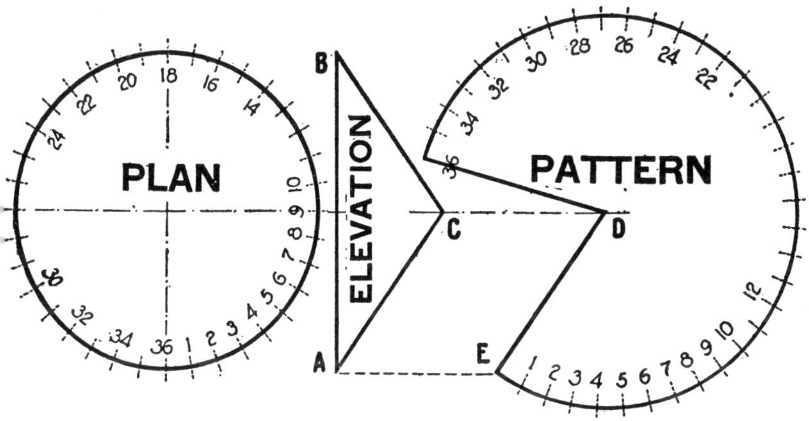

FIGS. 9,578 to 9,581.—*Problem 24.* Conical hood on a pipe and development of its pattern.

The pattern for a truncated pyramid is readily obtainable by the method used in the last problem on the development of a pyramid. In the case of a truncated pyramid, the latter is completed by suitable auxiliary lines so as to produce a whole pyramid which, then, is seen to consist of the given truncated pyramid and a superimposed small pyramid.

Both the large and the small pyramids are developed, with arcs equal to their oblique lengths, both arcs drawn from the same center, as shown in the illustration. Thus the two developments will appear one laid upon the other, the development of the smaller pyramid occupying the corner at the center of the two arcs. In the illustration, the dotted portion is that belonging to the smaller pyramid. The remaining part of the development of the large pyramid, designated by heavy lines, is the developed pattern for the truncated pyramid.

Problem 24.—Pattern for a conical hood on a pipe. Figs. 9,578 to 9,581.

The plan of the hood is a circle. The front view (also called elevation) is a triangle whose base is equal to the diameter of the circle in the plan and whose two other sides are equal to each other. These two equal sides of the triangle give the oblique height of the cone which forms the hood.

To develop the pattern for the cone, divide the base circle into any number of equal parts, say 36. Then, from any point D, as center, with the radius DE, equal to AC, that is equal to the oblique height of the cone, describe an arc upon which, starting at the point E, set off, in succession, all the parts into which the base circle is divided, in this case, 36, equal parts, so that the full length of the circumference of the base circle will be stretched out upon the arc.

The boundaries of the required pattern are the two radial lines from the point D, and the arc that is equal, in length, to the circumference of the base circle of the cone.

Problem 25.—Pattern for smoke stack with conical hood or base. Figs. 9,582 to 9,587.

Draw elevation and plans of top and bottom.

In the elevation, the base, as seen is a truncated cone, whose oblique outlines produced upward, intersect at J. This gives a large cone CJF, and a small cone DJE. Develop both of these cones from the common vertex P, and one common edge PR.

The circumference of the base circle A (in plan), divided into a number of equal parts, is stretched out upon the arc RS, whose radius PR, equals the oblique height JF of the large cone, thus the development of the entire cone CJF, is bounded by the two radial lines PR, and PS, and by the arc RS. Within the space of this development, at its corner, the space PQT, shows the development of the small cone that stands upon the truncated cone. The smaller development is made by the arc QT, which is drawn from P, as center, with the radius PQ, equal to the oblique height of the small cone.

All the space within the larger development PRS, unoccupied by the smaller development PQT, belongs to the pattern of the truncated cone.

For the laps, a circular strip should be added along the edge QT, and a straight one along TS.

To obtain the pattern for the pipe, divide its circumference into a number of parts as shown in plan. The larger the number of divisions, the

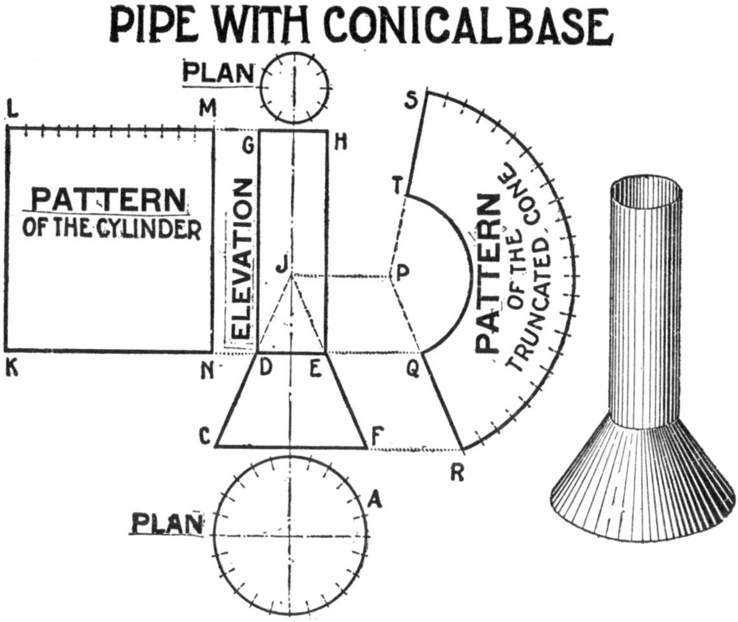

Figs. 9,582 to 9,587.—*Problem 25.* Smoke stack with conical hood or base and development of its patterns.

more nearly true will be the resulting pattern. Then, extend the lines DE, and GH, outward toward the desired pattern. Draw the line MN, parallel to the line DG, and stretch out on ML the circumference of the pipe that is all of its divisions laid off one after another.

Finally draw the edge LK, parallel to MN, completing the rectangle that is the pattern of the cylinder.

Of course, for the lap, a sufficiently wide strip should be added to the edge KL. The width of this strip will depend upon the thickness of the metal in the pipe.

Problem 29.—Pattern for conical roof hood or base connecting with stack. Figs. 9,591 to 9,594.

The problem calls, firstly, for the development of the cone ABC, of which the roof-hood DEFG, is the central part, and, secondly, for the develop-

MARINE STACKS

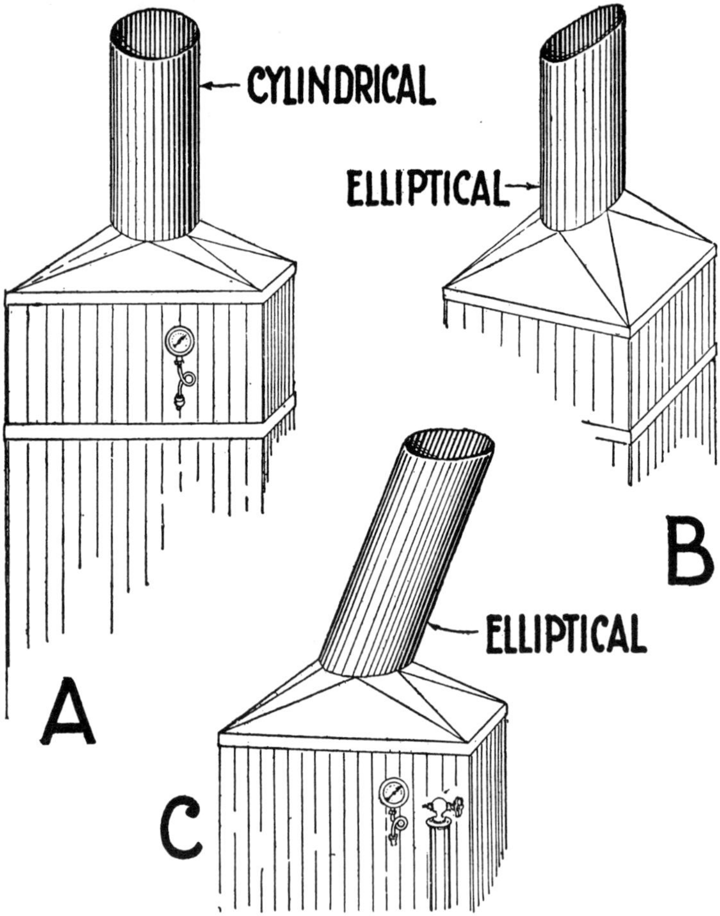

FIGS. 9,588 to 9,590.—*Problems 26 to 28, for practice.* **A**, cylindrical upright stack; **B**, elliptical upright stack; **C**, elliptical raked stack.

ment of the obliquely cut lowest portion of the cone, the part ADGC, shown in the elevation.

The base circle P, of the cone, is shown in plan. This whole circle is divided into 24 equal parts, marked by the points 1,2,3, etc. From these points on the circle, vertical lines are drawn cutting the base line AC, at

CONICAL ROOF CONNECTION FOR VENT PIPE

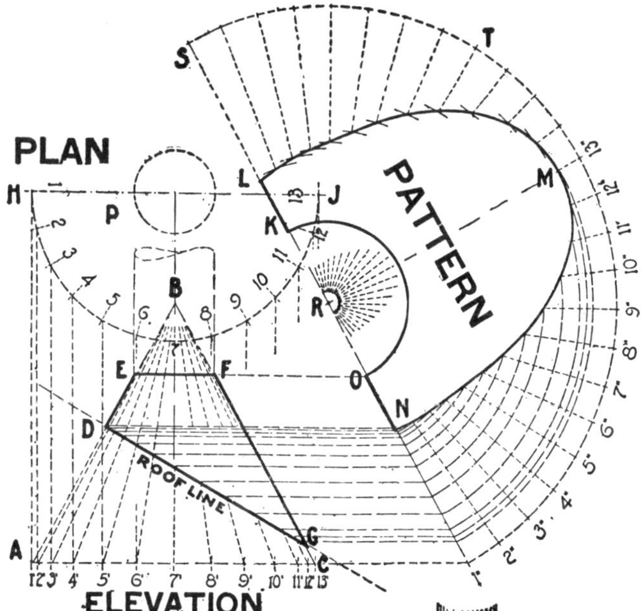

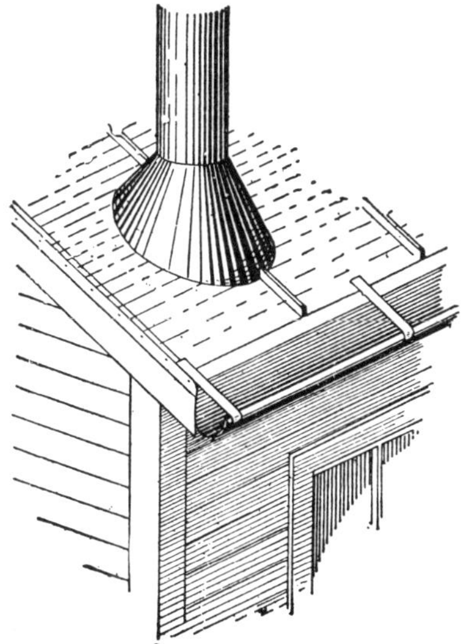

FIGS. 9,591 to 9,594.—***Problem 29.*** Conical roof hood or base connecting with stack and development of its pattern.

PATTERN FOR A CONICAL EAVE TROUGH OUTLET

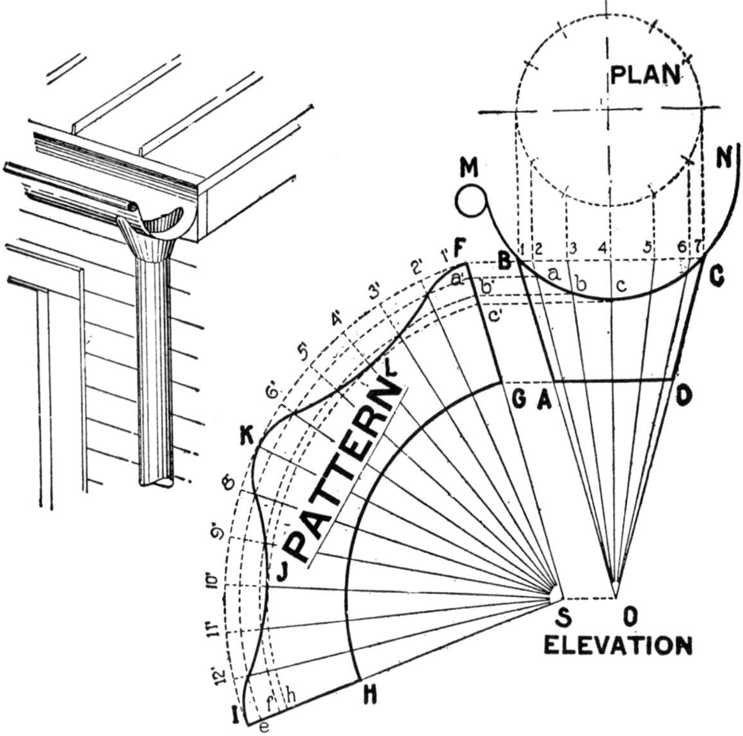

FIGS. 9,595 to 9,598.—*Problem 30.* Conical eave trough outlet and development of its pattern.

the points 1′,2′,3′, etc., from which points lines are drawn to the vertex B, namely the lines B1′,B2′,B3′, etc.

The development of the cone ABC, is laid out, with a radius equal to the oblique height of the cone, the distance BC, with the arc 1″TS, upon which all of the divisions of the base circle are set off, in succession, at the points 2″,3″,4″, etc., thus making the length of the arc 1″TS, equal to the whole circumference of the base circle. The development of the cone ABC is bounded by this arc and by the lines SR and RI″. Upon this development also is laid out the development of the hood DEFG. This development, is outlined with the heavy solid line ONMLK.

The roof line DG, giving the pitch of the roof, is intersected in 13 points by the lines 1′B,2′B,3′B, etc. From these points of intersection, horizontal lines are drawn to the line ON1″, giving starting points for a number of arcs upon the development. These arcs, at their meeting points with the radial lines R1″,R2″,R3″, etc., define one half of the curve.

TAPERING ELBOWS

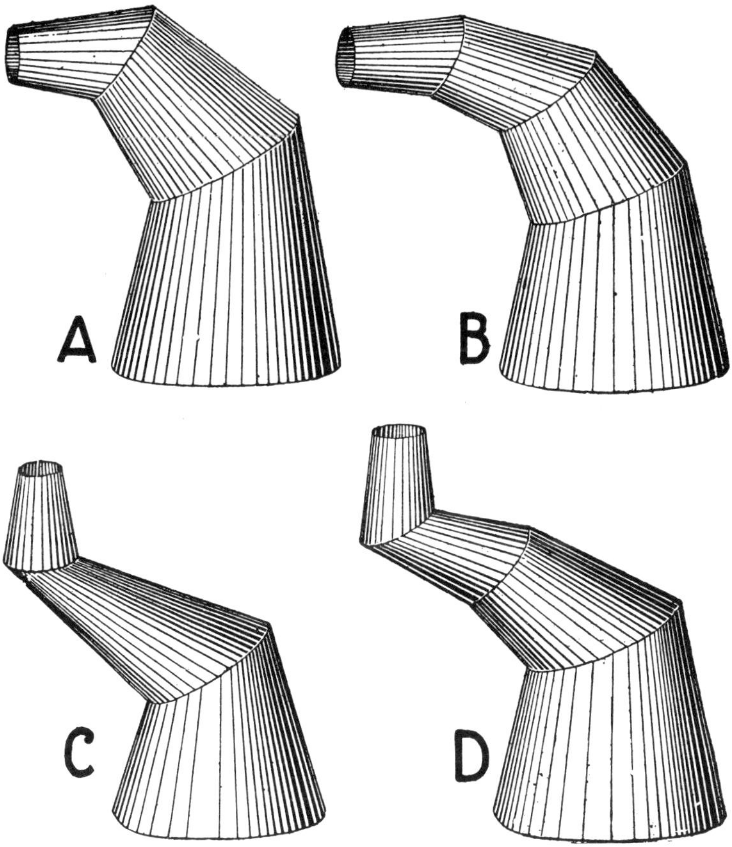

FIGS. 9,599 to 9,602.—***Problems 31 to 34, for practice.*** **A,** 3 piece elbow; **B,** 4 piece elbow; **C,** 3 piece offset; **D,** 4 piece offset.

The other half of this curve being exactly equal to the first, it can be laid out in a similar manner by means of these arcs and the radial lines in the second half of the development, the complete curve being NML.

Additional stock has to be allowed for laps, along both curved outlines and along one of the straight edges of the pattern of the connection piece.

Problem 30.—Pattern for a conical eave trough outlet. Figs. 9,595 to 9,598.

It does not matter what curve be given to the gutter, oval or circular. The method explained in this case is suitable for gutters of all curvatures.

In this case the cross section of the gutter is represented at MBCN in elevation. The conical outlet which is to form a connecting piece between the gutter and the pipe, is a truncated cone whose wider end is shaped so as to conform with the surface of the gutter. The cone of which the connection piece is a part, is represented as BOC. The dotted line BC, represents the diameter of the base of this cone; the circumference of it being shown in plan.

To develop the pattern, first develop the truncated cone ABCD, obtain the figure bounded by the radial lines HI, and GF (each equal to AB), and by the arc IKF and HG, the larger arc described with a radius equal to OB, while the smaller arc has a radius equal to OA.

The length of the larger arc is equal to the sum of all the divisions on the circumference of the circle shown in plan. These divisions are marked on

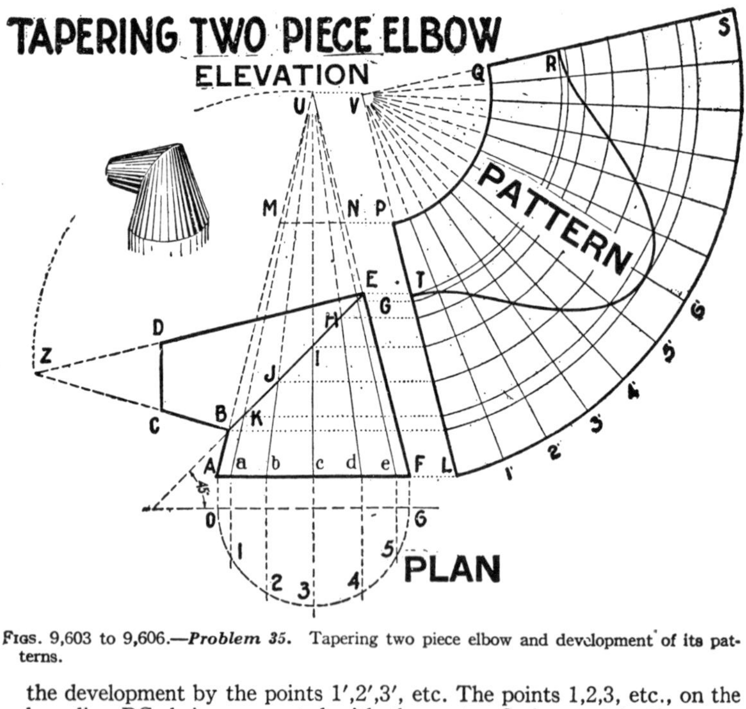

FIGS. 9,603 to 9,606.—*Problem 35.* Tapering two piece elbow and development of its patterns.

the development by the points 1′,2′,3′, etc. The points 1,2,3, etc., on the base line BC, being connected with the vertex O, by straight lines, the surface of the cone appears with a series of elements or lines converging in the vertex. On the development of the cone, these elements appear as the radiating lines S1′,S2′,S3′, etc.

Where the elements of the cone are intercepted by the cross section of the gutter, there are the points B,*a*,*b* and *c*, which are projected to the line FG, by projectors parallel to BF, thus marking off on the line GF, the points F,*a*′,*b*′ and *c*′. From these points, describe the arcs *a′e*,*b′f* and *c′h*. These arcs, in their intersections with the lines S1′,S2′,S3′, etc., define the curved outward boundary FLKJI, of the required pattern.

Of course, the actual pattern should be enlarged along the curved boundary by the addition of stock for laps or locks.

Problem 35.—Patterns for a tapering two piece elbow. Figs. 9,603 to 9,606.

The pattern for this elbow is made up of the two halves of a truncated cone, which is cut into two parts by a 45° miter angle.

In the elevation AMNF, is the truncated cone cut by the miter from B to E. The upper part BMNE, of the truncated cone, when joined to the lower part, gives the horizontal part BCDE, of the elbow.

To obtain the desired pattern, begin by developing the truncated cone, using the method explained for the problem of the conical base for a pipe (problem 25).

In the development, LPQS, represents the development of the truncated cone. What now remains to be done is to divide this pattern into two parts, corresponding to the two parts of the elbow. This is accomplished by laying out, upon the development, the curved line TR, which is the upper boundary of the development of the obliquely cut, lower part ABEF, of the cone.

The divisions upon the circumference of the base circle and the corresponding radiating lines upon the surface of the cone used for the development of the cone, serve also for the development of its oblique lowest portion. These lines AU,*a*U,*b*U, etc., cut the miter line in the points B,K,J, etc. From these points, horizontal lines are drawn to the boundary line TL of the development, from whence arcs are described upon the development. These arcs, at their meeting points with the radial lines VL,V1′, V2′, etc., upon the development, define one half of the curve TR. The other half of this curve is an exact counterpart of the first and is laid out in a similar manner.

Although the development of the two parts of the elbow are plotted with the common curved edge TR, in shop practice, when a strip of stock has to be provided on one of the patterns, for joining purposes, along this curved outline, the two patterns have to be cut detached, with the aid of templates each of which is identical with the corresponding part of the development.

Problem 36.—Patterns for a ventilator duct. Figs. 9,607 to 9,613.

VENTILATOR DUCT

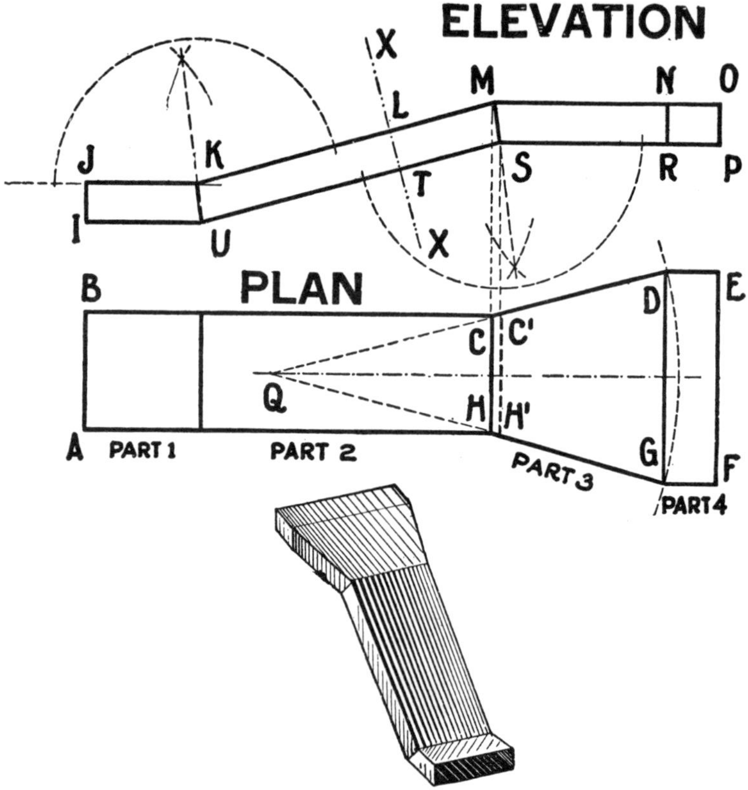

Figs. 9,607 to 9,613.—***Problem 36.*** Four part ventilation duct and development of its patterns.

First draw plan and elevation. The duct as shown in the plan has four parts. Horizontally, all four parts are lying in one direction, as is seen in the plan. This view also shows that part three is a tapering part.

The elevation shows that part two rises obliquely while all other parts are situated horizontally. The elevation, furthermore, shows that all four parts have the same height, namely, equal to IJ, or OP. To insure equal heights in the two adjoining horizontal and oblique parts, the angle between them is bisected so that the joint between them falls upon the bisecting line. Thus the joint KU, falls within the bisecting line of the angle JKM. For the pattern of part one, draw the vertical line 5,10, upon which measure consecutively the width 10,8, the height 8,1, the width once more (1,2)

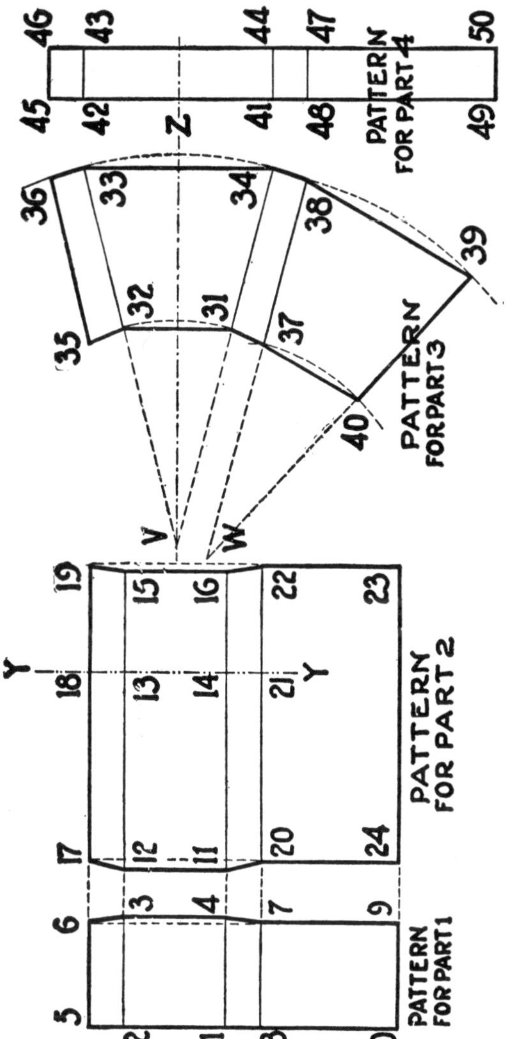

FIGS. 9,607 to 9,613.—*Problem 36.—Continued.*

and the height once more (2,5). At the points 10,8,1, etc., draw five indefinitely long horizontal lines, all at right angles to the line 10,5. Upon the horizontal lines, make 10, 9, 87 and 56 equal to JK, and measure the distances 14 and 23, equal to IU. Connect 6,3,4,7,9, completing the pattern for part 1.

The pattern of the second part of the duct may be laid out on the same horizontal lines, however, to begin with, cross part two, in the elevation, by a line at right angles to it, at any point, say at L. Let this "stretch-out" line be XX. Now, returning to the five horizontal lines extending from the pattern of part one, cross the upper four of these lines, at any point, say at point 18, by a vertical line YY.

Now, starting to measure from the vertical line YY, upon the horizontal lines, set off 18, 19 and 21,22, each equal to LM. Also, starting from line YY, measure 13,15 and 14,16, equal to TS. On the same lines, then, make 12,13 and 11, 14, each equal to the line UT, also, make 20,21 and 17,18, each equal to KL. To complete the pattern for part two, draw the lines 20, 24 and 22,23, at right angles to the line 20,22. If properly constructed, the points 24,20 and 17, will be found lying in one vertical line. Similarly, the points 23,22 and 19, should fall within one vertical line.

For the pattern of the tapering part, reproduce, to start with, the angle GQD, together with its center line, at any convenient place, say at 34,V,33. With a radius equal to QD, reproduce the arc DG, at 33,34, from V, as center. Make 33,34, equal to GD, one half of it on each side of the center line VZ. Then make 33,32, equal to DC; also equal to 31,34. The figure 31,32,33,34, forms the bottom part of the tapering section of the duct. Adjoining each side of this, construct the upright walls of this section, as follows:

At 34, draw line 34,38, at right angles to line 31,34. From point 38, then, pass the line 37,38, parallel to 31,34 and make the distance 37,38, equal to CD. Similarly, the line 33,36, is at right angles to 33,32, the line 35,36, parallel to 32,33, the length 35,36, equal to MN.

Now, for the upper (horizontal) wall of the tapering section, extend the line 37,38, and make 38,W, equal to D Q. With this distance as radius, describe the arc 38,39. Then, from point 38, with a radius equal to 33, 34, strike off upon the arc the distance 38,39. Then, having connected 39, and W, by a straight line, cut this line at the point 40, with the arc 37,40, whose center is at W. Note, the arcs 37,40, and 32,31, are not of equal radii.

The pattern for the fourth part of the duct, 49,45,46,50, is a rectangle wherein 45,46, is equal to PO, whereas 45,52, equals 36, 33; 42,41, equals 33,34; 41,48 equals to 34,38 and 48,49 equals 38,39.

All patterns have to be provided with laps for joints.

Problem 40.—Patterns for a scalene cone. Figs. 9,617 to 9,620.

In the plan, divide the circular base into any number of equal parts by the points 1,2,3, etc. Project these points up to the base line AI, of the cone, obtaining the points 1′,2′,3′, etc. Connect all these points with the vertex D, by means of straight lines giving the elementary lines D1′, D2′, D3′, etc.

Since these elementary lines (with exception of D1′, and D7′) appear shorter than their actual lengths, find their actual lengths by constructing a series of triangles, according to the following rule:

OBLIQUE TAPERING FLANGES

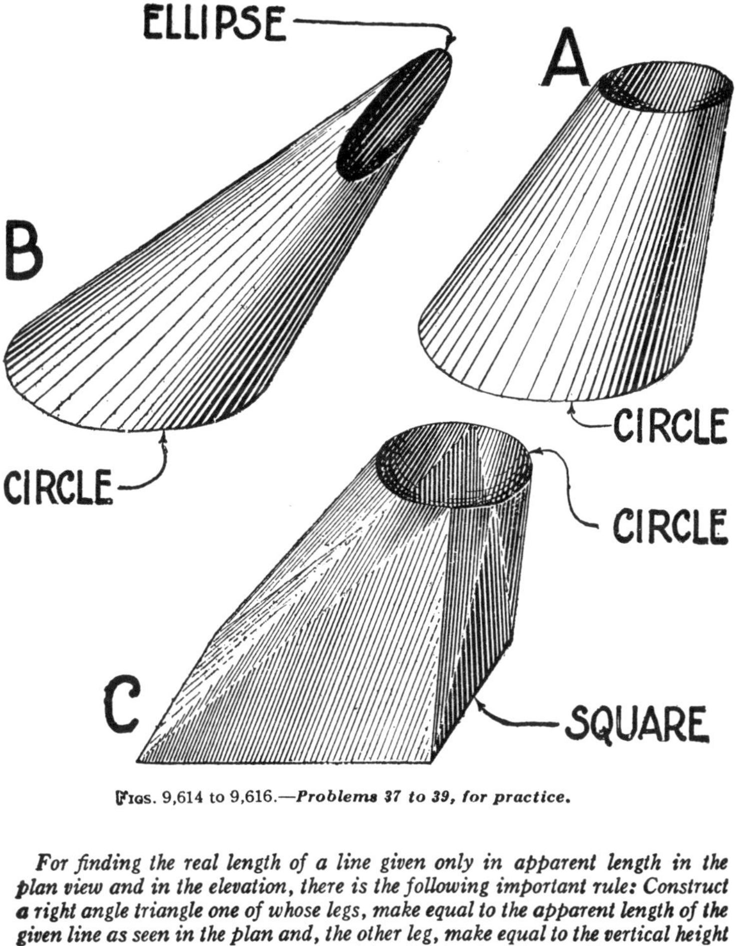

FIGS. 9,614 to 9,616.—***Problems 37 to 39, for practice.***

For finding the real length of a line given only in apparent length in the plan view and in the elevation, there is the following important rule: Construct a right angle triangle one of whose legs, make equal to the apparent length of the given line as seen in the plan and, the other leg, make equal to the vertical height which the given line rises in its position in the elevation. The hypotenuse is the real length of the given line.

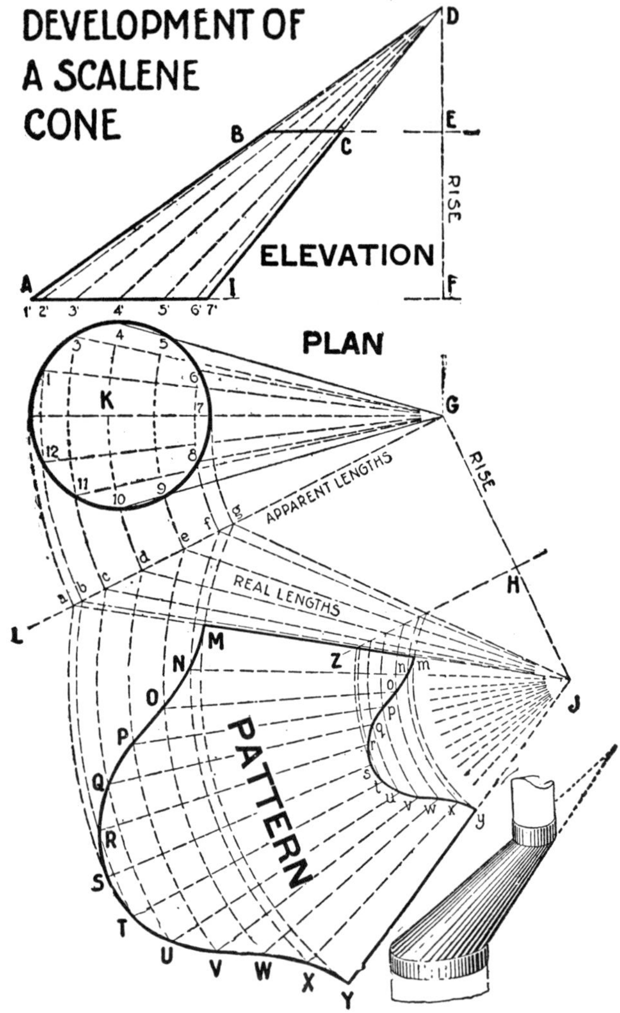

FIGS. 9,617 to 9,620.—***Problem 40.*** Scalene cone and development of its patterns

Accordingly, lay off the right angle LGJ, for the required right angle triangle. Upon the line LG, set off the distance Gf, equal to the apparent length 8G (in plan), as one leg of the triangle. The same line appears in elevation as 6′D; it rises in a vertical plane, from the level of the line 6′F, to the vertex D, that is, its vertical rise is equal to FD. This rise set off from G to J, to form the second leg of the triangle. The line joining the points *t* and J (the hypotenuse of the triangle *t*GJ) is the real length of the line 6′D (or 8G).

PATTERNS FOR A FLARING MEASURE

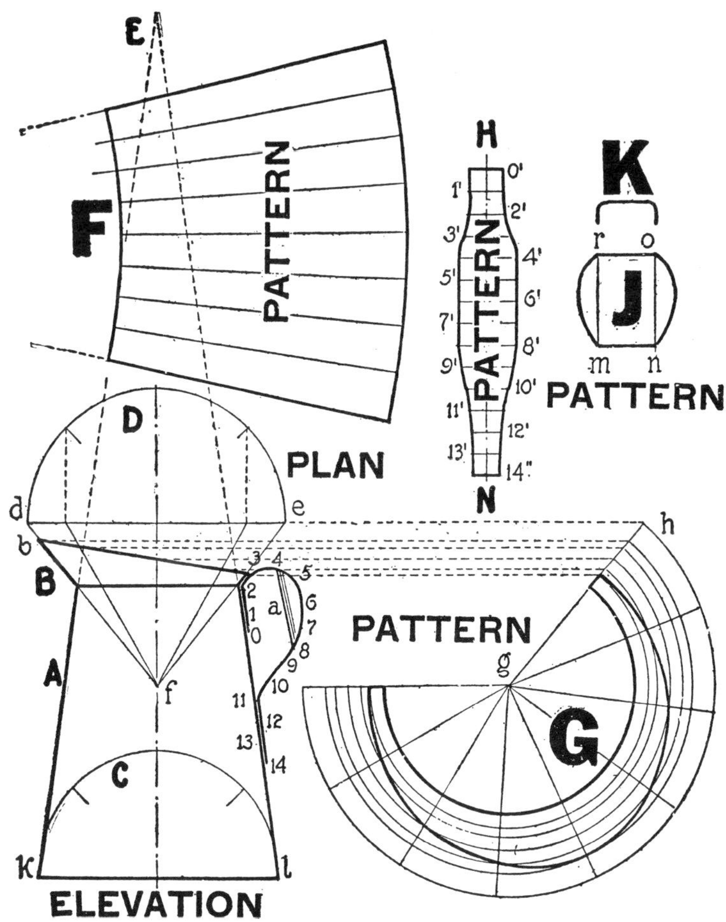

FIGS. 9,621 to 9,627.—*Problem 41.* Flaring measure and development of its patterns.

In a similar manner other real lengths as *e*J, *d*J, *c*J, etc., are obtained.

Describe from J, as center, arcs through *g,f,e,* etc.

To lay out the required pattern, describe, from J, as center, the arcs *g*M, *f*NX, *e*OW, etc. Then, from point M, as center, with a radius equal to *one of the divisions* on the base circle, step from arc through *g*, to the next arc, at the point N. Then, from the point N, with the radius equal to division of the base circle, step to the next arc at the point O, and in a similar way obtain the other points which define the curve.

The drawing plainly shows how the second half of this curve is plotted on the above arcs so as to form an exact counterpart to the first half of the development all the points gives the curve MSY.

The joining development of the top BC, is the curve *msy*, obtained in a similar manner. These two curves and the radial lines M*m* and Y*y*, complete the pattern, with exception of proper laps which must be added for the joints.

Problem 41.—Patterns for a flaring measure. Figs. 9,621 to 9,627.

To save space some of the drawings overlap, however, this need not cause confusion, and the student should become accustomed to reading drawings which overlap.

The patterns for the flaring measure are laid out according to the rules explained for the development of cones and truncated cones.

The cone of which the body of the measure forms a part is shown in the elevation as *k*E*l*, while its truncated part, the body of the measure, *k*B2*l* is developed at **F** (only a half of the development is here presented). The outer arc of this development has a radius equal to *k*E, the inner arc, a radius equal to *k*B. The length of the outer arc is made equal to the circumference of the base circle C, of the cone.

The lip B*b*32, is a part of an inverted cone *fde*, the development of which is given at **G.** In the development the outer arc is drawn with a radius *gh*, equal to *fd*, the length of this arc being equal to the circumference of the circle D, the base circle for the second cone. The equal divisions upon this circle projected to the edge of the lip *b*3, give a series of points which are in turn, projected to the line *gh*, there giving starting points for the arcs upon the development of the lip, the curve for which is obtained by means of intersections of these arcs with the radial lines drawn to the points upon the outer arc where the equal divisions taken from the circle D, were consecutively set off.

To plot the pattern for the handle, divide its profile into a number of equal parts, 1,2,3, etc. Then, set off these divisions one next to the other a line HN, which is to serve as a center line for the pattern of the handle. Through the divisions upon this center line draw, at right angles to it, a number of perpendiculars upon which lay off the widths of the handle at the different parts of it, always one half of the width to each side of the center line.

Thus the outline 0′,14,13,1 will furnish a symmetrical shape.

The boss, added to the handle at *a*, is made of a piece shown at **J** and and at **K.** In this piece, the rectangle *mron*, is made as wide as the handle at the place where the boss is attached. The height of this rectangle, *mr* is equal to the shortest distance between the points 8 and 4, on the profile of the handle. The irregularly shaped parts to each side of the rectangle *mron*, are exact reproductions of the shape enclosed by the line 9*a*4, and the curve 45678, in the elevation of the measure.

At **K**, is shown the manner in which this boss is to be bent. This is an easily obtained pattern for the boss. Where it is required to give the boss a wholly rounded bend, its pattern will take the shape of an oval whose length and width may be made equal to the above boss.

Measures are made in many different shapes and proportions, by different manufacturers. However, while there is no national standard for flaring measures, it might be helpful to keep in mind the proportions recommended by the U. S. federal government. According to these, the bottom and top diameters should be designed in a definite ratio to the perpendicular height of the measure, namely, the bottom base should be made two thirds of the vertical height (in diameter) while the top base should be made two thirds of the bottom base, in diameter. As to the vertical height, it should be 9.8 in. for a gallon measure; 7.78 in. for a half gallon measure; 6.17 in. for a quart measure; 4.90 in. for a pint measure; 3.89 ins. for a half pint measure and 3.09 in. for a one gill measure.

Development by Triangulation

There is a third class of elementary surfaces which contains neither parallel nor radial elements. If is found, however, that on such surfaces a series of two or more elements may be drawn in certain directions forming angles.

On such irregular surfaces it may happen that no two of the angles thus drawn on the solid, or represented, either correctly or foreshortened, in the projection draiwng, will lie in the same plane or be equal to each other. Since it is possible thus to project these angles, evidently they may be reproduced on the flat surface of the drawing paper in their correct size. If this can be done, it may be reasonably assumed that the surfaces thus represented will be the same as the corresponding surface of the solid surfaces belonging to the class and developed by the method of triangulation, and to the student who thoroughly understands the principles of projection, present no serious obstacles.

This process of triangulation depends on two operations.

1, Finding the true length of all lines, real or assumed, appearing on the surfaces of the solid, and 2, constructing triangles similar in form and relation to those shown on the solid.

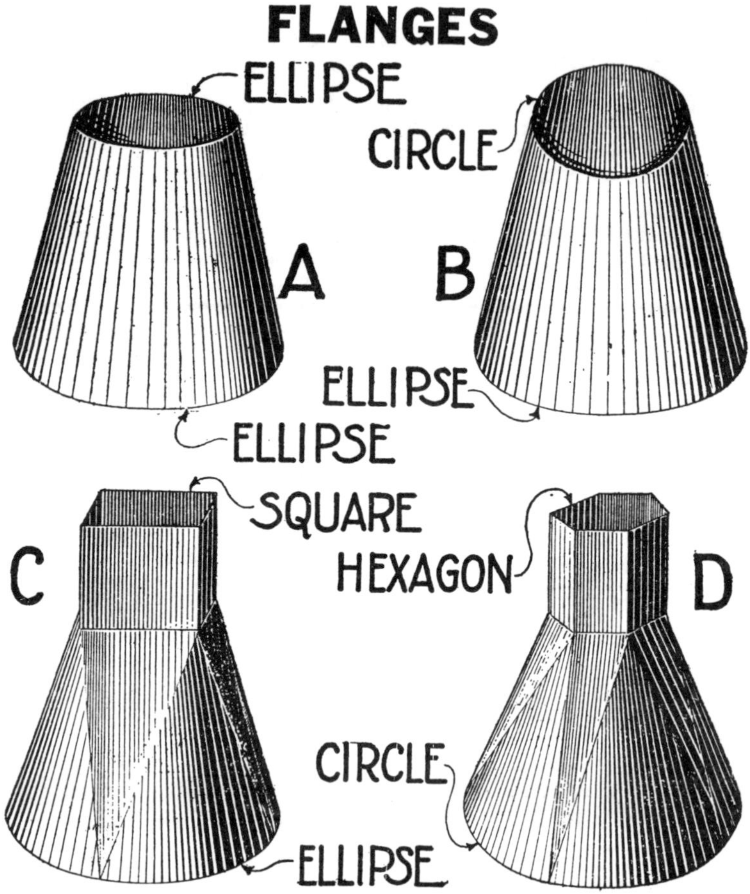

FIGS. 9,628 to 9,631.—*Problems 42 to 45, for practice.*

Problem 46.—Pattern for automobile engine hood. Figs. 9,632 to 9,645.

The *hood* or cover is made up of two rounded top parts and two flat parts. In the side view, IKLH, shows one of the flat parts, foreshortened. The real length of IK, is equal to FA (in the top view), the line IH, appears in true length as *a*O, and the line KL, as 1N, in the front view.

The pattern for the flat part is drawn by making the line RS, equal to AF, then erecting the lines RU (equal to *a*O), and ST (equal to 1N), perpendicular to RS, and finally joining the points U and T, to complete this pattern.

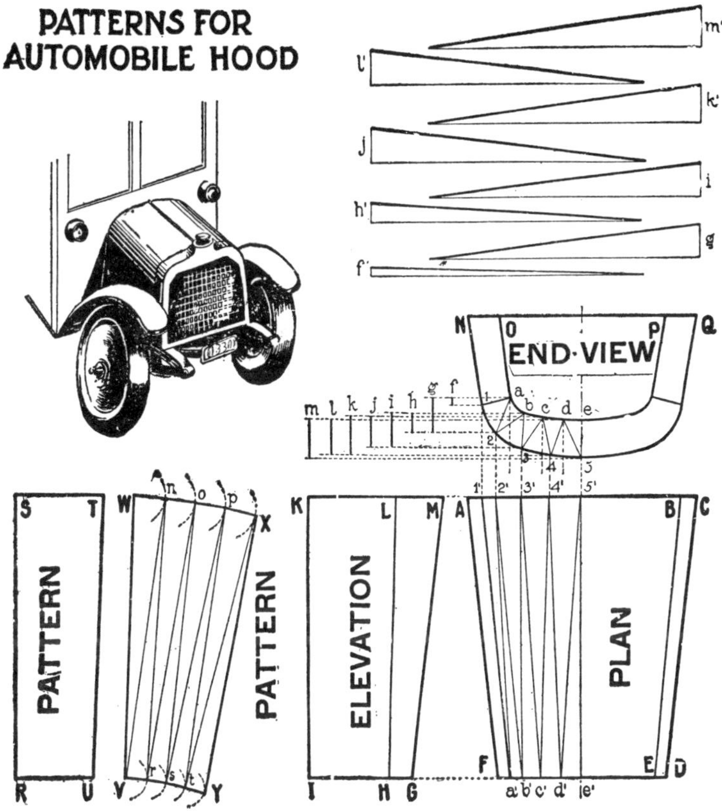

FIGS. 9,632 to 9,645.—***Problem 46.*** Automobile engine hood and development of its patterns

The rounded part designated in the front view of *abcde* 54321, and in the top view as *a′b′c′d′e′* 5′4′3′2′1′, is divided (applying the method of triangulation into a suitable number of triangles which, when put together in their true shape and proper order, will furnish the desired pattern. The smaller curved outline of the cover is divided into a number of equal parts, by the points *a,b,c,d,e*. The larger curve too is divided into the *same* number of parts, by the points 1,2,3,4,5. The divisions *a,b,c,d,e* are projected upon the edge FD, in the points *a′b′c′d′e′*, while the divisions 1,2,3,4,5 are projected to the edge AC, by the points 1′2′3′4′5′. Connecting these points, seven triangles are obtained upon the one half of the rounded part of the cover. These triangles appear in the front view greatly foreshortened, bounded by the lines 1*a*,*a*2,2*b*, etc.

For the purpose of obtaining the true lengths of the sides of all the triangles, determine the vertical rise of each. The end view shows that for the line 1*a*, the rise is *f*, for the line *a*2, the rise is *g*, etc.

A SQUARE BASE FOR A PIPE

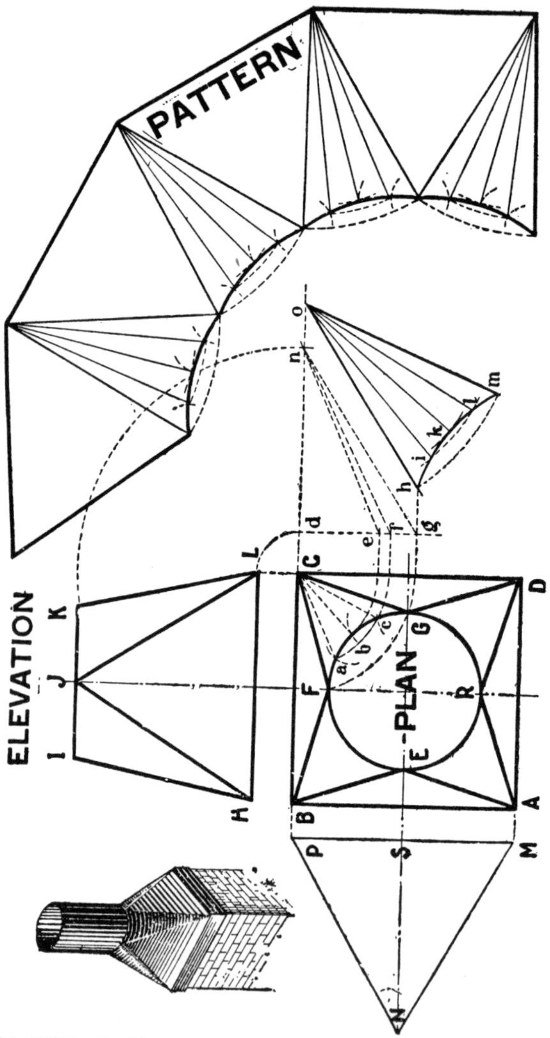

FIGS. 9,646 to 9,651.—***Problem 47.*** Square base of a pipe or transition piece between square and round pipes and development of its patterns.

Finally, the rise of *e*5, is equal to its length since the line is vertical. The true length of the line 1′*a*′, is equal to the hypotenuse of a triangle, one of whose legs is the apparent length of the line 1′*a*′, and the other leg, the rise of this line as appearing in the other view. The rise for this line is *f*. The true length of the line appearing as *a*'2', again, is the hypotenuse of a triangle one leg of which is the apparent length *a*'2' and the other leg, the rise of the same line as it appears in the other view, this time equal to *g*. Similarly, for all the remaining lines. The illustration shows the rectangular triangles built to obtain the true lengths of the eight lines, 1'*a*', *a*'2', 2'*b*', etc. The true length of the line 5'*e*' is equal to GM.

Now proceed to put together all the triangles with their sides drawn in their true lengths as follows: Make the line VW, equal to the true length of the first line, 1'*a*' (equal to the hypotenuse of triangle 8) describe an arc from W, as center, with a radius equal to 12, (measured on the larger curved profile in the front view), and intersect this arc, in the point *n*, by another arc, described from V, as center, with a radius equal to the length of the next line *a*'2', thus completing the first triangle. Then, from V, as center, describe an arc with the length of *ab* (measured on the smaller curved outline, in the front view) and intersect this arc at the point *r*, by means of another arc drawn from *n*, as center, with a radius equal to the third line 2'*b*', thus completing the second triangle.

The remaining triangles necessary to complete the pattern VWXY, are obtained in a similar manner. The pattern thus obtained forms one half of the rounded part of the hood.

Problem 47.—Pattern for square base of a pipe, or transition piece between square and round pipes. Figs. 9,646 to 9,651.

The square base is made up of four plane triangles whose adjoining base lines AB,BC,CD, and AD (in plan), form the square ABCD, at the bottom, and of the four irregularly shaped triangular corner pieces which occupy the spaces between the sides of the four plane triangles.

The four plane triangles appear in the plan as AEB,BFC,CGD and ARD. These triangles appear foreshortened because they stand oblique. HI (or LK), shows the actual height of these triangles. Hence, to obtain the actual shape of one of these triangles, make MP, equal to AB, and SN, equal to HI. The triangle MNP, then, gives the pattern for one of the four equal plane triangles.

The four irregularly shaped corner A pieces, EAR, EBF, FCG, and GDR, can be developed by triangulation. The pattern for one for one of them is shown as *ohiklm*. This pattern was obtained by triangulating the corner piece FCG, as follows: The arc FG, was divided into four equal parts at points *a*, *b*, and *c*, by means of which divisions the corner piece FCG, was subdivided into the four triangles FC*a*, *a*C*b*,*b*C*c* and *c*CG. To lay out these triangles in their real sizes, the real lengths of their sides were constructed, as *gn*, which is the real length of GC (also of FC); *fn*, which is the real length of *c*C, or of *a*C and, *n*, which is the real length of *b*C. The pattern of the corner piece *hom*, has the edge *oh*, equal to *gn* and the arcs *ohm*, with a radius equal to *oh*, the arc *il*, with a radius equal to *fn*, and the arc *k*, with a radius equal to *en*. Set off the length of the line *oh*, and then with a radius equal to F*a*, step from the point *h*, to the next arc, at the point *i*; then, from point *i*, with the same radius, step to the next arc at the point

AN OFFSET CONNECTING AN OBLONG PIPE WITH A ROUND PIPE

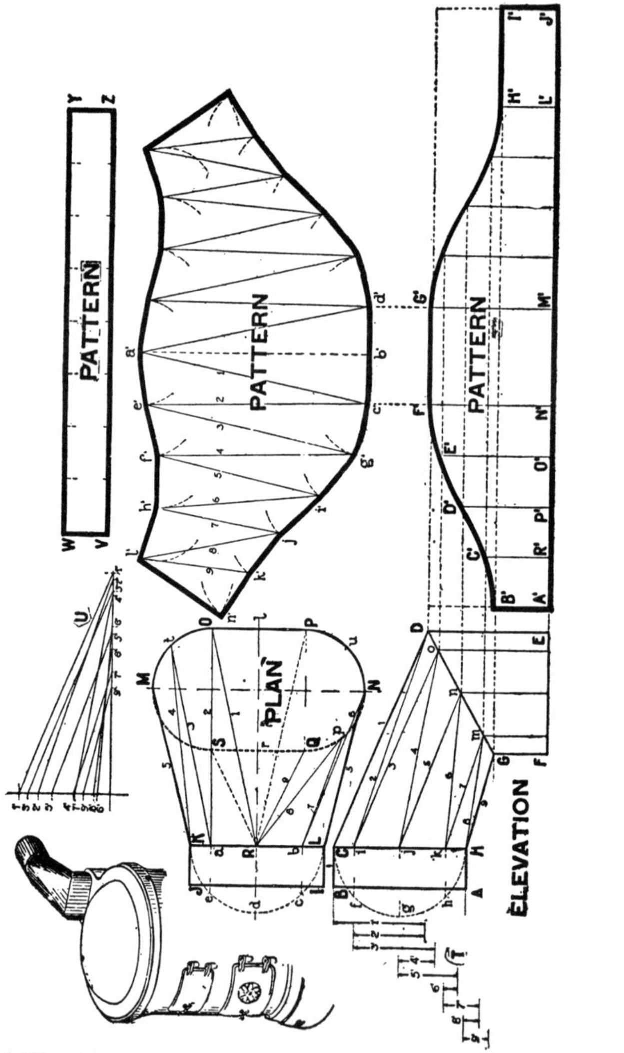

FIGS. 9,652 to 9,658.—*Problem 48* **Offset connecting an oblong pipe with a round pipe and development of its patterns.**

k; then, from point *k*, with the same radius, step to point *l* and thence, finally, to point *m*.

The illustration shows how the component four patterns of the plane triangles, and the patterns of the four corner pieces, are joined together so as to produce the entire pattern of the required square base for the round pipe. As a rule, however, the pattern is too large to be constructed of one piece of sheet metal. Hence it is usual in practice to plot the pattern for only a quarter or one-half of the entire pattern.

Problem 48.—Patterns for an offset connecting an oblong pipe with a round pipe. Figs. 9,652 to 9.658.

The offset consists of three quite different parts. First, the oblong part in the plan by the letters NQSMOP; secondly, the oblique part, indicated by NLKMOP, thirdly, the cylindrical collar IJKL.

The oblong part is made up of two half cylinders QNP and SMO, connected by the two flat sheets QS and PO. The first of these flat sheets QS, appears in the elevation as the line FG, the second, as ED. Between the lines FG and ED, there is the half cylindrical part FGDE. Its pattern is developed exactly as in the case of all cylindrical elbows, by taking the edge GD, as miter edge.

The circumference of one of these semi-cylinders (in plan) is then divided into a number of equal parts, as at Q,*p*,N,*u*,P. From these dividing points, projection lines are drawn upon the elevation of the semi-cylinder EGDE, cutting the miter line GD, in the points G,*m*,*n*,*o*,D.

Then, to develop the pattern of the oblong part of the offset, draw the line A′J′, as a continuation of the line FE, and upon it set off the circumference of one of the semi-circular bases of this part, from A′, to N′, consisting of all the equal divisions in the semi-circle QNP; next, upon the same line, lay off N′M′, equal to the length PO, of the flat part; next, once more all the parts of the semi-circle, from M′ to L′; finally, the distance L′J′, equal to N′M′. Then erect perpendiculars to the stretchout line A′J′, from all its division points, R′,P′,O′, etc. These perpendiculars, intersected by projection lines that start from the several points on the miter line, from the points G,*m*,*n*,*o*,D, define the pattern A′B′D′F′G′I′J′, for the oblong part of the offset.

The oblique part, at one end, conforms to the oblong part, while at the other end it becomes circular so as to be joined to the cylindrical collar.

The irregular shape of this oblique part is developed by triangulation.

Not all of the surface of the oblique part of the offset is rounded. Namely, the portions of it that adjoin the flat sections of the oblong pipe being by necessity flattened, there are two corresponding flat portions upon the surface: above, there is the large plane triangle PRO (see plan) and below, the smaller plane triangle QRS. Hence only the remaining two sections of the surface are rounded and subject to development by triangulation. These two portions are opposite each other and identical. One of these is bounded by the semi-circles QNP, of the oblong and the corresponding half of the circle of the collar which, seen edgewise, appears in plan as the line R*b*L. The other reversed, but identical in size and shape, is bounded by the semi-circle SMO, of the oblong and the corresponding half of the circle of the collar that is represented by the line R*a*K.

For purposes of triangulation, divide each half of the circumference of the collar into as many equal parts as there are in the semi-circles QNP and SMO. In the plan, the divisions of the circumference of the collar are marked by the points L,*c*,*d*,*e*,K, which divisions are projected upon the edge LK, at the points L,*b*,R,*a*,K. The same divisions in the elevation are indicated by the points H,*k*,*j*,*i*,C.

Now, proceed to join the divisions on the semi-circles of the oblong pipe with successive points of divisions upon the edge of the collar, by straight lines which will furnish sides for the triangles with which the entire rounded portion of the oblique pipe will be covered. Obviously, on the outer side of the oblique pipe there will be longer triangles while on the underside of it will be shorter triangles.

The drawing shows only one half of the upper side and only one half of the underside subdivided into triangles; this to avoid unnecessary crowding and consequent confusion.

The triangles on the outer side are: RO*a*,*a*O*t*,*ta*K and K*t*M. On the underside the triangles are: NL*b*,*b*N*p*,*pb*R and R*p*Q. The long sides of the triangles are numbered in the illustration 1,2,3, etc. The same lines, upon the elevation are marked by the same numbers as in the plan. Of course, all these lines are given upon these views only in apparent, foreshortened lengths. Before using them for the spreading out of the triangles upon the desired patterns, the lines have to be determined in their real lengths. For this purpose, the rise of each line in one view, and its apparent length in the other view, are taken to supply the two legs of a right angle triangle whose hypotenuse will be equal to the real length of the line. The rise for each line of the elevation is given, in the drawing at T, where 1′, stands for the rise of line 1; 2′, for the rise of line 2, etc. At U, all the rise lengths are laid off as the upright legs of nine triangles, while the apparent lengths of the lines 1,2,3, etc., are laid off as horizontal legs. Thus, the hypotenuse 11′, represents the real length of the line 1; the hypotenuse 2 2′,represents the real length of the line 2, etc.

With the real lengths thus obtained, triangles are constructed that cover the oblique pipe, in their actual sizes,one next to another, to which will have to be added the two plane triangles PRO and QRS, that complete the surface of the oblique pipe.

The altitude of the larger of the plane triangles is reproduced at *b′a′*, upon the pattern. It is taken from the elevation where the edge CD, is equal to the true length of the altitude R1 (in plan).

The triangle *c′a′d′*, upon the pattern, represents the actual size of the triangle PRO (in plan), *c′d′*, being equal to the real length of PO, *d′a′* being equal to the real length of RO, each equal to line 1.

Now, with a radius equal to B′C′ (upon the pattern of the oblong pipe), and *a′*, as center, describe an arc which is intersected at *e′*, by another arc, drawn from *c′*, as center with a radius equal to the true length of the line 2. Now from *c′*,ascenter, with a radius equal to one of the equal divisions upon the circumference of the collar, say *cd*, describe an arc which intersects at the point *g′*, by means of another arc, drawn from *e′*, as center, with a radius

PATTERNS FOR TAPERING SQUARE ELBOW

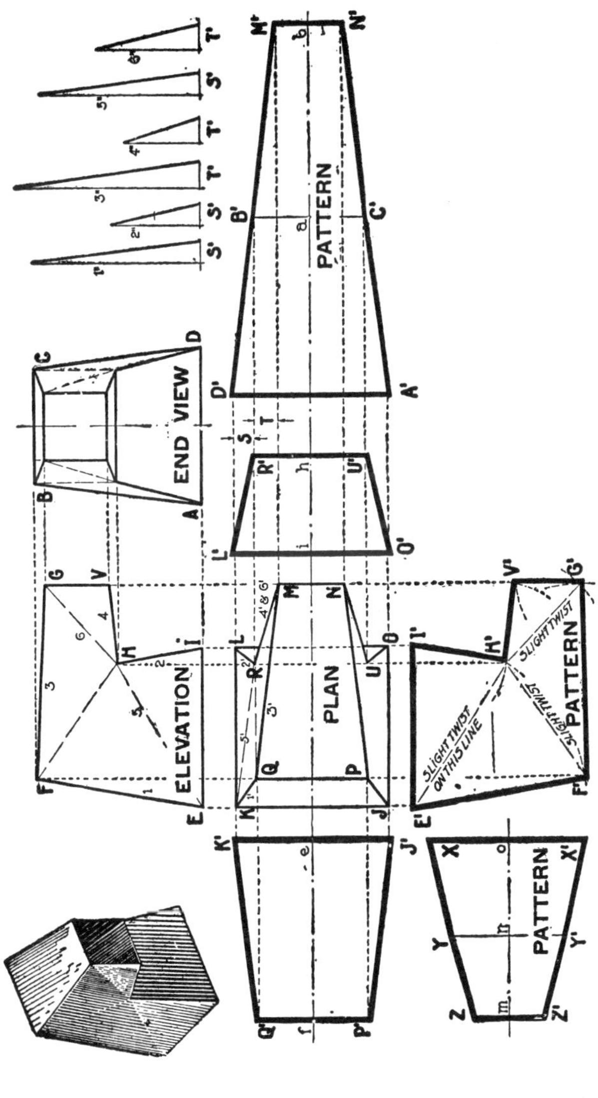

FIGS. 9,659 to 9,673.—*Problem 49.* Tapering square elbow and development of its patterns.

equal to the line 3. In a like manner lay out all the other triangles of the surface of the oblique pipe. In the drawing the long sides of the different triangles are numbered the same as in the plan and elevation.

The distances *g'c'*,*g'i'*,*i'j'*, and *j'k'*, are each equal to one of the divisions of the circumference of the collar; on the opposite side of the pattern, the divisions *a'e'*,*e'f'*,*f'h'* and *h'l'*, are equal in the same order, to B'C', C'D', D'E' and E'F'. The triangle *n'l'k'*, is equal to the actual size of one half of the triangle RQS; the point *n'*, is obtained by the intersection of two arcs, one, with a radius *k'n'*, equal to Q*r*, the other with a radius equal to the altitude of the triangle QRS, which is given in the elevation, in its true length, by the line HG. The other half of the pattern is a counterpart of the first and is constructed exactly like it.

The pattern of the collar is a straight band VWYZ, whose width VW, is equal to AH, and whose length is equal to the circumference of the collar.

The method here employed for the development of the parts of the offset may be applied to a great many similar problems, for patterns of irregular elbows, transition pieces and, in general, for the plotting of patterns of irregularly rounded bodies or parts of bodies. Of course, the patterns must, in all cases, be enlarged by the addition of stock for flaps and locks, where such are required.

Problem 49.—Patterns for tapering square elbow. Figs. 9,659 to 9,673.

For shop purposes it will suffice to draw the side elevation and plan. Here, the front elevation is added to give a better idea, to the beginner, of the peculiar shape of the elbow.

The elevation shows, mainly, the large face EFGVHI, which is divided into triangles by the lines EH,FH and HG along which the metal has to be slightly twisted so as to enable the fitting of this one piece face to the adjoining sides and top. The pattern for this face is best constructed by means of triangulation, the method used for the development of the scalene cone, radiator cover, etc., in previous problems.

For purposes of triangulation, it is necessary to determine the true length of the lines in the triangles that are to be used for making up the required patterns. For this, take the vertical rise of each line, in one of the views, as one leg, and the apparent length of each line, as it appears in the other view, for the other leg; the hypotenuse gives the true length of the line.

For convenience, the lines, whose true length must be found, are numbered, in the side view, as 1,2,3,4,5,6 and in the plan, as 1',2',3',4',5',6'.

The rise of line 3', is the same as that of the line RM, which latter represents two lines 4' and 6', which coincide in the plan. The rise for each one of these lines is equal to the distance marked T. The rise of 1', is the same as that of line 5', and also the same as the rise of line 2'. Each of these three lines rises by the amount indicated as S. These rises are used as the shorter legs for six right angled triangles, the other legs of which are made

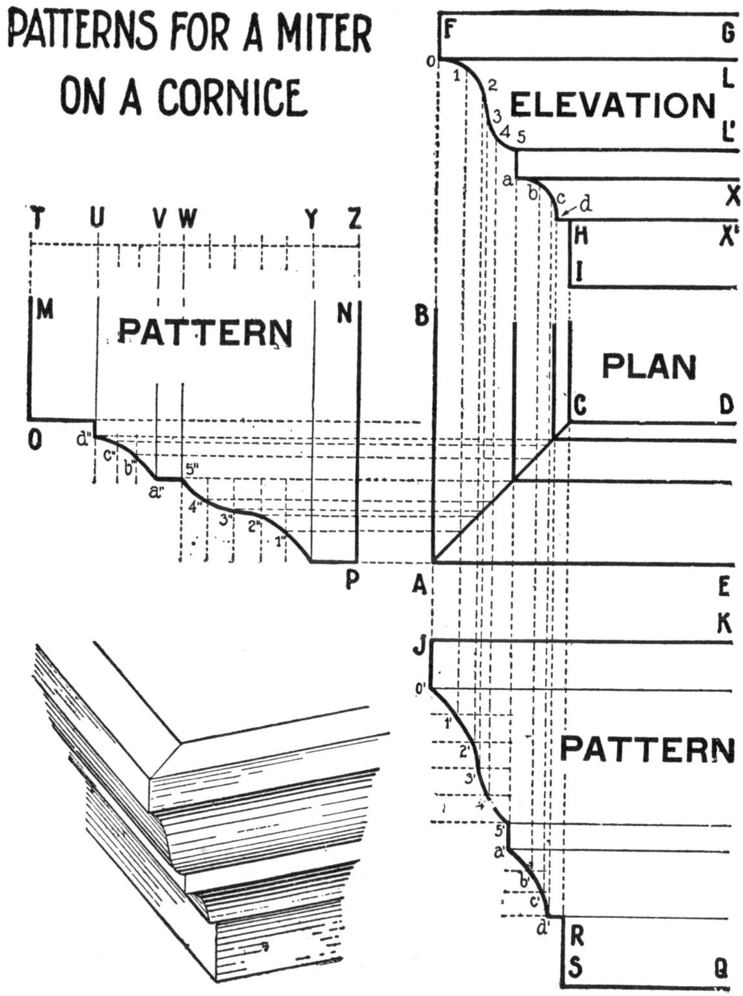

FIGS. 9,674 to 9,678.—*Problem 50.* Miter on a cornice and development of its patterns.

equal to the apparent lengths of the corresponding lines upon the side view, namely, the legs 1″,2″,3″,4″,5″,6″.

The hypotenuses of these triangles give the true lengths of the six lines that are to be used for triangulation. Now proceed to lay out the pattern F′E′I′H′V′G′.

Make E′I′, parallel and equal to JO, from plan. Draw two vertical projection lines from P and U, downward, across the desired pattern. Then, with a radius equal to the true length of line 1, from E′, as center, draw an arc intersecting the projecting line PF′, in the point F′. Then, with a radius equal to the true length of line 2, from I′, as center, draw an arc intersecting the projection line UH′, at H′. To test the result, describe an arc from E′, as center, with a radius equal to the true length of line 5, which arc should pass through the point H′.

Next, draw an arc from F′, as center, with a radius equal to the true length of line 3, which arc is intersected at the point G′, by another arc whose radius is equal to line 6, and whose center is at H′.

Now, from the newly obtained point G′, as center, with a radius equal to the length VG (of the side elevation) describe an arc which is intersected at the point V′, by an arc drawn from H′, as center, with a radius equal to line 4. This completes the pattern of the part F′E′I′H′V′G′. The lines E′H′,F′H′ and H′G′, upon this pattern indicate the direction of the slight twists required.

The face JKQP, is represented in its true shape as J′K′Q′P′. The projection lines drawn across this latter figure, from the plan, show how J′K′, is made equal to JK, and P′Q′, is made equal to PQ, the perpendicular distance between the lines J′K′ and P′Q′, marked as *ef*, is taken from the side elevation, equal to EF. This figure was laid out separately so as to make it easier to understand its derivation. It really should form a part of a larger pattern that includes the top PQMN. This larger pattern is shown at A′D′M′N′, wherein the wider end is a reproduction of the figure J′K′Q′P′, while the narrower end C′B′M′N′, is the true shape of the top face PQMN.

Projection lines drawn from the elevation show that M′N′, is equal to MN. The perpendicular distance between the lines B′C′ and M′N′, marked as *ab*, is taken from the elevation view, equal to line FG. Furthermore, the face represented in the elevation, edgewise, by the single line HI, and, in plan, as face OURL, is shown in its true shape at O′U′R′L′, wherein the line O′L′, equals OL, and line U′R′, equals UR, while the perpendicular distance, *ih*, between the lines O′L′ and U′R′, is equal to line 2. This figure too was drawn separately merely to better explain its derivation. Again, it is made a part of a larger pattern which is shown at Z′ZXX′, wherein the wider portion Y′YXX′, is a repetition of O′L′R′U′, and the narrower portion Z′ZYY′, is the true shape of the face that is represented edgewise, in the elevation, by the single line HV, and in plan, by the hidden portion URMN. Z′Z, is made equal to MN, and Y′Y, is equal to U′R′, or UR. The perpendicular distance *mn*, is equal to line 4.

It is obvious that the patterns here shown separately may be combined by the mechanic, into larger sections, to be cut out of one piece of stock, whenever that may prove desirable, as, especially, in elbows of small size. If cut separately, for large sized elbows, the different parts will have to be provided with additional strips of liberal size, for laps or locks.

Miscellaneous Problems.

Problem 50.—Pattern for a miter on a cornice. Figs. 9,674 to 9,678.

The curved parts in the elevation of the cornice IHF, are divided into equal parts by the points 1,2,3,4,5, and *a,b,c,d,* from which projecting lines are drawn downward, toward *o*′,1′,2′,3′,4′,5′,′*a,b*′,*c*′,*d*′ and also to right, toward 1″,2″,3″,4″,5″,*a*″,*b*″,*c*″,*d*″. YZ, is made equal to F*o*′,YW, to the distance between the lines *o*L and 5L′. Then the distance WY, is divided into as many parts as there are equal parts in the curve *o*5, and from these divisions upon the line WY, vertical projection lines are drawn to meet the previously drawn horizontal projection lines in the points 1″,2″,3″,4″,5″, which define the corresponding outline of the pattern.

The distance WV, is made equal to that of 5*a*′, the distance VU, to the distance between the lines *a*X and *d*X′, and is divided into as many equal parts as are in the curve *ad*.

PATTERNS FOR COMMON AND RAKED BRACKETS

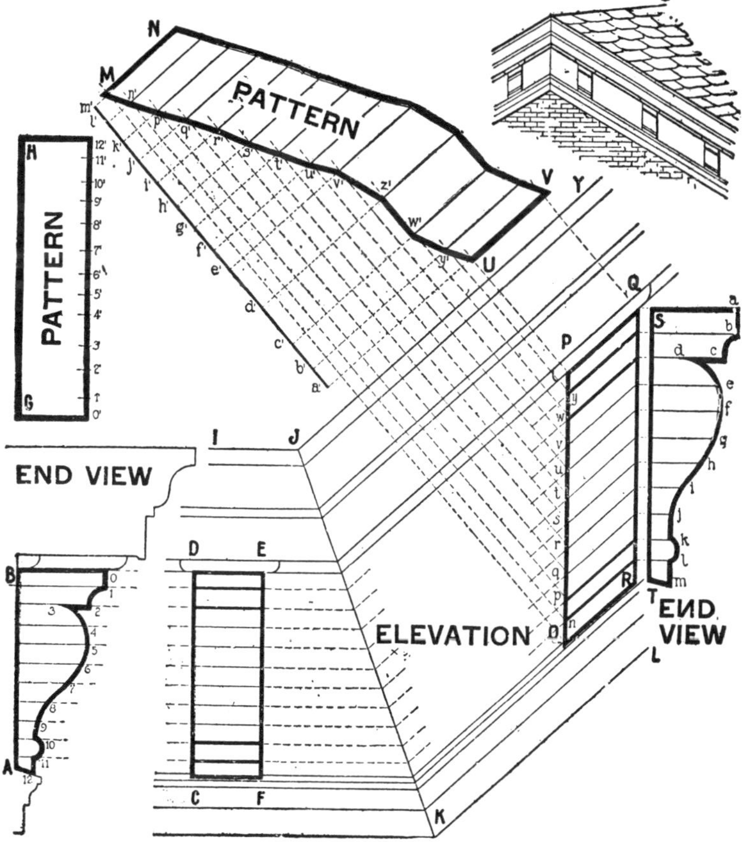

FIGS. 9,679 to 9,684.—***Problem 51.*** Common and raked brackets on a cornice and development of its patterns.

Vertical projection lines from the dividing points upon VU, at their intersections with the previously drawn horizontal projection lines, give the points for the corresponding part of the pattern. The line TU, is made equal to HI. The other pattern QSRJK, is an exact counterpart of the first pattern.

Problem 51.—Patterns for common and raked brackets on a cornice. Figs. 9,679 to 9,684.

The common bracket is frequently used as ornament on a horizontal cornice. When such a bracket ornament is to be used on a gable cornice, it is called a raked bracket or gable bracket. The patterns for each bracket consist of two flat sides exactly shaped like the profile and of the development of the front part of the bracket that is to be bent so as to truly join to the flat sides.

In the drawing AB*o*12, represents the side view of the common bracket and CDEF, represents the front view.

The pattern for the flat sides of the bracket must be cut identically to the size and shape of the end view. To reproduce the end view, for the pattern, the curved profile is, first divided into a number of parts, preferably but not necessarily equal, in the points 1,2,3, etc., and then its surface divided by parallel lines drawn through the several points of division on the end view.

For the drawing of the pattern of the flat side, draw a straight line of the length AB, upon which set off the divisions made on it by the parallels.

Then draw these parallels and, in proper order, make these equal to the lengths of the parallels on the end view. The end points of these produced parallels are then joined by straight lines and curves, to conform with the shape in the end view. The pattern for the front of the common bracket, shown at GH12′,O′, is a rectangular strip whose width is equal to the width CF (in the elevation), and whose length is equal to the stretchout of the profile 0,1,2. . . . . .12, that is, to the combined length of all the divisions from *o*, to 12, in the end view. The numbers 0′,1,2,3′, etc., of the pattern GH12′0′, correspond to 0,1,2,3, in the end view.

In the illustration, JK, is the miter line between the horizontal and the raking parts of the cornice. This miter line should be made so as to bisect the angle IJY. Only then the identically shaped horizontal and gable parts of the cornice can be made to meet each other exactly.

The raked bracket is made as wide as the common bracket in a horizontal direction, both in its elevation and end views. However, in the direction of the gable line, the raked bracket will show a greater length than that of its horizontal width. Thus the distances PQ, and its counterpart OR, are greater than the shortest distance between the lines OP and RQ, which shortest distance, as stated, is equal to CF. The height of the raked bracket is longer than that of the common bracket since it is the oblique distance between two parallels in the cornice while the common bracket occupies the perpendicular, shortest distance between them. As to the side view of the raked bracket, while it is longer than the side view of the common bracket, since it is equal to the vertical height of the raked bracket, horizontally, in every part, it is exactly equal to the corresponding part in the common bracket. Thus S*a*, and all the parallels in the end view of the raked bracket TS*am*, are equal to the corresponding parallels upon the end view AB*o*,12.

The pattern for the flat side of the raked bracket is omitted in the drawings because it is an exact reproduction of the side view and is obtained in the same way as for the pattern for the flat side of the common bracket.

As to the pattern for the front part of the raked bracket, it has an irregularly oblique shape, shown at UVNM, whose width UV, is equal to PQ, and whose surface is made up lengthwise, of obliquely terminating parallel strips that are the variously shaped sections upon the front of the raked bracket produced by the parallels. The profile of the raked bracket is divided into a number of parts by the points *a*,*b*,*c*,. . . .*m*, which are derived from the parallels of the common bracket produced to their intersections

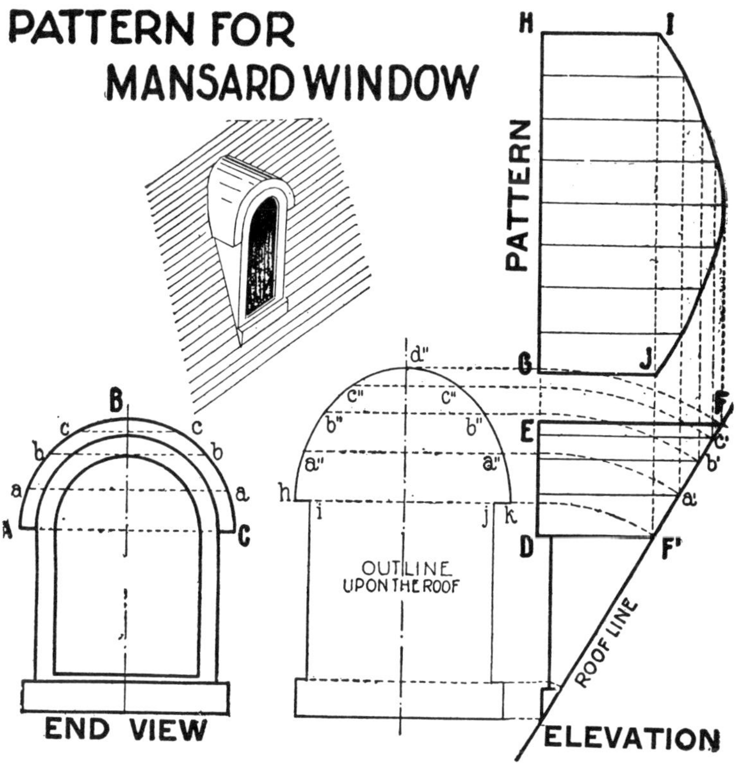

FIGS. 9.685 to 9.689. —*Problem 52.* Mansard window and development of its patterns.

with the miter line KJ, and from there, projected upon the front view of the raked bracket, by parallels to the roof line; thence to the edge ST, of its side view whence the projection lines run horizontally, parallel to *Sa*, to the points *a,b,c,*. . . .*m*.

To plot the pattern UVNM, draw the stretchout line *a'm'*, perpendicular to the gable line JY, and lay off upon it all the distances into which the profile is divided by the parallels. Thus the combined parts of the stretchout line, *a'b',b'c',c'd'*, etc., will be equal to the total length of the combined parts *ab,bc,cd*, etc., of the profile in the end view. From the division points on the stretchout line, draw the lines *a'*U',*b'y',c'w'*, etc., parallel to the gable line JY. Then, from the several points of intersection of the parallels with the bracket edge PO, draw a series of projection lines at right angles to the gable line, up to the parallels drawn from the divisions of the stretched line, thus obtaining the points U,*y',w'*, etc., as one of the longitudinal outlines of the pattern which is then completed by drawing the opposite longitudinal outline, VN, every part of which being drawn parallel to corresponding, parallel to the first part of UM.

A sufficient amount of stock should be added to the patterns of the flat side pieces for required laps. Along the curved parts of the outline, the laps will have to be notched to facilitate bending.

The smaller the radius of a curve, the closer the notches should be placed, along the outline of the pattern.

Problem 52.—Pattern for mansard window. Figs. 9,685 to 9,689.

The circumference of the hood ABC, is divided into a number of equal parts, by points *a,b,c*, etc., which are projected to the points F',*a',b'*, etc., upon the roof line. The pattern GHIJ, is obtained like the development of a cylinder. The line GH, is equal to the total length of all the divisions in the semi-circumference ABC. If it be required to plot the outline upon the roof where the window booth is joined to the roof, it may be obtained by arcs drawn from the points F',*a',b'*, etc., which locate the distance between the parallels *hk,a''a'',b''b''*, etc. The distance *hk*, is equal to AC, and *a''a''*, to *aa*, etc.

The other parts of this outline are too obviously derived to call for an explanation.

Problem 53.—Pattern for dormer. Figs. 9,690 to 9,697.

The hood over the dormer is developed after the manner used for the pattern of the slope sheet of a boiler. Here, KLM, in the elevation is divided into a number of equal parts in the points 1,2,3, etc., which are projected to the edge ON, of the end view, obtaining points 1'2'3', etc. Thence, the lines 1'8,2'7, etc., are drawn parallel to the line NP. Now the true profile of the rounded hood of the dormer has to be laid out. For this, on the re-

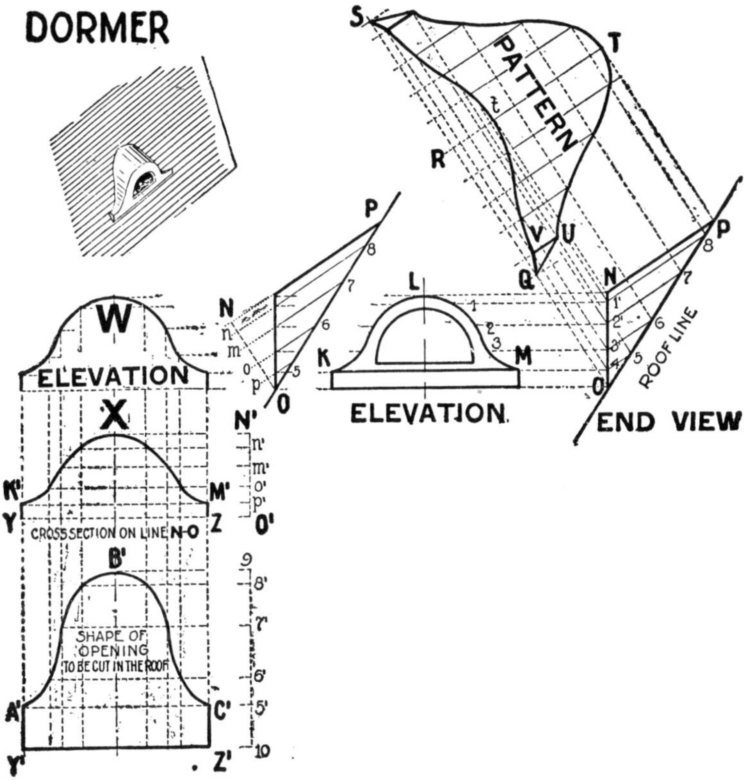

FIGS. 9,690 to 9,697.—*Problem 53*. Dormer and development of its patterns.

produced end view, to the left of the elevation, the line NP, and all the parallels to it are extended outward, just as if the hood itself were enlarged forward. The extended hood is then cut by line NO, at right angles to NP, and intersecting the oblique parallels in the points *m,n,o,p*. At W, is shown an exact reproduction of the outline of the elevation. Beneath this, at X, is the cross section on the line NO.

Its defining points, as is plain from the drawing, are derived from the intersection of the vertical projection lines (from the dividing points upon the circular edge of the elevation) with the horizontal projection lines coming from the points N'*n*'M'*o*'*p*'O', which are exactly equal to the corresponding distances on the cutting line NO.

Now, to plot the pattern for the hood of the dormer, a stretchout OQS, is drawn at a right angle to NP, which line is crossed at *t*, by a line RT, perpendicular to it (hence parallel to NP). To either side of the line

RT, which is intended as a center line for the pattern, set off upon SQ, the divisions which the defining points made upon the profile of the cross section YK'M'Z, in proper order, so that the central division X, should fall upon R, and the extreme divisions Y and Z to fall upon S and Q respectively.

From the divisions thus obtained on the stretchout line, perpendiculars are erected across the desired pattern.

These perpendiculars being cut by the projecting lines which start from the points P,8,7, etc. (in the end view), give the defining points of the pattern, which is then completed by the additions of the triangle QVU, and the one at S, to represent the little triangle 045, of the side view.

The shape of the opening to be cut in the roof for the dormer is plotted by means of vertical projection lines from the cross section and the horizontal lines which are drawn from the points 9,8',7', etc., which are equal to the corresponding lengths P8,87,76, etc., upon the roof line.

CHAPTER 4

Development of Air Conditioning Ducts

Rectangular Offset Mitered on Flat or Long Side.—A manner of laying out the patterns which makes it possible to cut out the entire offset from one piece of metal. Since this offset is mitered on the long side of the rectangle, obviously the short side will appear in elevation.

Draw the plan as shown and designate the corners as *A,B,C,D*, or by numbers if desired. Extend the line *CB*, and erect the line *EFGH*, in elevation, representing the angles of offset elbows required. Bisect the angle *EFG*, and obtain the miter line *FK*, as follows: With *F*, as center and any convenient radius, draw arcs, cutting the angles at *d* and *e*. Now, with a slightly greater radius and *d*, and *e*, as centers, draw arcs intersecting each other at *f*. A line drawn from *f*, through *F*, indefinitely, will be the required miter line. Bisect the angle *FGH*, in like manner, as indicated by *a*, *b*, and *c*, thus obtaining miter line *GJ*. With the miter lines established, complete the elevation indicated by *EFGHIJKL*, as shown.

The two angles of this offset are alike since both upper and lower arms run parallel; hence the pattern for one angle will suffice for all.

For the patterns, lay off, consecutively, on the line *EL*, extended as *EM*, the four sides of the plan *ABCD*, beginning and ending at the seam *X*, as shown by *XABCDX*, from which points erect vertical lines indefinitely as shown and interersect them by lines drawn parallel to *L'M*, from intersections on miter line *FK*. Trace a line from *K'* to *N*, reproducing miter line *FK*, from *A* to *B* and from *C* to *D*, as shown.

Then *K′L′MN*, will be the pattern for the lower arm or section. Take length of *FG*, or *JK*, and set it off on pattern lines from *K′* to *J′* and from *N*, to *O*. Draw line from *J′* to *O*, reproducing miter cut shown by *K′N*. Now take the length of *IJ*, in elevation and set it off on pattern lines from *J′* to *I′* and from *O* to *P*, and draw a line from *I′* to *P*; thus completing the pattern shapes for the three sections of the offset shown in elevation, with the seams occurring in center of long sides as indicated by the arrows.

A Twisted Rectangular Duct.—This is a fitting having two pipes identical in shape placed centrally one above the other. The letters *a,b,c,d* and *e,f,g,h* in plan, represent the profiles of the upper and lower pipes respectively.

It is obvious that since the projection of all four sides is the same, it is necessary to construct but one elevation as shown by *A,B,C,D*; making its required height as *XY*, and thus determining the true length of the sides *AB* and *DC*.

To lay out the one pattern which serves for all four sides: Through the center of the profile *a,b,c,d*, in plan, draw a horizontal line to the right indefinitely. Take the length of slant height *AB*, in elevation and set it off on this line as shown by *A′B′*. Through these points draw the measuring lines *EF* and *GH*, at right angles to *A′B′*, and intersect them by lines drawn parallel to *A′B′* from the profiles in plan, as shown; add the collars indicated by "*n*", and complete pattern *EFGH*. Allow sufficient lap for seaming on all four patterns.

A Square to Square Taper.—It is readily apparent from the perspective view that this fitting is a simple reducing joint, and that it has the shape of a four sided pyramid. The principles of the method used to develop the patterns for such figures, is fully explained in chapter No. 3 under the heading Development by Radial Lines.

Lay out the plan by drawing the profile of the large pipe represented by *ABCD* in plan, draw in the diagonal lines intersecting at *O*, and put in the profile of the small pipe as shown by *EFGH*. The vertical height of the joint is indicated by the distance *a,b*, in the elevation.

To develop the pattern proceed as follows: From the points *O* and *D*, in plan and at right angles to the line *OD*, draw lines indefinitely and intersect them by the line *O′D′* drawn parallel to *OD* in plan.

Take the distance of the vertical height *a,b*, in elevation and place it on the line from *O′* to *b′*. From *b′* draw a line at right

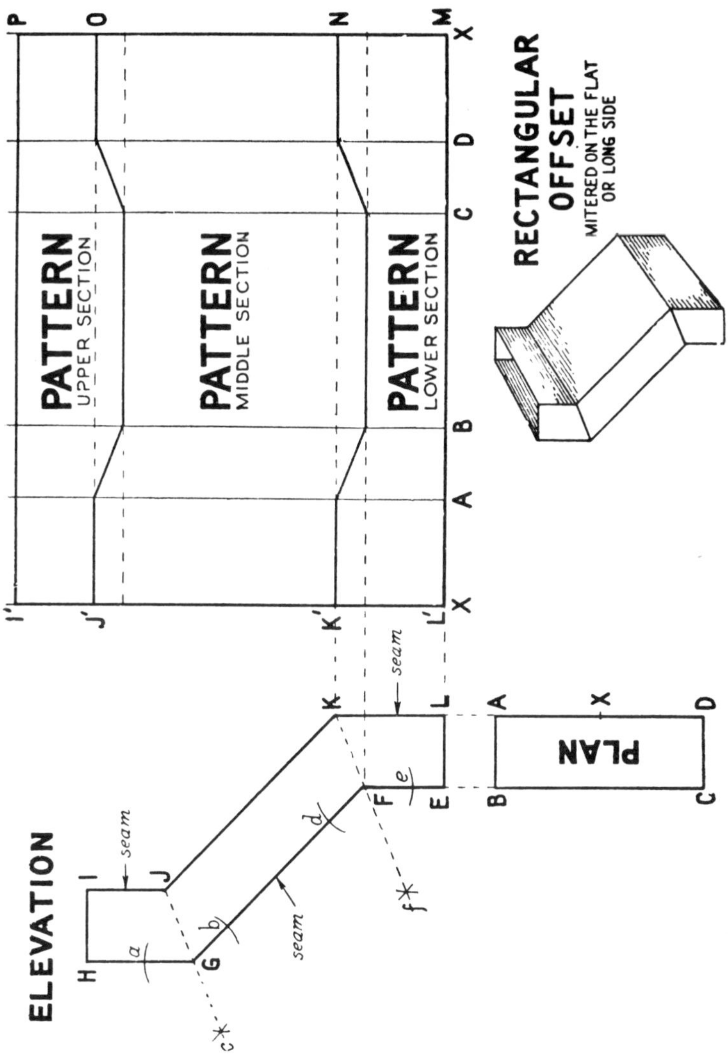

angles to O-O', and intersect it by line drawn from H, parallel to O-O', and establish the point H'. Now from D' draw a line through H' to the line O-O' at P. With P' as center and PH' and PD' as radii, strike the arcs EG and AC, respectively. Now with the dividers set at AB, and BC, in plan, point off these two distances on outer arc as indicated by similar letters.

A TWISTED RECTANGULAR DUCT

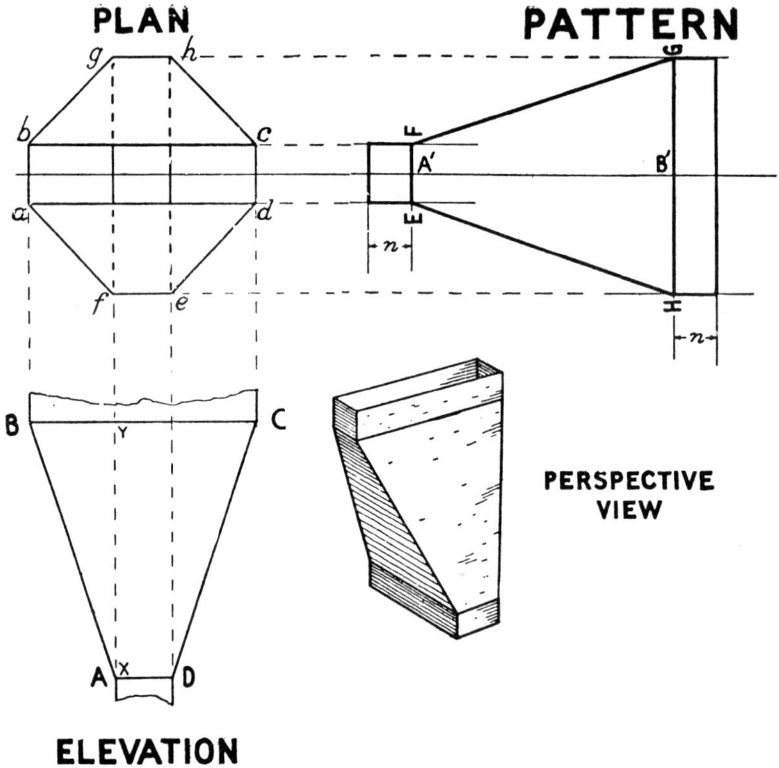

From the points A, B, and C, on outer arc draw lines to point P', cutting the inner arc at E, F, and G. Trace a line through ABC $GFEA$ and complete the pattern for one half the taper, E,F,G, in pattern equals E,F,G, in plan. Laps for seams are not shown in these patterns.

Curved Offset Having Rectangular Profile.—Laying out an elbow to a given dimension. Begin by laying out the space according to dimension in which the offset is to be drawn, as represented by $WXYZ$; and place the plan a,b,c,d in position as indicated.

If, however, it be desired to draw the offset with the long

A SQUARE TO SQUARE TAPER

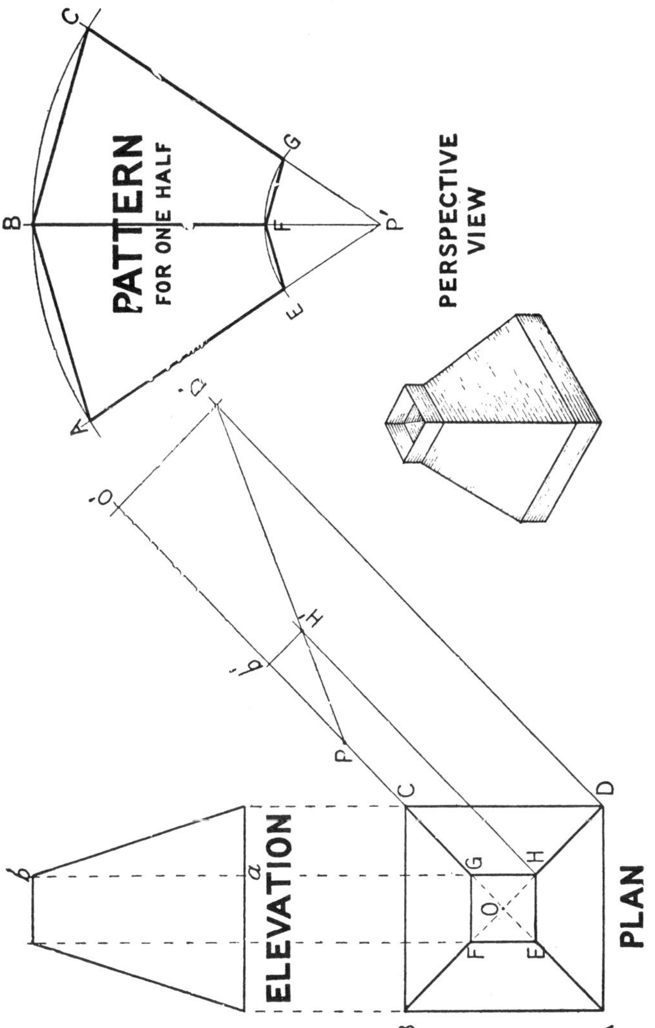

side of the profile in elevation, place the plan so that its length *a*,*d*, will be in a vertical position parallel with *WX*, and proceed as follows: Take the width *c,d*, of plan and set it off on the line *WX*, as indicated by *WA*. Take the distance *a,d*, the widest part of the profile, and set it off from *A* as *AB*; with *B* as center and *BW*, as radius, strike the arc *WC*, indefinitely.

On the line *YZ*, measure off the distances *WA* and *AB*, as indicated by *YE* and *ED*. Using *D*, as center and *DE*, as radius, draw arc *EF*, indefinitely. Connect the arcs *WC*, and *EF*, by drawing a line tangent to them as shown by *a'b'*; and from the points of tangency draw radial lines *c'* to *B*, and from *D*, through *d'* indefinitely. Then with *B*, as center and *BA*, as radius describe an arc intersecting the radial line at *f*; in like manner with *D*, as center and *DY*,as radius,draw an arc passing through the radial line at *e*. Connect *e* and *f*, and the form *W,E,Y,A* is the pattern for the offset.

When the dimensions are such that the offset cannot be made in one piece, the elbows are made separately and connected by a straight piece of pipe joined at *c'f* and *d'e*.

Compound Curved Reducing Elbow.—An elbow of this sort is not infrequently found where high class work prevails. This one makes a quarter turn in both front and side elevations. In the elevations as presented, the two profiles are plainly shown by their heavy outlines; the small profile *1-2-2-1* in the side elevation is shown in the front elevation by the line *1-2*; and the line *9-X* in side elevation, shows the profile *9-X-X-9* in front elevation.

Draw the elevations as shown, according to given dimensions; the throat and heel curves in side elevation being drawn from centers *a* and *b*, respectively, and in the front elevation, from *d* and *c*, respectively. Bisect the profiles in both elevations by perpendicular lines as *A-B* and *C-D*. Point off an equal number of spaces on the back curve as indicated, and from these points draw horizontal lines to the right, cutting across both sides of front elevation as shown by corresponding numbers. Add the points *e*, on the left side, and *f,g*, on the right side to insure

CURVED OFFSET HAVING RECTANGULAR PROFILE

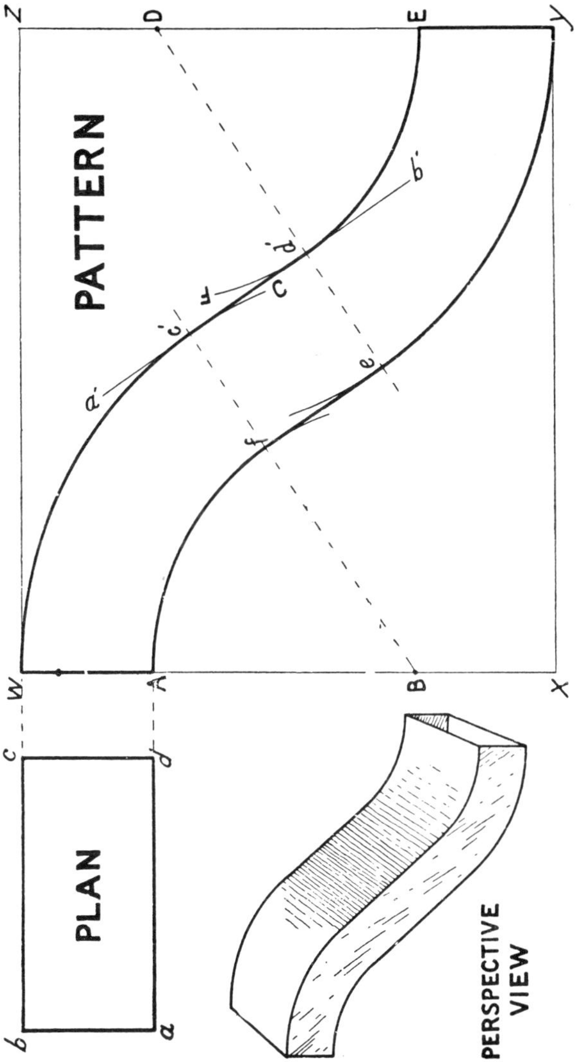

greater accuracy in taking the girth.

Beginning with the pattern for left side: With the dividers,

COMPOUND CURVED REDUCING ELBOW

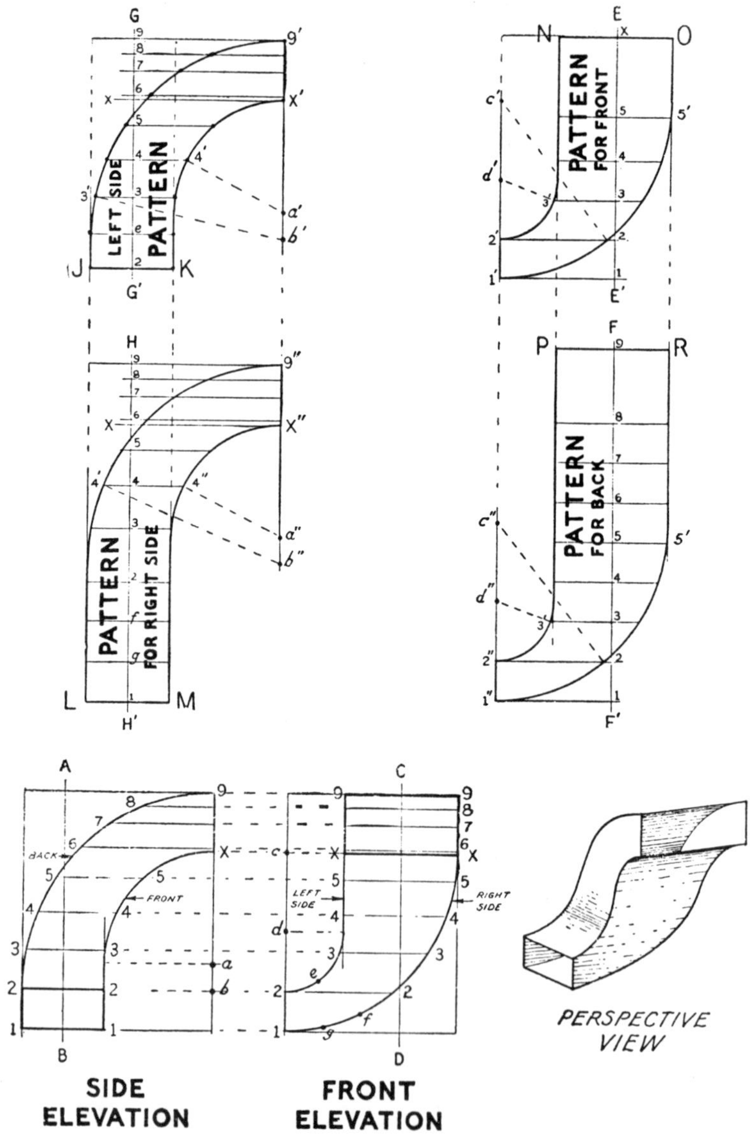

take the girth of the left side in front elevation and lay it off on a vertical line, as *G-G′* and through these points draw horizontal lines indefinitely. Then from the line *AB* in side elevation, take the distances to the numbered points at the right and left of line, and put them to the right and left of the line *G-G′* on like numbered lines as shown. Drop a perpendicular line from 9′ through *X′* indefinitely. Take the distance from *9* in side elevation to points *a* and *b*, and place them on the line from *9′* as centers *a′* and *b′*. Using *a′* as center, draw arc *X′-4′* and with *b′* as center, draw arc *9′-3′*; complete the outline and *J-9′-X′-K* is the pattern for left side.

For the pattern for the right side, extend the vertical line *G-G′* as *H-H′*, and place on it the girth of the right side in the front elevation, and draw the usual measuring lines through the points, at right angles to *H-H′*. Take the numbered points to the right and left of the line *AB*, in side elevation, and put them to the right and left of the line *H-H′* as in the pattern for left side. The arc for the throat is drawn from *a″* as center, and the arc for the heel from *b″* as center. The completed outline shows *L-9″-X″-M* as the pattern for the right side.

The patterns for the front and back are arrived at in a similar manner. For the front, place the distances *1-X* in side elevation, on the vertical line *E-E′* as shown, and draw horizontal measuring lines through them. From the line *C-D* in front elevation, measure the distances from the points *1* to *X*, on the right side, and *2-X* on the left side, placing them to the right and left of the line *E-E′* as indicated. To locate the centers *c′* and *d′*, measure the distance from *1* to *d* and *c*, in front elevation, and place them on a line drawn from *1′* through *2′* as shown. From *c′* as center, describe the arc *1′-5′*, and using *d′* as center, draw the arc *2′-3′*, complete the outline and *1′-2′-N-O* shows the pattern for the front.

For the back pattern, set out the girth on vertical line *F-F′* as indicated by the numbers, and draw the usual measuring lines. Locate the intersection points by taking the distances of

COMPOUND CURVED OFFSET

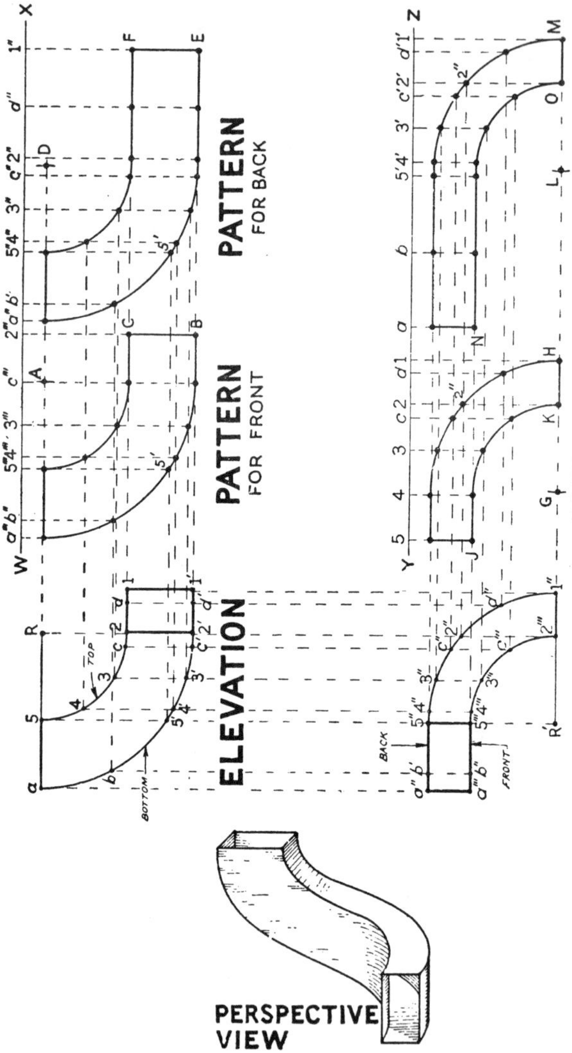

the projections each side of the line *C-D* in front elevation, and placing them to the right and left of the line *F-F′*. Fix center points *c″* and *d″* as in front pattern; describe the arcs *1″-5′* and *2″-3′* respectively, draw the straight parts and finish the outline; then *P-R-1″-2″* will be the pattern for the back. No allowance has been made for seaming.

Compound Curved Offset.— Here is an offset which curves in both the front and the side elevations, with the two profiles of the same dimensions, but in different positions. The upper one occupies a horizontal position and the lower one a vertical position.

To develop the patterns begin by drawing the elevation and plan in the proper relative position as follows: Draw the profile of the opening represented by the heavy lines *1-2-2′-1′* in elevation and with *R-2* as the given radius of throat, draw the quadrant *2-5*; using *R*, as center and *R-2′* as radius, draw the heel *2′-a*. Then *1-1′-a-5* shows the elevation.

Below the elevation and in direct line with it, construct the plan. Drop perpendicular lines from points *1,2,5* and *a*, in elevation, and on the line from point *5*, locate the point *R′*. With *R′* as center and *R′-2‴* as radius, describe the arc *2‴-5‴*, take the width of the narrow side of the rectangle and set it off on the perpendicular line *5-R′* from *5‴* to *5″*, and with *R′* as center and *R′-5″* as radius draw the arc *5″-1″* for the heel; complete the rectangle *5″-a″-a‴-5‴* as shown, thus completing the plan and elevation.

The curves in both views of this offset are the same, being quarter circles; sometimes they are elliptical or irregular in shape, but in any case the same principles of pattern development as here employed, would be applicable. As a means of quick identification, the four sides of this offset shown in elevation and plan are designated by their names in small letters.

Divide the curve of throat in elevation into any number of equal spaces, as shown by the figures *2,3,4* and *5*. (Use more divisions in actual work to facilitate measuring the girth), from which points drop perpendicular lines intersecting the heel in elevation, and the heel and throat in place at points indicated by similar figures. As the space in the heel in elevation between the points *a* and *5′* is too great, fix an extra point as at *b*, and from *b*, drop a perpendicular line intersecting the profile in plan at *b′* and *b″*.

As the space in the throat in plan between *2‴* and 3‴ is

too great, therefore an extra point is established as at c''', and from there a perpendicular line is erected to intersect the heel in plan at c'' and the heel and throat in elevation at c' and c respectively. For the same reason an extra point is fixed in the heel in plan between $1''$ and $2''$ as at d'' from which a perpendicular line is erected to intersect the rectangle in elevation at d' and d.

Having located the measuring points on all four curves in both views proceed to develop the patterns.

Parallel to R-a in elevation draw a line as W-X. Beginning with the pattern for the front a'''-$2'''$ in plan, take this girth and set it off on the line W-X as shown by similar numbers and letters. From these numbers and letters drop perpendicular lines indefinitely, and intersect them by lines drawn parallel to W-X from similar numbers and letters in the top and bottom curves in elevation.

Since the space $5'''$-a''' in plan is in a horizontal plane, it is apparent that the pattern for that space for both front and back, will be a replica of the segment 5-$5'$-a in elevation. Therefore this sequence must be reproduced in both patterns.

For the front pattern, take the radius of R-a in elevation and set it off in pattern for the front from a''' to A. With the same radius and A, as center describe the arc a'''-$5'$. Continue the line through points of intersection to B, and draw line from $5'''$ through intersecting points to C, as shown; join points B and C and $5'''$ and a''' thus completing pattern.

For the pattern for the back, take the girth of the back $1''$-a'' in plan, and lay it out on the line W-X as shown by corresponding numbers and letters. From these points drop perpendicular lines cutting through horizontal lines drawn from corresponding letters and numbers in elevation.

Take the radius R-a in elevation and set it off from a'' to D, in pattern for the back and, with D as center, draw the curve a''-$5'$, thus again reproducing the segment 5-$5'$-a; continue the line through points of intersection to E. Trace a line from $5''$ through intersecting points to F, connect E and F, and a''-$5''$ thus defining the pattern. Draw the stretch outline Y-Z parallel to R'-$1''$ in plan, and proceed as before.

RECTANGULAR DOUBLE OFFSET

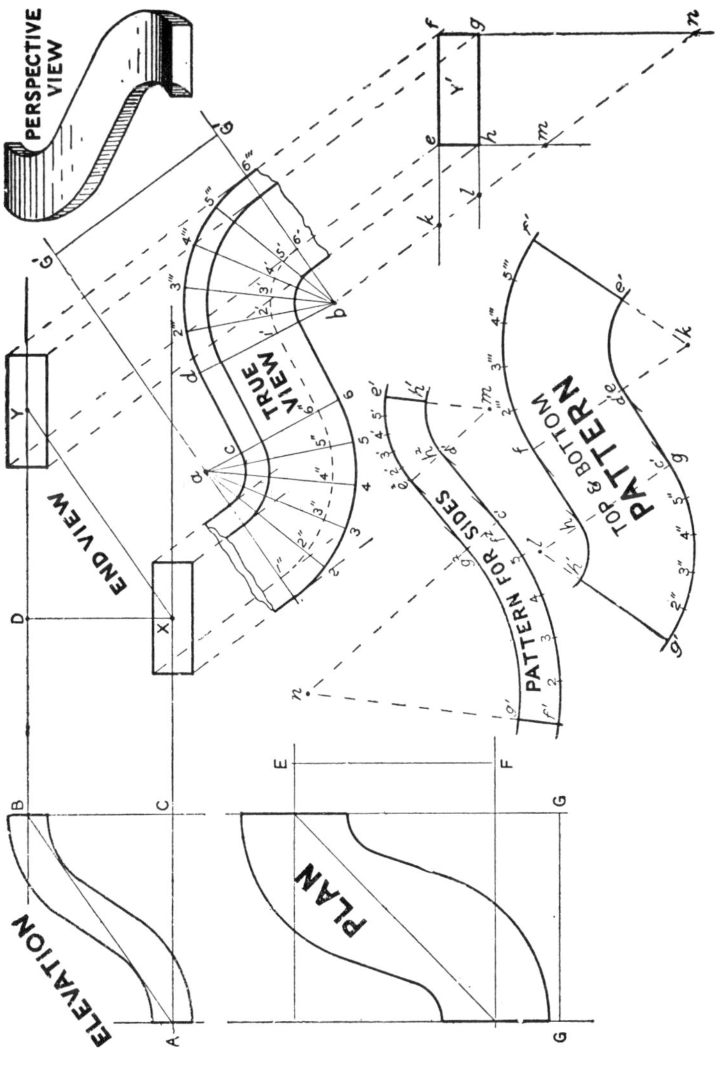

For the pattern for the top *1-5* in elevation, take this girth and lay it out on the line *Y-Z* as indicated by its numbers and letters. Draw horizontal and perpendicular lines obtaining the points of intersection as shown. The space *1-2* in elevation being in a horizontal plane, its pattern shape would be a replica of the segment *1″-2-″-2‴* in plan. Reproduce this segment in pattern thus; take the radius *R′-1″* in plan and with *H*, as center, establish the intersecting point *G*. With *G*, as center and *G-H* as radius, draw the arc *H-2″*. Continue the line from *2″* through points of intersection to *5* and *J-K-H* thus completing the pattern.

For the bottom pattern, take its girth *1′-a* in elevation and stretch it out on the line *Y-Z* as shown, and obtain the intersecting points of perpendicular and horizontal lines as previously explained. Reproduce the segment *1″-2″-2‴* in plan by taking the radius *R′-1″* in plan and with *M* in pattern as center locate the point *L*. With *L* as center and *L-M* as radius, describe the arc *M-2″*. Through the established points of intersection, continue the line from *2″* to *a*, and to *N-O-M* which completes the pattern.

In these patterns no allowance has been made for seaming; hence proper laps should be added. Be careful to roll the patterns so that the elbow will turn in the right direction.

Rectangular Double Offset.—The requirement of this offset is that it shall retain the same sectional area throughout. Therefore the development of the patterns is such as to achieve that result.

In addition to the plan and elevation there is shown an end view, from which it should be observed that center lines is the basis of the pattern development.

Draw in the plan and elevation as shown, the vertical height of elevation between centers being as indicated by *C-B*. Extend the center lines *AC* and *BY* in elevation to the right indefinitely, and draw in profile of pipe as shown at *X*. From *X* draw vertical to interset *B-Y*, and locate point *D*; Set off to the right from *D*, the distance *E-F*, in plan and establish point *Y*; draw profile with *Y* as center; and connect profiles by center line *XY*; thus completing the end view.

As a means of simplifying the method of developing the patterns, and for ascertaining the curves for heel and throat, a true view is erected in line with the end view as shown, according to the following procedure: From the angles of the profiles *X* and *Y*, draw lines at right angles to the center line *X-Y*, indefinitely; draw the lines *aG′* and *bG′* horizontal to the perpendicular from the profiles, and separated by a distance equal to *GG* in plan as indicated by *G′G′*, locate the points *a* and *b*, as centers for throat and heel curves. Complete the true view as shown.

Extend the lines from profile *Y*, sufficiently far below the true view and duplicate the profile as shown by *Y′*. Extend the top and bottom lines of *Y′* to the left and the side lines downward to intersect the line *bn*, drawn at right angles to *XY*, from *b*. From the point thus obtained, we get the length of radaii, for the patterns.

Beginning with the pattern for the sides: With *n*, as center, and *n,f*, and *n,g*, as radaii, draw arcs f'-f^2 and g'-g^2 in pattern, making g'-g^2 equal to the curve *1* to *6* in true view, as shown by similar figures. Add the straight parts *c′-d′*, which should equal *c-d* in true view; then with *m*, as center and *m,e*, and *m,h*, as radaii, strike the arcs as indicated by similar figures, making e^2-e' equal to *1′* to *6′* of true view as shown, thus completing pattern.

The pattern for the top and bottom is drawn in the same manner. From *l* in pattern as center, with *l-g* and *l-h* as radaii describe the arcs, making *g′-g* equal to *1″* to *6″* of true view as shown, add straight part *c′-d′* equal to *c-d* of true view. Complete the pattern by striking arcs *e-e′* and *f-f′*, from *k*, as center, making *f-f′* the same length as *d-6‴* in true view, as is shown by corresponding figures.

It should be remembered when putting this fitting together that though the patterns for both sides are alike, they should be reversed to each other. This also applies to patterns for top and bottom.

Offset Transition—Rectangular or Square Having One Side Flat.—This fitting is not an uncommon one. While the two

OFFSET TRANSITION

RECTANGULAR TO SQUARE HAVING ONE SIDE FLAT

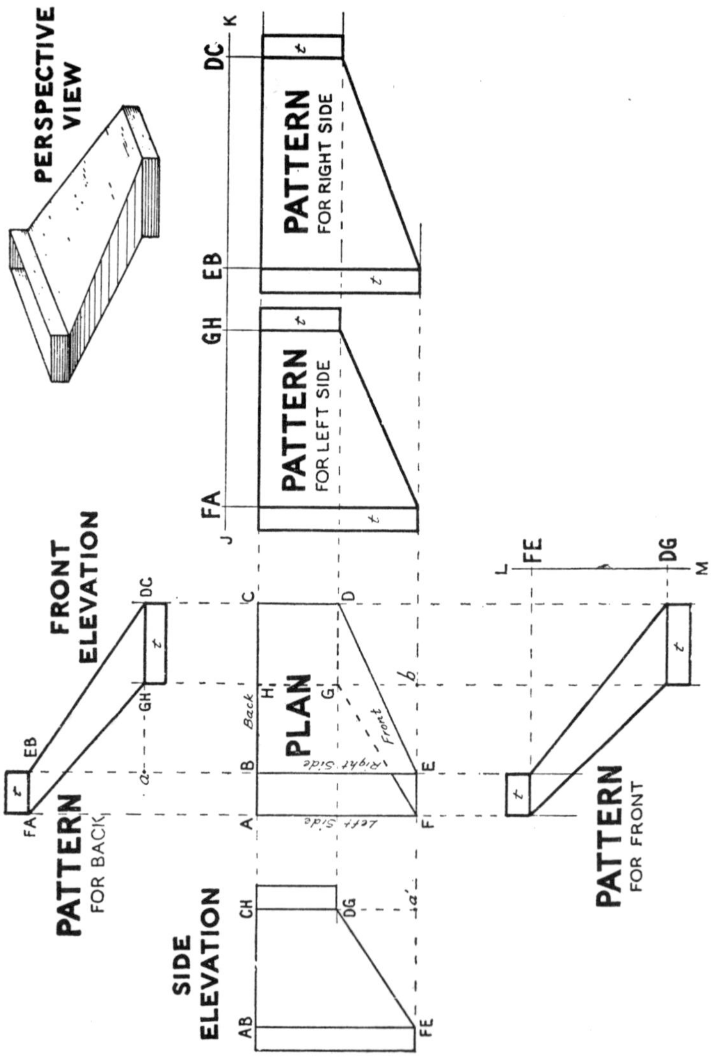

TWISTED OFFSET
FOR RECTANGULAR PIPE

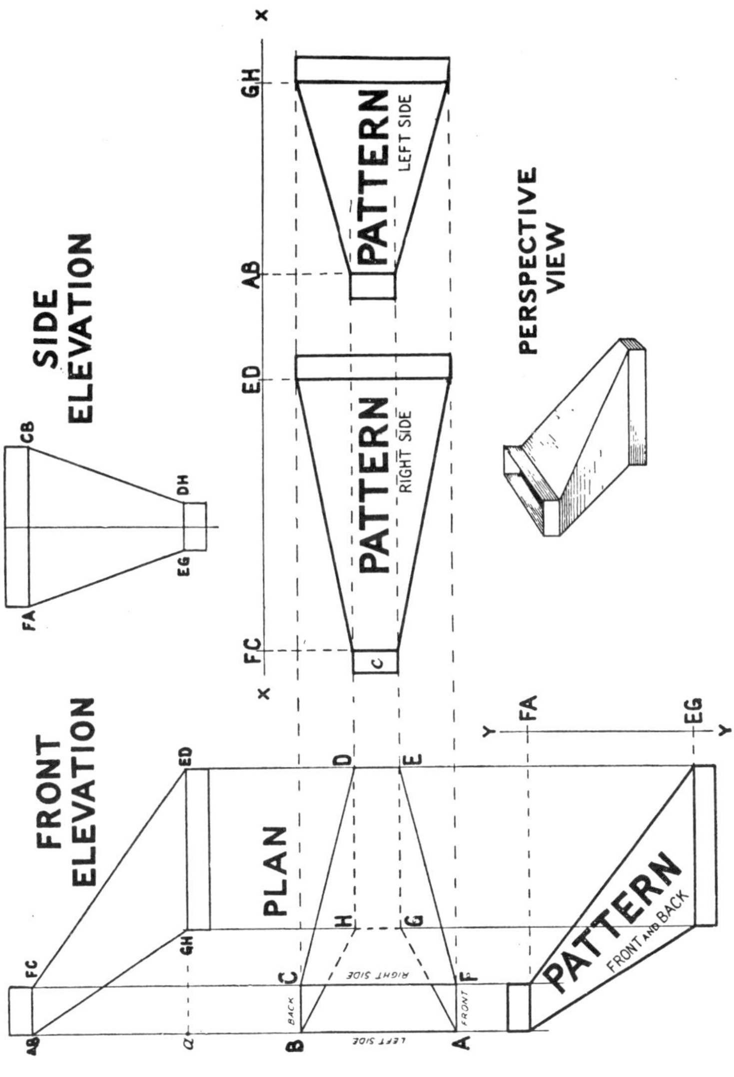

profiles are different in contour, they are equal in area. As the upper and lower profiles of this offset run flush in the back, therefore the back when shown in elevation appears as the front elevation and also becomes the pattern for the back.

Draw in the plan as shown with *ABEF* representing the profile of the rectangular pipe at the top, and *HCDG* the profile of the square pipe at the bottom which offsets a distance equal to *Eb*, in plan. Construct the front elevation with its required vertical height equal to *a-EB* in elevation, as shown, and *FA*, *EB*, *DC*, *GH*, will be the pattern for the back.

Erect the side elevation as shown, having its vertical height *a'*-*FE* equal to *a-EB*, in front elevation. Take the distance *FE-DG* in side elevation and stretch it out on a vertical line as *LM* below the plan; from these points draw the horizontal lines to intersect the perpendicular lines drawn from corresponding letters in plan. Draw a line through the points of intersection and these form the pattern for the front *FEDG* in plan.

Extend the horizontal lines *AC*, *GD*, and *FE* in plan to the right indefinitely; and above them draw another horizontal line as *JK*, on which set off the distances *FA-GH* and *EB-DC* in the front elevation as shown.

Draw the usual measuring lines vertically from these points to intersect the horizontal lines from similar letters in plan, connect the points thus obtained and complete the patterns for the two sides as shown. Add the collar "*i*" to top and bottom of all four patterns, also laps for seams.

Twisted Offset For Rectangular Pipe.—This fitting is a variation of Fig. page 124 having both profiles alike and centrally placed in relation to each other, but offset a distance equal to *a-GH* in front elevation.

Lay out the plan as shown, with *ABCF* representing the upper profile and *GHDE* the lower profile. Erect the front elevation from the plan as indicated, its required vertical height being the distance from *a* to *AB*, as shown in elevation. In line with the front elevation erect the side elevation giving it the same vertical height.

Then *FC-ED* and *AB-GH* in the front elevation show the act-

A TWO WAY OFFSET TWISTED DUCT

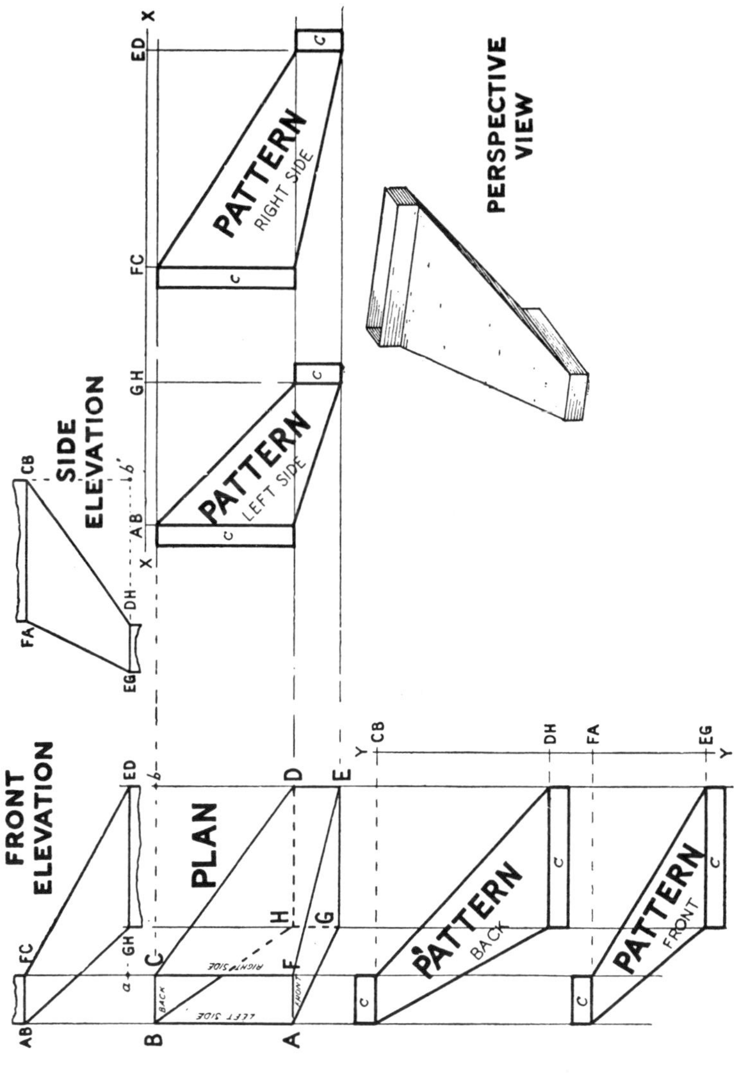

ual lengths of material required for the right and left sides shown in plan. Set these lengths off on the horizontal line *XX* as shown; and from these points drop perpendicular lines indefinitely, and intersect them by horizontal lines drawn from similarly lettered points in plan. Draw the outlines connecting the intersecting points and the result will be the patterns for the right and left sides.

Drop vertical lines from the points *A*,*F*,*G* and *E*, in plan, and the vertical line *YY*, at the right; on which place the distance *FA-EG* in side elevation, and draw lines from these points at right angles to *YY* cutting the vertical lines from the plan. Draw the outline connecting the points and complete the pattern; as both sides in the side elevation are alike, this will be the pattern for the front and back. Add to top and bottom of the patterns the collars "*c*" and sufficient laps for seams.

A Two-Way Offset Twisted Duct.—This fitting like the previous one, has two profiles identical in shape, making a quarter turn, but differs in that it is offset both ways.

The patterns for this fitting may be developed as follows: Assuming that *ABCF* in plan is the required profile of the upper pipe, and *GHDE* the profile of the lower pipe, and that *b* to *D*, and *b* to *C*, is the amount of offset both ways, draw in the plan as shown.

Erect vertical lines from profiles in plan as shown, and construct the front elevation according to the given height as represented by *a* to *FC*, in elevation, and in line with the front elevation, construct the side elevation as shown with the offset *b′* to *DH* equal to *b* to *D*, in plan.

Draw any horizontal line as *XX*, and set off on it the lengths *AB-GH* and *FC-ED* which are the actual lengths of the right and left sides. From these points drop perpendicular lines and intersect them by horizontal lines drawn from similarly lettered points in plan as shown; connect the points by the diagonal lines and complete the patterns for the right and left sides.

Now draw any perpendicular line as *YY*, and set off on it the actual lengths *CB-DH* and *FA-EG*. From these points

TRANSITION ELBOW

RECTANGLE TO RECTANGLE

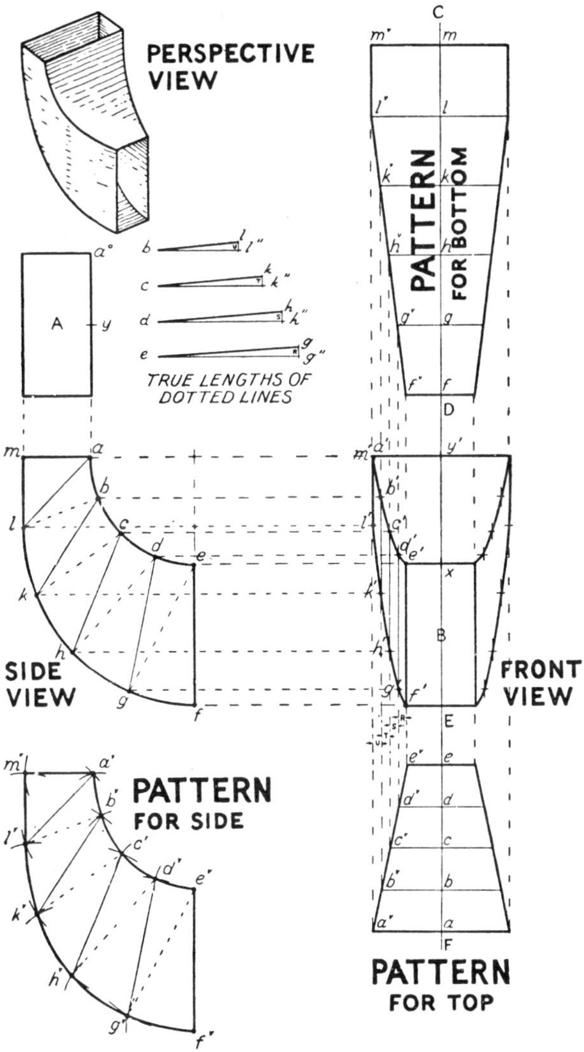

draw horizontal lines to the left intersecting the perpendicular lines drawn from similarly lettered points in plan; connect the points by the diagonal lines and complete the patterns for the back and front of the offset as shown. Add the collars indicated by "*c*", at top and bottom to all patterns. Allow laps for seaming as required.

Transitional Elbow—Rectangle to Rectangle.—In the planning of heating and ventilating work, a transitional elbow of this type is frequently used. It is a transition of the same profile from a horizontal to a vertical position, making a quarter turn.

Sometimes a twist in the checks or sides of the elbow is developed by the way in which the patterns are laid out; therefore to obviate that, the present method is employed.

Construct the side view with the given radaii for throat and heel, as shown by *a-e-f-m*. Above it draw the profile of the pipe as indicated by rectangle *A*; to the right and in line with *e-f* draw the other profile as represented by rectangle *B*.

Locate the center of the narrow side of rectangle *B*, as at *x*, and through it draw the perpendicular line *C-F*. Likewise bisect the wide side of rectangle *A*, as at *y*. Extend line *m-a* in side view to right indefinitely, and to the left from center line *C-F*, set off on it the distance y'-a' equal to y-a° in profile *A*.

Divide the throat and heel curves of side view each into the same number of spaces, as shown by *a-e*, and *f-l*. Transfer the girth of the upper or throat curve, *a* to *e*, to the line *EF*, as shown by similar letters. Through these points draw horizontal lines as indicated, making the distance from *e* to e^v equal to x-e' in profile *B*, and from *a* to a^v equal to y-a^o in profile *A*; draw a line connecting points a^v and e^v which will show the half pattern; reproduce this to the right of center line and complete the whole pattern for the top.

From the points of intersection a^v, b^v, c^v, d^v, e^v, on the line a^v-e^v, in pattern for the top, draw perpendicular lines indefinitely as shown. Now from the letters *a* to *e*, and *f* to *l*, in side view, draw horizontal lines to the right to intersect the verticals from top pattern and locate the points $a', b', c', d', e', f', g', h', k', l', m'$, in front view; draw lines through the point of intersection and obtain the miter lines in the front view as shown.

Take the girth of the heel, *f* to *m*, in side view, and place it on the vertical line *C-D* as shown by similar letters; at right angles to *C-D*, draw lines through these points cutting through the verticals from the top pattern and resulting in the points f^v, g^v, h^v, k^v, l^v and m^v draw a line through these points from f^v to m^v and trace the half pattern thus produced to the right of center line, and complete the pattern for the bottom.

The pattern for the sides is developed by triangulation; hence it is necessary to determine the true lengths of the lines in the triangles that are to be used for making up the required patterns. To do this, take the vertical rise of each line, in one of the views as one leg, and the apparent length of each line, as it appears in the other view for the other leg; the hypotenuse gives the true length of the line.

Draw the solid and the dotted lines which form the angles in the side view, as shown. Since the solid lines in the side view show as vertical lines in the front elevation, as shown by similar letters, it may be assumed that the solid lines in the side view show their true length.

To find the true lengths of the dotted lines in side view; Take the line *e-g*, and set it down as one leg of a right angle triangle as shown by *e-g''*; for the other leg erect the vertical line *g''-g* to equal the horizontal distance between the lines *e'-f'* and *d'-g* in front view and indicated by *R*. A line drawn from *e* to *g* in the true lengths, will be the length sought. Proceed in this manner to find the remaining true length on dotted lines, using the horizontal distances marked *S*, *T*, and *U* in front view, for the shorter legs of the right angled triangles.

Having established the true lengths of the lines of measurement, proceed to lay out the pattern for the sides. Take the distance *e-f* in side view, this being its true length, and set it off as shown by e^v-f^v in side pattern. With f^v-g^v in bottom pattern as radius, and f^v in side pattern as center, draw the arc g^v

and intersect it by arc drawn from e^v as center, and *e-g* in true lengths of dotted lines, as radius. Now, with e^v in side pattern as center, and e^v-d^v in top pattern as radius, describe an arc as at d^v and intersect it by an arc drawn from g^v as center, and *g-d* in side view (this being its true length) as radius.

Follow in this alternate manner, using a radaii, first the divisions on the miter line in the bottom pattern, then the true lengths of the dotted lines; next the distances on the miter line in top pattern, then the lengths of the solid lines in side view (which are their true lengths) until all the points a^v-e^v-f^v-l^v, in the side pattern are established, the triangle a^v-l^v-m^v being a duplicate of *a-l-m* in the side view. Draw a line through the points and complete the pattern. Allow laps for seaming.

CHAPTER 5

Sheet Metal Machines

Sheet metal work is a large and growing industry. It is closely connected with the building trades in turning out such work as cornices, roofing, skylights, ornamental ceilings, ventilating and heating pipes, etc. The various machines and tools forming the equipment of a sheet metal working shop may be classified as

1. Floor machines
2. Bench machines
3. Stakes
4. Hand tools
5. Soldering equipment.

There is a great multiplicity of machines of each class available and the proper selection of these tools for any particular shop will depend upon the kind of work to be done and special care should be taken to choose such machines and tools to properly turn out the work required of the shop.

The layout of a small sheet metal shop is shown in fig. 9,751. This plan is for a general shop in which sheet metal work is combined with some occupation, as for instance, the sheet metal department of a plumber's shop. The layout is based on the idea of using one corner of a room approximately 25×30. The following equipment will be sufficient for 4 to 6 workmen.

Bench Machines

1 30 in. Improved sheet iron folder;

1 Bench machine with stand, with one pair each 3/16 turning rolls, burring rolls, wiring rolls, crimping rolls, 3/16 in. single beading rolls and 3/4 in. O. G. beading rolls.

1 2×30 in. Niagara slip roll former.

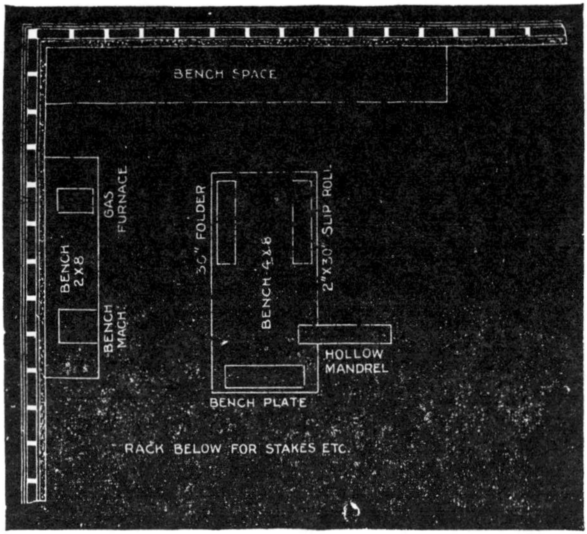

FIG. 9,751. —Layout for small sheet metal shop.

Stakes

1 Beakhorn stake, No. 2
1 Blowhorn stake
1 Needle case stake
1 Common square stake
1 Hatchet stake, No. 5
1 Hollow mandrel, No. 0
1 Bench plates—8×37 in. No. 1.

Hand Tools

4 Straight snips
1 Circle snips
1 Bench shear
1 Set of 3 rivet sets: Nos. 0, 2 and 5
1 Set of 2 grooving tools: Nos. 3 and 5
1 Set of 2 hollow punches, 3/8 and 1/2 in.
1 Set of solid punches, 9/32, 7/32 and 9/64
1 Prick punch
4 Scratch awls
4 Riveting hammers, No. 3

FIGS. 9,752 and 9,753.—Niagara lever punches, intended for punching small holes. Adjustable back gauge regulates the distance from the holes to the edge of the sheet and a stripper removes the stock from the punch at the up stroke. Fig. 9,572 shows a small punch having three punches and one three hole die for 5/32, 7/32 and 9/32 through 1/4 in. iron or equivalent. The lever works both ways, forward and backward for direct action, backward only when using the ratchet. Fig. 9,753 shows a small one-hand punch used for small punching operations in the sheet metal shop or on construction jobs. An adjustable gauge in the throat keeps successive holes equidistant from the edge. It will punch 1/4 in. holes through No. 18 gauge sheets. An adjustable gauge is also furnished for equally spacing a row of holes.

1-36 in. plain circumference rule
4-2½ in. hickory mallets
1 Cutting nipper
1-10 in. wing dividers
1-2 ft. steel squares
1-6 in. flat nose pliers
1-5/8 in. cold chisels.

Soldering Equipment

1 Double burner gas furnace
1 Pair soldering coppers, 3 lb. per pair
2 Soldering copper handles.

The machines ordinarily used are described in the following paragraphs:

Squaring Shears.—All stock for articles made of sheet metal must first be cut from the original sheets. The machine used for this work is the squaring shear. It may also be used for

FIG. 9,754.—Niagara double end machine for shearing and hole punching, suitable for cutting apart and slitting sheets of any length and width. Frame is off-set to provide clearance for the several edges of long sheets during slitting. In operation, the lever works down toward the operator whether he faces the punching or the shearing end of the machine. The hold down attachment prevents material from raising while being cut and an adjustable gauge is provided.

much straight cutting in trimming stock to exact sizes. The cut is made in straight lines only and the machine has guides and gauges to be used in squaring and cutting to the required widths and lengths.

FIGS. 9,755 to 9,763.—Pexto "Utility" slitting shears and samples of work cut from blanks to line. These shears will do the work of ordinary bench shears, over which they have the following advantages: The length of cut is longer to the same movement of the hand; the same pressure of the hand will cut thicker stock; they cut with the same ease at all points of the cut, while ordinary bench shears cut harder near the point than near the bolt. The lower blades of these shears are stationary, so that when cutting to line, the mark may easily be followed with accuracy. The blades are so constructed that the line drawn is always exposed to the view of the operator.

Usually the work is inserted from the front of the machine, but long sheets may be worked from either front or back. The side guides should always be used in squaring work. Care must always be taken to keep fingers from under the blade. The work should be held down firmly on the bed of the machine while cutting. It will be advisable not to depend on the scale marked on the bed for accurate work. A steel rule should be used to check the setting. Keep the blades well oiled.

The method of operating a foot power squaring shears is shown in fig. 9,764 and the combination of a power squaring shears in fig. 9,765 and the combination of a power squaring shears in fig. 9,765.

Circular Shears.--The ring and circular shears has inclinable cutters for cutting circles for the bottoms of vessels, etc., as well as cutting a circle inside of a circle or cutting a circle from a square sheet. A perfect burring operation demands that the bottom to be burred should be cut with extreme accuracy. Circles can be cut with the hand circular snips, but not with the same accuracy as the circular shears will produce.

FIG. 9,764.—Squaring shears. They are used for trimming and squaring sheet iron. ***In selection,*** decide on the length of shears that will answer all requirements and the heaviest sheet metal that will be cut. Then select the type of shears best suited for the requirements.

The shears consist of a base and cutting head, with parallel cuttershafts driven with gears by means of a hand crank, or with a belt when the machine is arranged with pulleys for power drive. A sliding circle arm is fitted to the base and is adjustable for different diameter circles by sliding on the base to and from the cutters. The blank from which the circle is to be cut must be squared previous to cutting the circle. After being squared true and of a correct size nearest to the size circle to be cut, providing for as little waste as possible, the square blank is inserted between two clamping discs in the circle arm. Through an eccentric clamping lever, or with

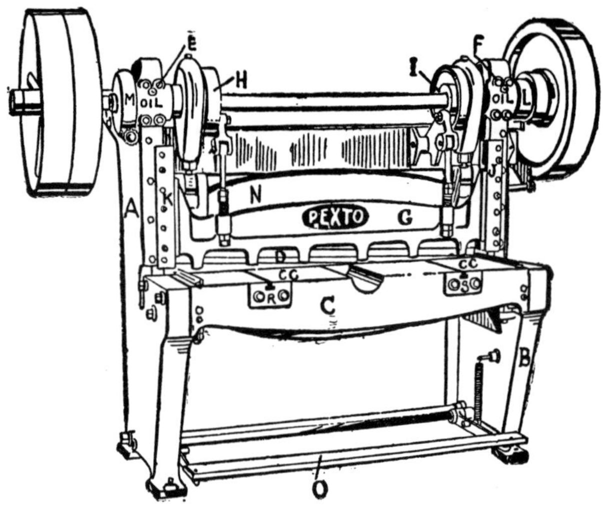

Fig. 9,765.—Pexto power squaring shears with overhead drive and gap in housing for permitting the trimming of sheets before squaring. ***In construction,*** the bed C, is accurately planed and bolted to the housings A and B, and it can be shifted towards or away from the upper cutting blade D. Graduations in fractions of an inch are cut in the bed along gauge slots CC. Gauge extension arms are fastened to the bed C, at R and S, and are held with two bolts for each arm.

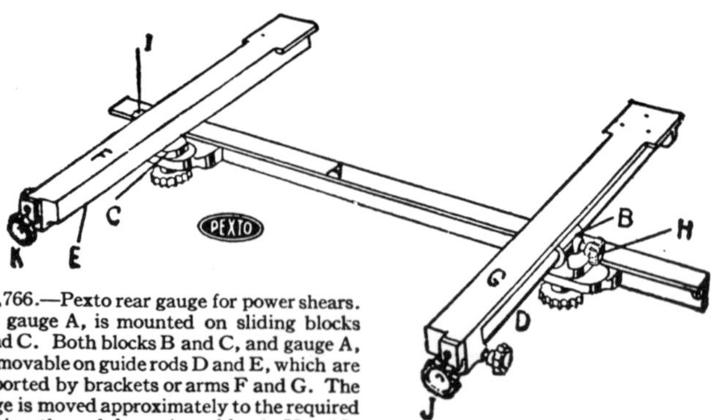

Fig. 9,766.—Pexto rear gauge for power shears. The gauge A, is mounted on sliding blocks B and C. Both blocks B and C, and gauge A, are movable on guide rods D and E, which are supported by brackets or arms F and G. The gauge is moved approximately to the required position through loosening of knobs H and I, and are again tightened after the gauge A, is approximately set. The final adjustment is then easily secured by turning adjusting knob screws J and K, to position desired.

a crank screw or a hand wheel, according to the design of the machine, the blank is securely clamped and the rotating of the cutters draws the material through the machine, cutting a true circle.

Figs. 9,767 to 9,770 shows a typical circular shears and work done by same.

Ring and Circular Shears.—This type shears is intended for cutting both outside and inside circles and irregular curves.

FIGS. 9,767 to 9,770.—Waugh's circular shears and work done by same. A, eccentric lever to quickly clamp material between discs. A graduated scale is attached to the bed C, for the purpose of securing quick and accurate adjustment of the sliding arm D. The clamping discs B, are rubber covered which prevents slipping of the blanks. ***In operation,*** if for instance, a circle 5 ins. in diameter is to be cut, proceed to loosen the lock nut on the circle arm and set the sliding circle arm, drawing it to or from the cutters until the correct position of the cirlce arm is secured. With indicator on circle arm pointing to 5 ins. on the scale, lock the circle arm securely through the lock nut. Place the square blank between the clamping discs in a central position of equal distance on all sides of the blank from the outside diameter of the clamping discs. If many of the same size of circles are to be cut, after the correct first setting, adjust the swinging gauge so it will strike the edge of the square blank while remaining in the clamping discs. The machine properly set, any number of the same size circles can be cut accurately. Care should be exercised not to allow the cutters to have too much clearance or to rub against each other too hard; and when adjustment of the cutters for more or less cutter clearance is necessary, the lower shaft to which the lower rotary cutter is attached can be drawn away or to the upper cutter by means of the adjusting clasp nut on the lower shaft next to the gear. After the correct adjustment is secured, be sure to fasten the adjusting clasp nut securely. The upper cutter shaft bearing is adjustable and will raise or lower, allowing for the taking up of the wear of the cutters; and this adjustment should never be tampered with while the cutters are in good condition.

For ordinary outside circle cutting, the ring and circular shears does the same work as already described under circular shears and owing to the cutter raising feature in the ring and circular shears, it is not so limited in the work performed. In addition to doing the work already described, the ring and circular shears offers a suitable means for cutting irregular curves when following a scribed line, and for such work the sliding circle arm is not used. The rotary cutters being of small diameter, measuring approximately 1⅝ ins. in diameter, and set angular in position, they will cut as true and clean on the inside as on the outside of a sheet of metal.

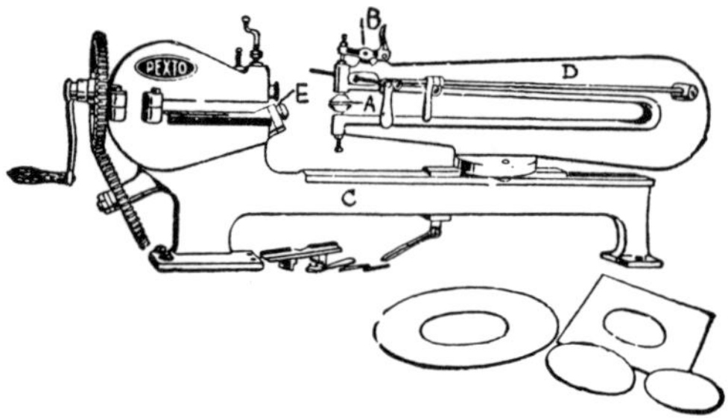

FIGS. 9,771 to 9,775.—Pexto ring and circular shears and work done by same. ***In construction,*** rubber covered clamps A, are provided for securing the material to the cut. Clamps operated by quick acting lever B. The bed C, on which the tail piece D, slides, is graduated in ins. and fractions of ins. for quick setting of the tail piece D, for the size circle to be cut. The angular position of cutters E, allows for as clean cutting on the inside as on the outside of circle. ***In operation,*** to cut an "internal" circle or circle out of a circle, for making a ring, or to cut a circle from the inside of a square sheet of metal, the prepared blank is clamped between the clamping discs in the usual way, and the sliding circle arm with the sheet metal blank inserted is brought toward the cutters, permitting as much of the edge of the sheet metal blank as necessary, according to the size of the inside circle that is to be cut, to slide between the cutters. With the proper alignment of the blank in the machine secured, bring the upper cutter down on the material by turning the crank screw hard enough, so that the cutters will cut the material without burring or buckling the edge. The ring and circular shears has in addition to the regular swinging gauge a ring gauge which slides on a rod along the circle arm for facilitating the cutting of rings, and through the proper setting of both gauges many quantities of the same kind and size of circles and rings can be cut alike with accuracy.

In using this machine never allow the cutters to have too much clearance or to rub against each other too hard; and when adjustment of the cutters for more or less cutter clearance is necessary, a satisfactory adjustment of the cutters is secured through the adjusting clamp nuts next to the gears on both the upper and lower shafts.

In cutting circles of a very small diameter, and sometimes when cutting a small circle inside of a circle for making a ring, if on the edge of the material there should appear a burr or a buckle, this indicates that the circle arm is in too straight a line with the center of the cutters; and in such case the circle arm must be set a trifle out of line with the center of the cutter, just enough for cutting a true and clean edge. This adjustment is secured through loosening the bolts marked A, in the sliding circle arm plate and moving the circle arm as necessary until a proper adjustment is secured for assuring a clean cut.

Forming Machine.—This machine is used for bending sheet metal or wire to a curved form. All articles made in the shape of a cylinder and others which are to be bent to a radius greater than 1 in. can be quickly formed in this machine. The machine consists of two geared rolls and one loose roll which serves to bend the work which passes between the first two. The distances between the geared rolls can be regulated by thumb screws. The rear roll can be raised to bend the sheet to a smaller radius.

The general type of all forming machines is similar, though there are many variations in their mechanism. In principle the forming is done by three rolls. The two front rolls grip the sheet of metal and force is against the rear roll, which bend it around the front upper roll, forming a cylinder. The size of circle that can be formed on a forming machine depends on the nearness of the rear or forming roll to the front upper roll in the machine. The pressure of the gripping rolls, which are the two rolls looking from the front of the machine, is regulated by thumb screws, and the rear or forming roll is regulated in the same manner. Fig. 9,778 shows method of operating a forming machine.

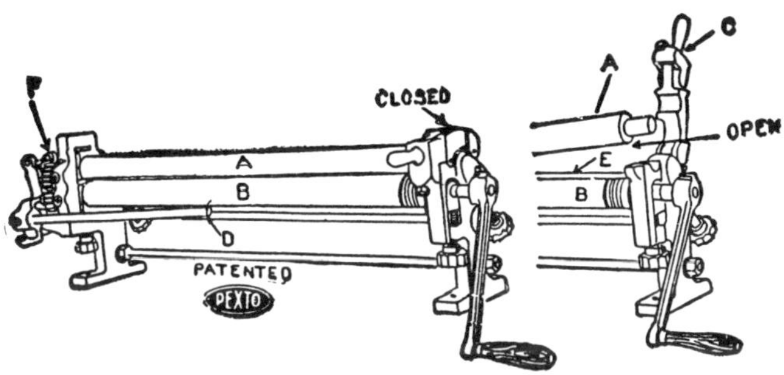

FIGS. 9,776 and 9,777.—Pexto slip roll formers (forming machine). *In operation*, the sheet to be formed is placed between the gripping rolls A and B. The cylinder formed, open latch C, then by means of lever D, raise upper gripping roll A and slip the formed cylinder from the roll A. It will be readily seen that all operations are performed at the crank end of the machine. Rolls B and E, are adjusted with knurled screws, and once adjusted they cannot slip. The roll raising mechanism in these machines is so balanced that the roll A, is easily lifted with a slight pressure on the lever D. The latch C, is released and closed with one movement and is self locking.

Folding Machines.—In order to turn a hem or lock on the edge of a piece of sheet metal some means must be found to hold the metal firmly while the edge is being turned, or the edge may be gripped and the fold turned by moving the piece itself. The bar folder does the latter. These machines will also prepare the edge of sheet metal to receive a wire. All bar folders have adjustments to regulate the width of the fold and also the sharpness of the bend, so that they may also be used to prepare the edge of sheets to receive a wire. It is always necessary to know just how the machine is adjusted before attempting to make a fold or to wire an edge.

Fig. 9,779 shows method of operating a folding machine. The most popular pattern of folding machines are the bar folder, sheet iron folder and pipe folder, the bar folder being the most important.

FIG. 9,778.—Forming machine and method of operation. ***In operation,*** the sheet of metal to be formed is inserted from the front of the machine and passed through the two front gripping rolls, securing uniform pressure of the rolls on the material through thumb screws, so that the sheets will ride through the rolls freely. The sheet, in passing through the gripping rolls, will strike the rear or forming roll, if adjusted high enough, thereby forming a circle. If the circle secured in the first experiment should not prove of correct diameter, the lowering of the rear or forming roll will increase the diameter of the cylinder, or by raising the forming roll the diameter of the cylinder will be decreased. In forming cylinders of very small diameters the blank as soon as entered between the gripping rolls must be given enough curvature with the hands by means of bending the material upward, otherwise it will not strike the rear forming roll for the proper shaping of the cylinder to be formed. When the operator becomes accustomed to the forming machine and its mechanism, accurate settings are easily made after the first few trials, according to the judgment used.

Grooving Machine.—After a lock seam has been folded on the folder, it should be closed down with a grooved wheel on the grooving machine. Hand grooving tools are also used for this, but in all cases where it is possible to use the machine it is better to do so. The grooving rolls are made to fit several widths of seam and the proper roll should be used. A closed lock is first turned on the sheet of metal by the use of the folding machine. The sheet is then rolled into a cylinder by the forming machine, when the corresponding edges, as prepared in the cylinder are snapped together and laid on the grooving horn in the grooving machine. The grooving rolls in the grooving machine run over the seam lengthwise, effecting an operation called "grooving" or "seam closing" and completing the lock.

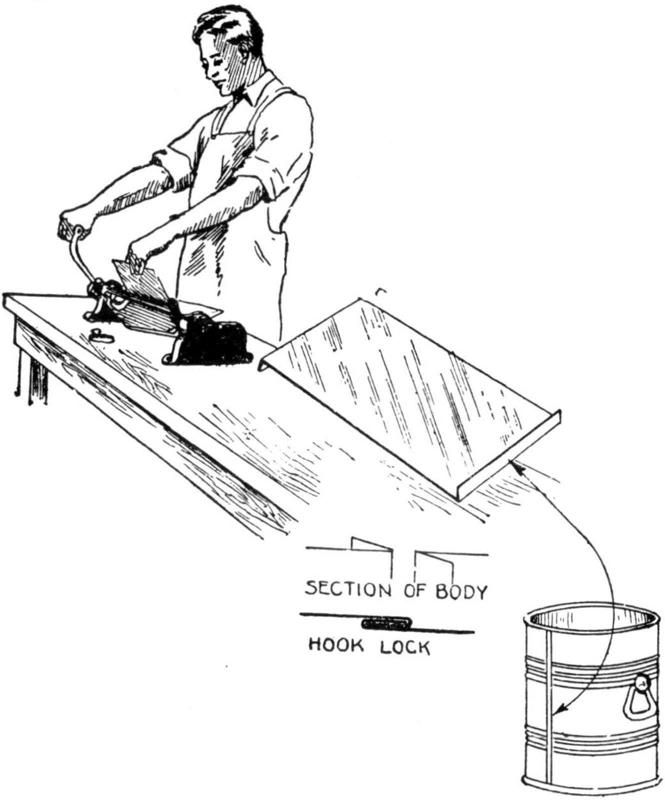

FIGS. 9,779 to 9,782.—Folding machine and method of operation, with illustrations of work done by the machine

Fig. 9,784 shows the grooving operation and figs. 9,786 and 9,787 the appearance of lock before and after grooving.

Burring Machine.—This machine is used for turning an edge on cylinders of metal or on discs such as can bottoms. In preparing vessels for double seaming, a burr is first turned at a right angle on the body and then one of the same width on the edge of the bottom. This last operation is quite difficult and takes considerable practice.

In using the burring machine, it should be noted that only a narrow burr about ⅛ in. wide can be turned. The burring machine is the hardest

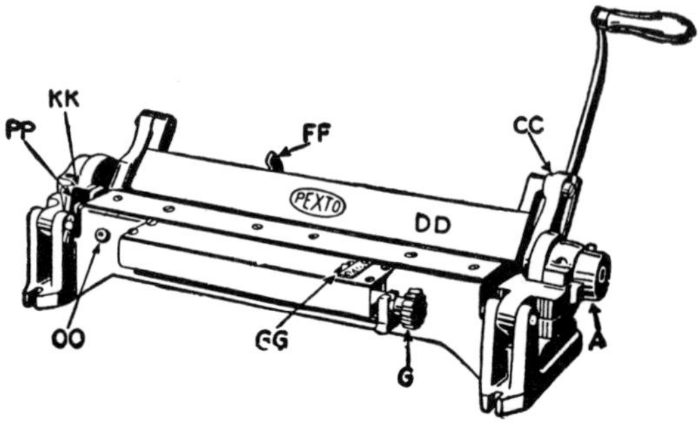

FIG. 9,783.—Pexto "Arrow" bar folder. It is intended for forming the edges of sheet metal at various angles. They will produce closed locks as well as open or round locks for inserting a wire in the flat sheet. Open or round locks for wiring are made by raising the folding bar CC, at right angle. The wing DD, is then adjusted for the size of wire to be used through the setting of wedge FF, that moves to left and right on folding bar. An improvement, consisting of a pin in the frame, prevents the dropping of the wing below the gripping jaw, the wing DD, dropping in a proper position automatically in the process of folding, producing accurate and uniform round locks. Gauge GG, is adjustable, moved by a screw and is adjusted by turning gauge knob G. The width to which the gauge is adjusted is indicated on a graduated brass plate, and after set is firmly secured through lock screw OO. Gauge is so designed that it cannot twist, insuring accurate lock forming. Adjustable stop A, is provided to permit the forming of any desired angle, in addition to regular square and bevel stops KK and PP.

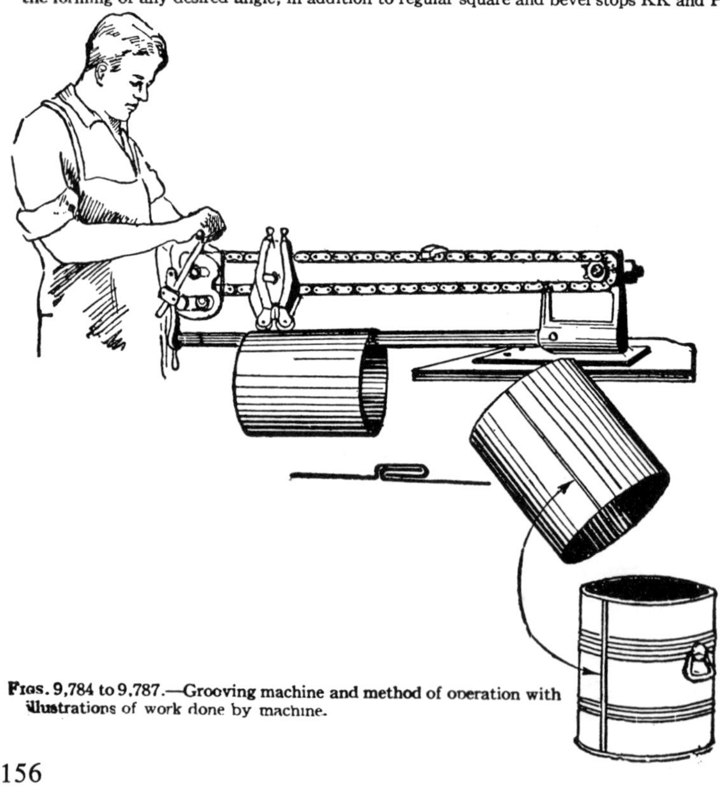

FIGS. 9,784 to 9,787.—Grooving machine and method of operation with illustrations of work done by machine.

FIGS. 9,788 to 9,790.—Niagara grooving machine. This type of machine is used for stove pipe manufacture, tinware and general sheet metal work. It produces both inside and outside seams. Its principal parts are the frame, guide bar and horn socket, which are cast in one piece, together with the carriage and rolls.

In construction, the horn can be rotated and locked to hold the desired groove on flat area in working position, a swinging latch supports the outer end. A rotation of the pinion by the crank handle from either side, imparts motion to the carriage. Interchangeable pinions provide two speeds to suit the metal being worked. The crank may be removed and a rod inserted through the carriage for punching it rapidly over very light work. The rolls are carried in swinging cradles, one operating on the outward carriage travel, the other on the return. Change takes place automatically. Two adjusting screws regulate the pressure. The setting is locked by a yoke which fits over the adjusting screw heads. A clamp at the inner end of the horn holds down the work on the outward travel of the carriage. It raises automatically when carriage is in extreme return position. An adjustable hinged stop holds work at outer end. An adjustable guide on the carriage facilitates locating the work in the grooves.

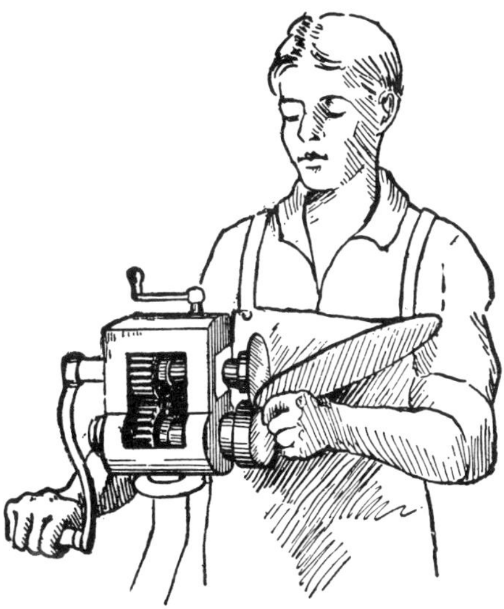

FIG. 9,791.—Burring machine and method of turning a burr. This is a difficult operation to master but practice will produce uniform flanges on sheet metal bodies and prepares the burr for bottoms preparatory to setting down and double seaming.

machine for beginners to use. The pupil should avoid spoiling good material until he has had careful instruction.

Fig. 9,791 shows method of operating a burring machine.

Setting Down Machine.—This machine is used to close the seams left by the burring machine making ready the seams for double seaming. The machine is very simple and may be turned in either direction. It has no adjustments except for thickness of material.

FIGS. 9,792 to 9.794.—Setting down machine and its operation.

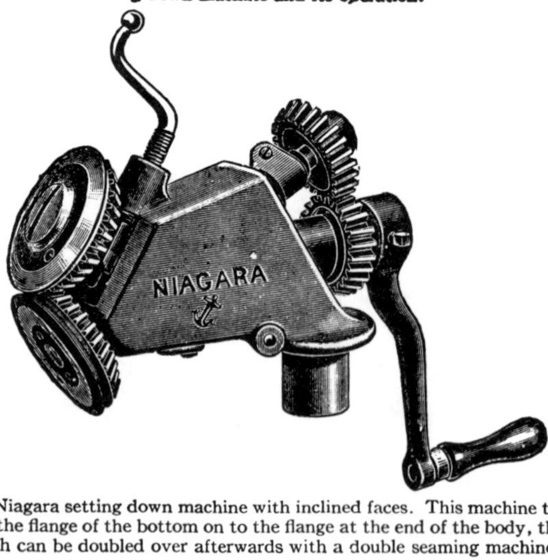

FIG. 9,795.—Niagara setting down machine with inclined faces. This machine turns down and compresses the flange of the bottom on to the flange at the end of the body, thereby forming a joint which can be doubled over afterwards with a double seaming machine.

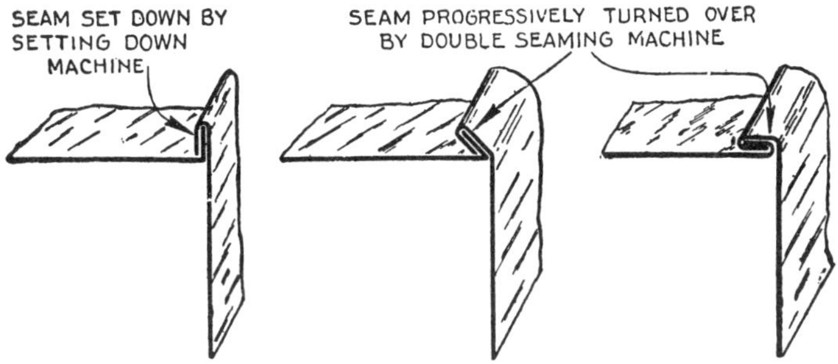

FIGS. 9,796 to 9,798.—Progressive operations of setting down machine and double seaming machine in forming a double seam.

Figs. 9,792 to 9,794 illustrate operation of a setting down machine. ***In operation,*** the vessel is held bottom upward and the edge A, of the bottom runs between two rolls, when the corresponding edges are pressed or closed tight ready for double seaming as shown at B.

Double Seaming Machine.—After the seams have been set

FIG. 9,799.—Double seaming machine and its operation. Fig. 9,796 shows a seam as set down with a setting down machine, and figs. 9,797 and 9,798, the progressive turning over of the seam with the double seaming machine making a tight joint.

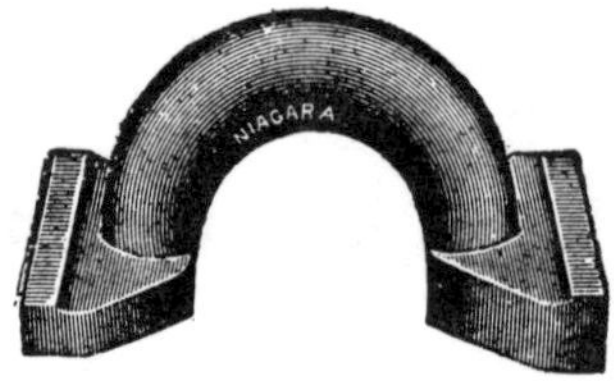

FIG. 9,800.—Niagara hand roofing double seamer. The standing seam of sheet metal roofing finished on these blocks after the edges were turned with tongs.

down properly with the setting down machine the flange as left must be turned against the body of the vessel. This double seaming operation can be effected with the double seaming stake and mallet. A double seaming machine will double seam the work with a greater degree of neatness and accuracy

FIGS. 9,801 to 9,803.—Niagara crimper and beader. This machine is not backgeared, being intended for quick operation on light work. It is adjustable for depth of crimp and bead.

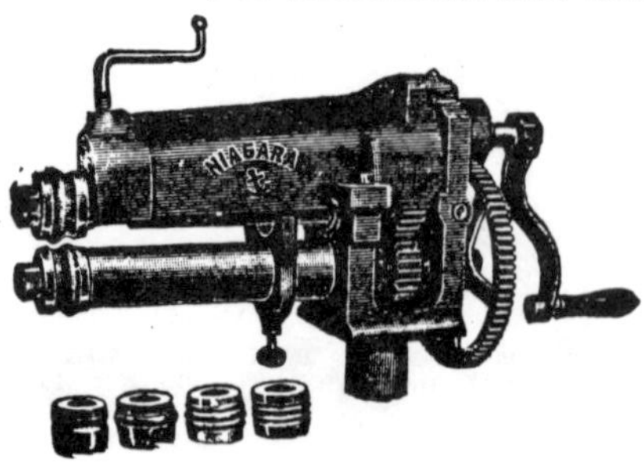

FIGS. 9,804 to 9,808.—Niagara beading machine and rolls. This machine is used for ornamenting and stiffening tinware and other sheet metal goods by forming beads corresponding with the shape of the rolls.

and more often saves the loss of a good job by unattractive stake double seaming.

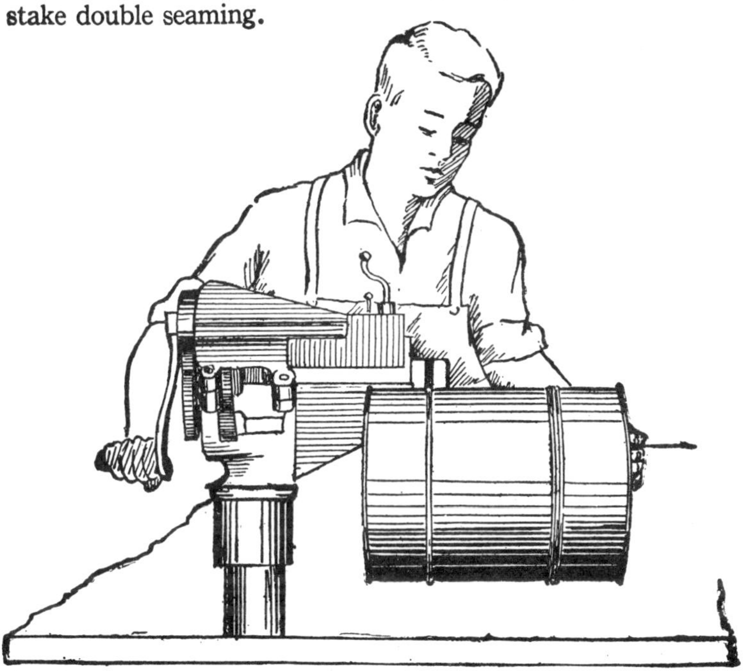

FIG. 9,809.—Beading machine and its operation.

FIGS. 9,810 and 9,811.—Niagara adjustable gutter beader and example of work. *It is intended* to round and stiffen the edge of gutter by forming a so called bead as shown in fig. 9,811. Rods from ⅜ to ¾ in. diameter can be used in the frame of adjustable gutter beaders. The two jaws can be spread apart to facilitate removing the work and rod. Stops fix the workng position of the moveable jaw.

Fig. 9,799 illustrates the operation of a double seaming machine.

Beading Machine.—This machine is a very simple one to operate. It is furnished with a series of rolls of different

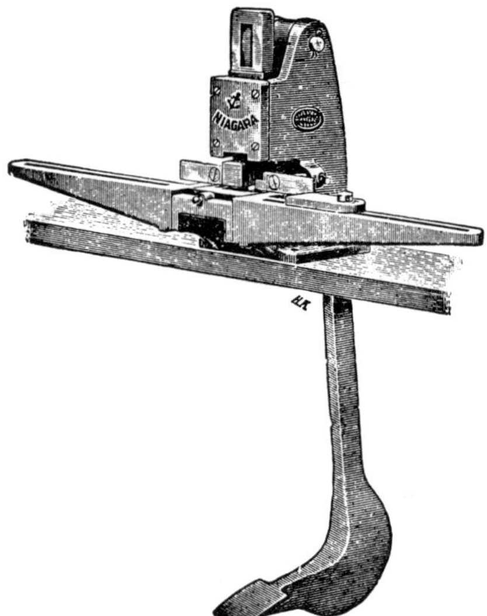

shapes. The impression made in the bodies of vessels will correspond with the shape of bead in the rolls used. Making these impressions is called beading which serves to ornament and strengthen the bodies of vessels or on any other work where the machine is applied.

FIG. 9,812.—Niagara notching machine for notching sheet metal for wiring and grooving, for cutting corner and hinge notches, etc. The rectangular die is suitable for notches up to 1¼ ins. wide and 2 ins. deep. Adjustable side and back gauges are provided, and several thicknesses of light tin can be cut at the same time.

Fig. 9,806 illustrates the operation of a beading machine.

Crimping and Beading Machine.—The crimp and bead so much in evidence on the edge of the common stove pipe are made with this machine.

Crimping and beading machines are intended to facilitate the making and putting together of sheet iron pipe of different diameters, by contracting the edge of the pipe so that one joint of pipe will enter another. In putting together the pipe the ogee bead next to the crimp prevents the joints slipping beyond the impression made with the beading rolls.

Turning Machine.—This machine is used to form a rounded edge into which a wire is inserted, the edge being closed by a wiring machine. The operation of a turning machine is shown in fig. 9,813 and the shape of the edge formed by the machine in fig. 9,814.

Wiring Machine.—This machine finishes the operation begun

FIG. 9,813.—Turning machine and its operation.

by the turning machine. Depending upon the shape of the work, the seals to receive the wire are sometimes prepared on the folder or brake instead of the turning machine.

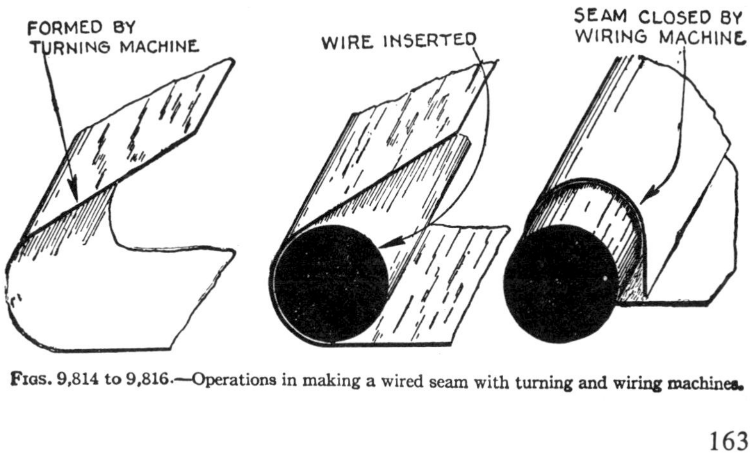

FIGS. 9,814 to 9,816.—Operations in making a wired seam with turning and wiring machines.

FIG. 9,817.—Wiring machine and its operation.

Brake.—This machine has a wider range of usefulness than the folder. It may be used to turn hems or folds and to make

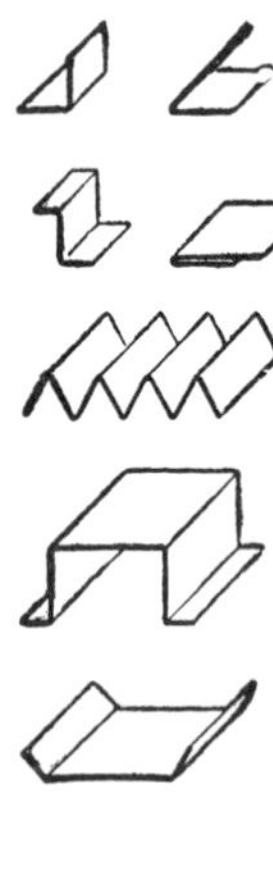

FIGS. 9,818 to 9,825.—Niagara foot brake and various bends produced by same. This machine is designed for dies or blades intended to perform straight bending operations on light sheets.

bends at all angles up to nearly 180 deg., and at any distance from the edge. It is supplied with moulds that permit the forming of moulded shapes to almost any pattern for cornices. These machines in their lengths up to 8 ft. are in common use. Brakes have a capacity for forming locks and angles in a wide range of sizes and of unusually large lengths. The folding

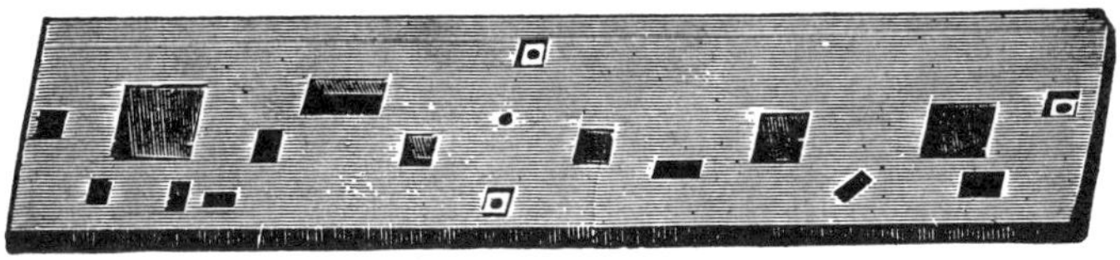

FIG. 9,826.—Niagara bench plate for holding stakes and bench shears.

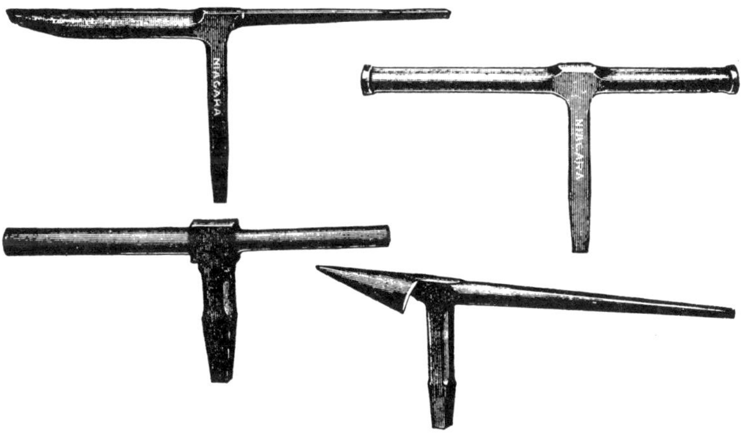

FIGS. 9,827 to 9,830.—Various tinner's stakes **1.** Fig. 9,827, beakhorn; fig. 9,828, double seaming; fig. 9,829, conductor; fig. 9,830, blowhorn.

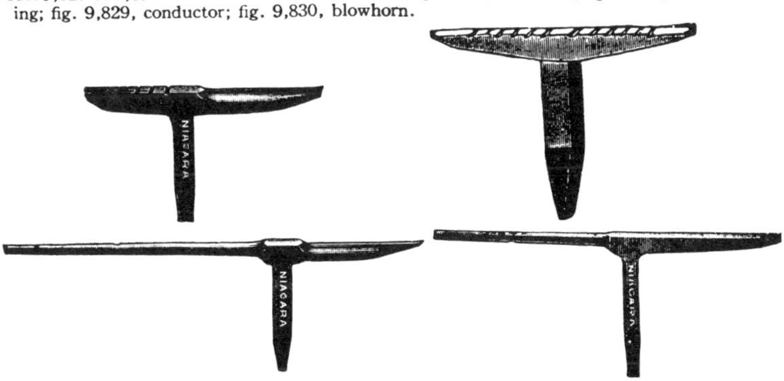

FIGS. 9,831 to 9,834.—Various tinner's stakes **2.** Fig. 9,831, creasing with horn; fig. 9,832, common creasing; fig. 9,833 candle mould; fig. 9,934, needle case.

machine can only form a lock or edge as wide as the depth of the jaw in the machine will permit, whereas, the brake

FIGS. 9,835 to 9,837.—Various tinner's stakes **3.** Fig. 9,835, common square; fig. 9,836, bevel edge; fig. 9,837 coppersmith's square.

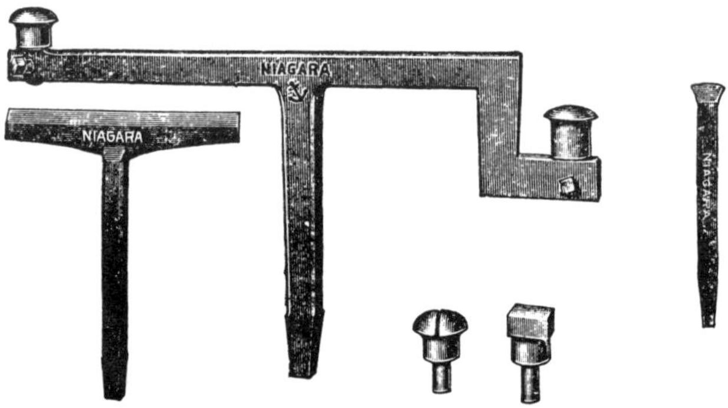

FIGS. 9,838 to 9,842.—Various tinner's stakes 4. Figs. 9,838 to 9,840 tea kettle with four heads; fig. 9,841, hatchet; fig. 9,842, bottom.

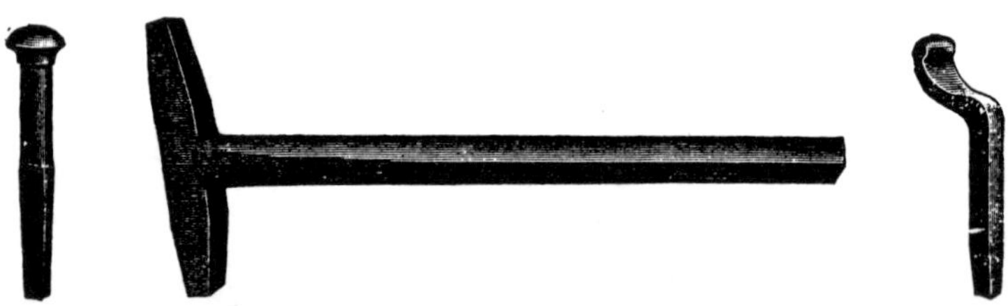

FIGS. 9,843 to 9,845.—Various tinner's stakes 5. Fig. 9,843 round head; fig. 9,844 solid mandrel; fig. 9,845, bath tub.

allows the sheet of metal that is to be edged or formed to pass through the jaws from front to back without obstruction.

Fig. 9,818 shows a typical foot brake, and figs. 9,819 to 9,825 some of the shapes produced by this machine.

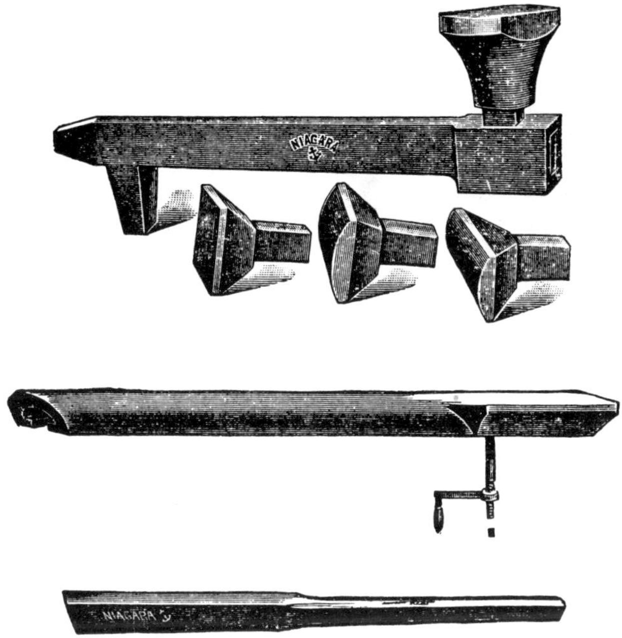

FIGS. 9,846 to 9,851.—Various tinner's stakes **6.** Figs. 9,846 to 9,849. double seaming stake with four heads; fig. 9,850, hollow mandrel; fig. 9,851 solid mandrel.

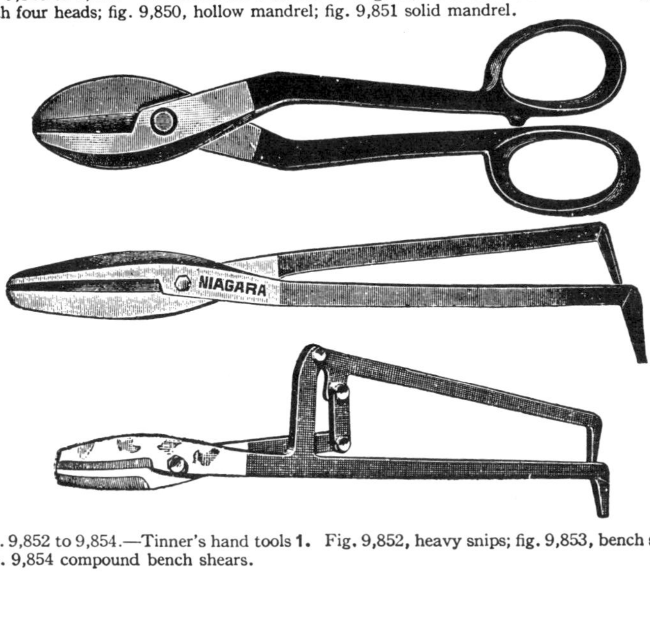

FIGS. 9,852 to 9,854.—Tinner's hand tools **1.** Fig. 9,852, heavy snips; fig. 9,853, bench shears fig. 9,854 compound bench shears.

FIGS. 9,855 to 9,859.—Tinner's hand tools **2.** Fig. 9,855, rivet set and headers; fig. **9,856**, grooving tool; fig. 9,857, hollow punch; figs. 98,58 and 9,859. solid punches.

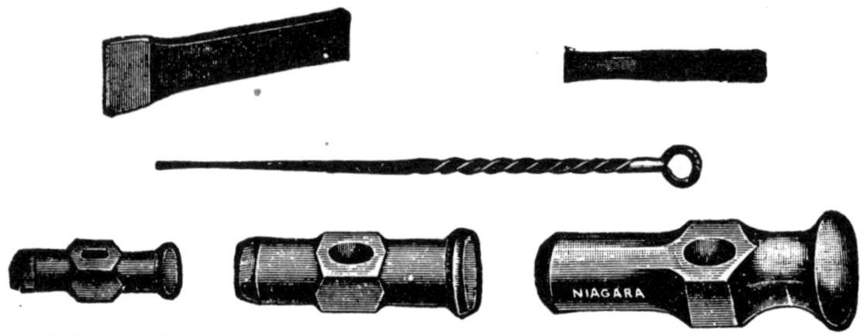

FIGS. 9,860 to 9,865.—Tinner's hand tools **3.** Fig. 9,860, wire chisel; fig. 9,861, lantern chisel; fig. 9,862, scratch awl; figs. 9,863 to 9,865 raising hammers.

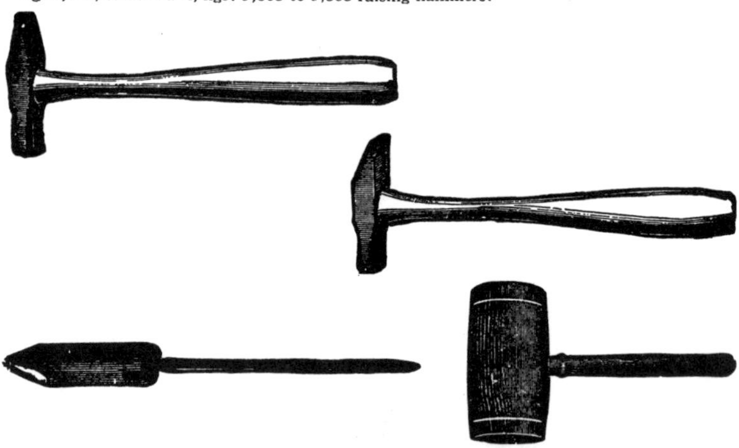

FIGS. 9,866 to 9,869.—Tinners' hand tools **4.** Fig. 9,866, riveting hammer; fig. 9,867 setting hammer; fig. 9,868 soldering copper; fig. 9,869 mallet.

Stakes.—Sheet metal is in many instances shaped by being bent over anvils of peculiar forms known as *stakes*. These fit into holes cut in the bench, the stakes being held rigid by a

FIG. 9,870.—Steel circumference rule. This rule gives the circumference of circles by simply measuring the diameters. The top edge is a rule graduated in sixteenths inch. The bottom indicates the circumference of circles, which are equal in diameter to the measurement given directly opposite on the other edge. The reverse side gives useful table of measurements.

bench plate as shown in fig. 9,826. The plate is fastened to the bench. The plate is also used to hold bench shears. There is a great variety of stakes as shown in the accompanying illustrations and the use of each will be inferred by its shape without a detailed description.

CHAPTER 6

Welding Sheet Metal

The distinction between sheet metal and metal plates is based upon *thickness*, that is up to a certain thickness it is called *sheet* and above that, *plate**.

Sheet metal of thicknesses from 22 gauge up to approximately 8 gauge is used in a great many applications, such as duct lines of all kinds, for heating, ventilating and air conditioning, and for making package conveyors, hoppers, bins, door and window casings, kitchen equipment, parts of buses, trucks, fenders, fan housings, furnace casings, metal furniture, pans, etc.

The welding of sheet metal can be done by various methods, such as:

1. Electric arc
   a. Carbon
   b. Metallic
2. Oxy-acetylene torch
3. Bronze welding (brazing)

Sheet Metal Arc Welding.—On the thinner sheets, according to Bainbridge, the carbon arc process has been used with considerable success. Sometimes the material is flanged at the edges, so that the weld is made without the use of filler rod.

Special electrodes have been made for the carbon arc welding of thin sheets and a special operating technique is recommended. With this technique the end of the electrode is placed right on the joint, and the electrode is held at a small angle extending in the direction of the weld.

*NOTE.—The dividing line as given by different authorities varies, for instance, Hawkins Mechanical Dictionary gives 3/16 (.1875); Lincoln 8, gauge (.1719) and The Linde Air Products Co. 11 gauge (1/8 or .125) inch.

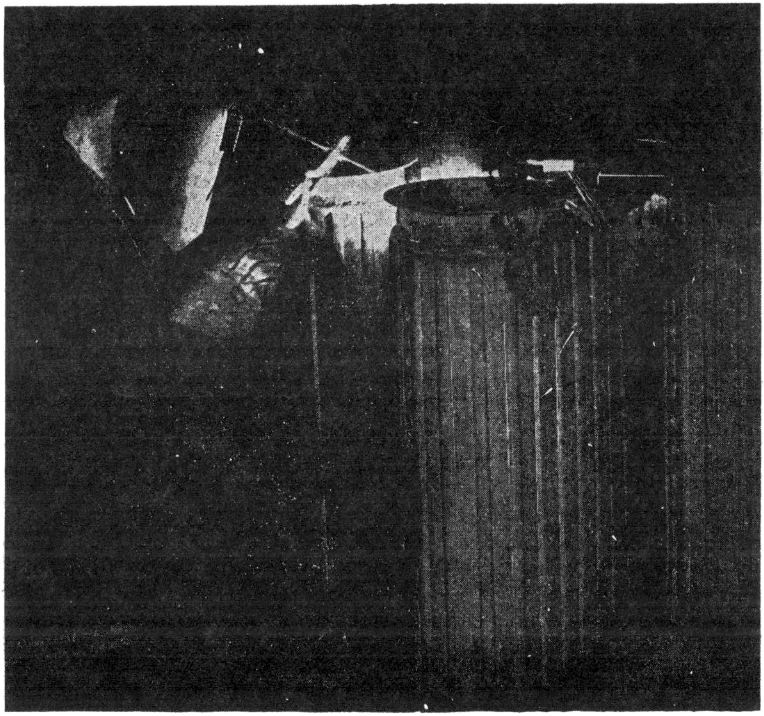

FIG. 1.—Arc welding light gauge sheet metal.

The arc is struck and kept just long enough so that the metal from the electrode flows steadily and evenly into the weld. The process is adaptable to lap welds, corner welds, and butt welds.

Welding by the metallic arc does not differ materially from the procedure used in welding ordinary light plate. The same precautions must be observed relative to expansion and contraction and securing full penetration.

In the case of very light sheets it is, of course, necessary to use low current and small diameter electrodes. Even with these adjustments, the operator who is accustomed to working on heavier stock will find it very difficult at first to keep from over-heating the seam.

Very thin materials have been welded in the laboratories by hand, but anything thinner than 12 gauge, is usually beyond the capacity of the ordinary operator.

In welding sheet metal it is advisable to use some sort of jigs or fixtures to hold the parts in place for welding and take care of the stresses which are put into the metal during rolling and which are released during welding.

These jigs or fixtures must be so designed to hold the parts in shape, without permitting them to warp. Copper clamps may be used.

Ques. When clamping devices are not used, how is the alignment of edges maintained?

Ans. By tacks spaced at regular intervals along the joint.

For steel, tacks should be about 2 in. apart, about 1½ in. to 2 in. apart for copper, brass, Monel metal, etc., and 1 in. to 1½ in. apart for aluminum.

Ques. Can welds in sheet steel be made without tacking the edges or without using welding wire?

Ans. This can only be done when at least one of the edges is free and capable of being manipulated.

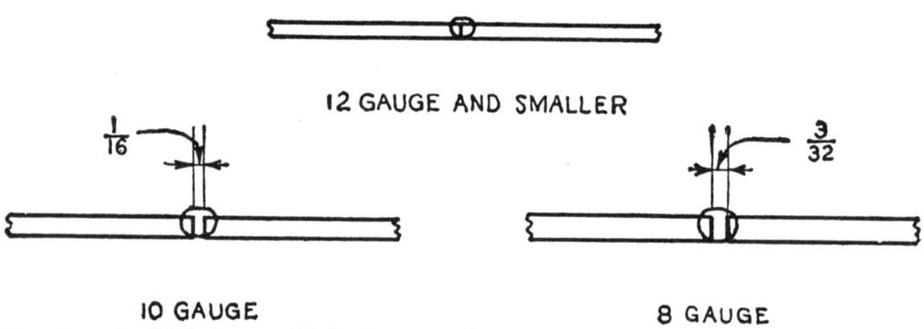

FIGS. 2 to 4.—Plain butt welds in sheet metal. ***In flat position,*** for 8 gauge and thinner sheets the work is fitted up as shown. Penetration will be 85% or better in 8-10-12 gauge sheets and 100% in 14 gauge and thinner.

The natural spring of the metal enables the welder to hold the free side with one hand while melting the edges together.

Joints.—The selection of the type of joint to use in joining sheet metal is determined by the service conditions to be met.

For example, a butt weld provides a smooth even surface at the joint. A lap weld, made in one bead only, has an opening or separation at the joint and substances may collect between the plates and cause corrosion.

Since both butt and lap joints are used extensively, it is evident that selection depends on the service to be met and becomes a question of methods of welding and cost. Various sheet metal joints with spacing as suggested by Lincoln are shown in figs. 2 to 14.

Ques. What is the simplest joint in sheet metal?

Ans. The butt joint.

Ques. What use is made of the corner joint?

Ans. It is largely used in the production welding of sheet metal products, where the parts to be joined are held in correct alignment by means of properly designed jigs or fixtures.

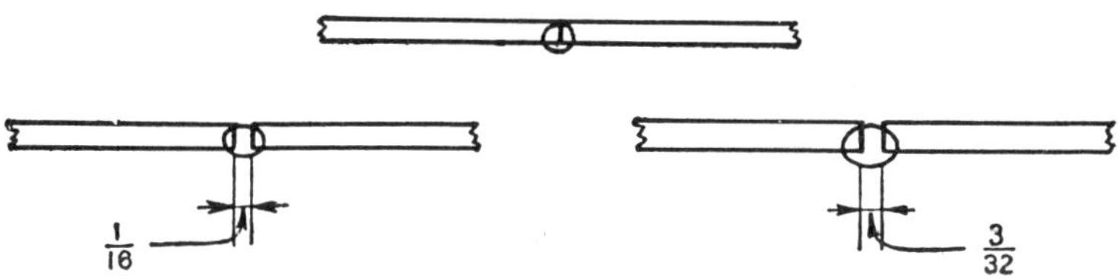

FIGS. 5 to 7.—Overhead butt welds.

Ques. When is the flange joint used?

Ans. It is particularly adapted for sheet lighter than 20 gauge.

Ques. How is the flange joint prepared?

Ans. By simply turning up a flange.

The upstanding portion of the flange should extend above the upper surface of the sheet a distance equal to the thickness of the sheet.

Ques. What should be noted about the lap joint?

Ans. It is not recommended and should be avoided.

The single lap weld has little resistance to bending and is consequently quite unsatisfactory for flat sheet. It may occasionally be used where, for example, it is necessary to join one cylindrical shell fitting inside another. The double lap weld has better strength characteristics than the single lap, but it requires nearly twice as much welding as the simpler and more satisfactory butt weld.

Procedure.—It takes a lot of practice to learn to do a good

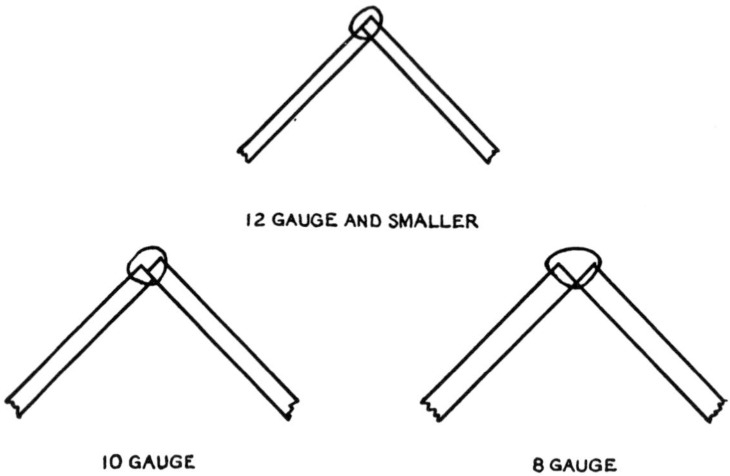

FIGS. 8 to 10.—Corner welds for various gauges.

job on light gauge sheet metal, although some operators with a lot of experience in this sort of work are able to weld successfully with bare electrodes, as small as 1/16 diameter.

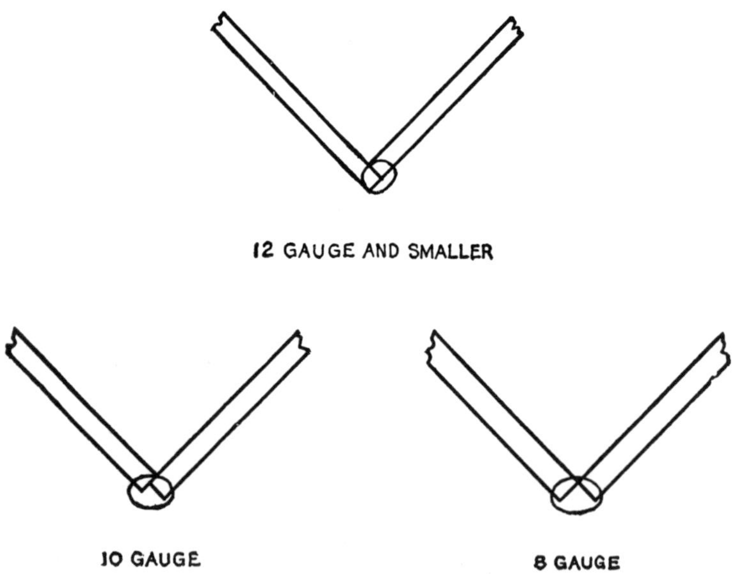

FIGS. 11 to 13.—Overhead corner welds.

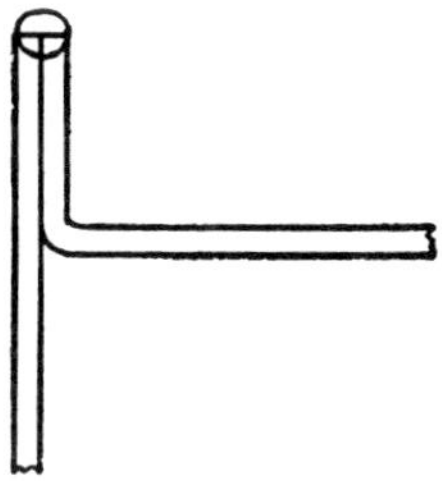

FIG. 14.—Edge weld flat position. Edges should be fitted up close. The electrode to be held perpendicular and drawn along seam with arc only long enough to obtain a smooth rounded bead.

The average operator, however, will find coated electrodes such as "Thinweld" for instance highly desirable. With it many operators have welded 22 and 24 gauge sheet steel and there are cases on record where experts have welded metal as light as 26 gauge in special applications, using these special electrodes.

It is more difficult to strike and maintain the arc with low current values. At the same time, if the current value be too high, it is difficult to work fast enough to keep the intense heat from melting or burning through the light sheets.

Warpage, too, due to expansion and contraction is more of a problem when welding light gauge metal.

Ques. How is the proper current adjustment of the welder determined?

Ans. This is best determined by experience.

It can be determined fairly well in advance of actually welding a joint by laying practice beads on scrap metal of the same thickness.

In practical work it will be necessary to adjust the current value, speed of travel and position of electrode to suit the particular conditions of backing, fit-up, etc.

Ques. Should backing be used on thin metal?

Ans. Yes.

Copper should be used if possible, as it helps dissipate excess heat and keep the molten metal from running through the joint.

Ques. How should an electrode of small diameter be held?

Ans. It should be gripped at the middle.

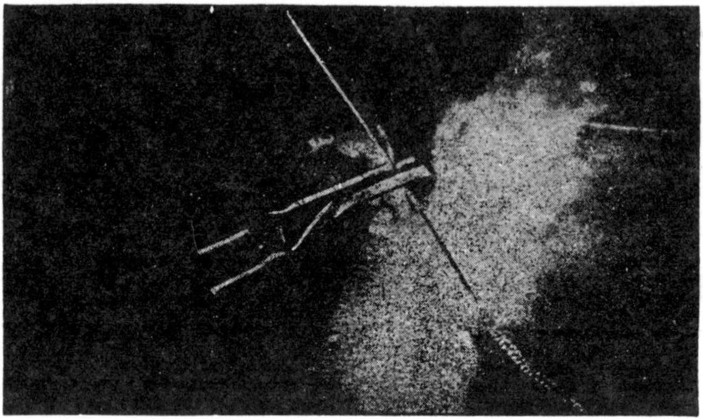

FIG. 15.—Best method of gripping 1/16 in. or 3/32 in. electrodes.—*Courtesy of Hobart Bros.*

Be sure that the coating is rubbed off at the gripping point so that the metal jaws of the electrode holder make a good electrical contact with the wire of the electrode.

Ques. How should light gauge metal be welded in vertical plane?

Ans. Start at the top and weld down.

This is the reverse of the direction recommended on heavier plates. It is better on light gauge because of the speed of travel that is necessary.

Ques. How is a lap weld made?

Ans. Tack the lap joint at both ends. Clamp the assembly flat on the bench so that the lower sheet is flat on the bench top for the full length of the joint. Lay a straight bead along the joint. Remove slag and inspect for thorough penetration, fusion in both plates, freedom from undercutting or burning through and absence of craters.

Break the sheets apart and inspect for freedom from porosity, entrapped slag and gas pockets as well as for penetration and fusion uniformly along the full length of the weld.

Ques. How is a butt weld made?

Ans. Tack the two sheets together with proper spacing. Lay the assembly flat on the bench and clamp it firmly to the

FIG. 16.—***Lesson 1.*** *Practice corner weld.* Begin by setting up two pieces of light material (perhaps 20 gauge metal) over a firebrick in the form of a tent (about a 60° angle) with the top edges just touching. These pieces should be about 6 by 8 in. Weld along the long side. Do not space the edges at the far ends; simply place them so the sheets are just touching all along the corner. Fit the welding blow pipe with the tip or welding head recommended for work on that thickness of sheet. Put on goggles, adjust the flame to neutral and begin at one end, holding the blow pipe as explained. Merely melt the two edges together. To do this, apply the flame to the edges, moving it along fast enough to fuse the edges without burning a hole between. Work carefully and steadily to see that the sheets are melted to the bottom and are fused together all the way to the under side. It may take some practice to avoid burning holes between the edges, but if the metal be melting too rapidly, simply flip the flame away momentarily and allow the puddle to solidify slightly before going ahead. Be sure not to work so fast that melted metal is blown ahead on cold edges.

bench so that the joint is in contact with the bench for its full length. Lay a straight bead rapidly along the joint, being sure to get thorough penetration to the bottom of the seam, with the top at least flush with the top of the seam.

NOTE.—***Testing the joint*** (fig. 16). After welding, set work on a flat surface; flatten by hammering on top. Examine joint from underneath. If the weld crack open and exhibit places where the original edge of the sheet is still to be seen, it indicates insufficient penetration or that the molten metal has been blown ahead. Defective weld may be due to "burned" metal with excess oxygen—indicated by coarse glittering crystals, frequently having peacock colors.

Ques. What should be noted about welding light gauge metal in vertical or inclined plane?

Ans. Always weld "downhill".

Oxy-Acetylene Welding Procedure.—Before starting to make actual welds, the learner should become familiar with

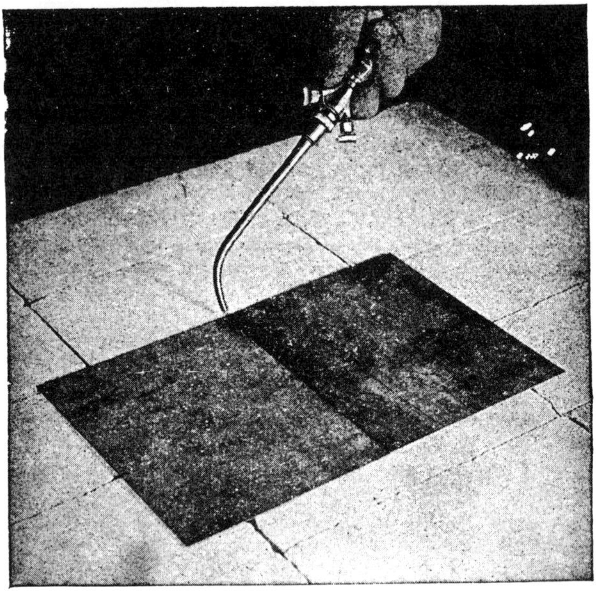

FIG. 17.—***Lesson 2.*** *Practice butt weld.* Take two pieces of 1/16 in. (16 gauge) sheet about 8 in. long and place them flat on the welding table with the edges just touching as shown. A firebrick or a piece of steel bar may be placed on each piece far enough from the edge so as not to interfere with the work. Adjust the welding blow pipe with the proper head and oxygen pressure and weld the seam from end to end. Move the blow pipe along in a straight line at just sufficient speed to fuse the edges together thoroughly. Be sure to fuse the metal through to the bottom of the sheet. Do not be disturbed if the finished weld be depressed a little below the surface of the two sheets. Even on a very good weld it probably will be; but in this exercise, the principal thing is to secure complete penetration between the edges. ***In testing,*** examine under side. There should be some small rounded beads, but all traces of the crack between the edges should be gone. Clamp work in a vise or along the edge of an anvil and hammer the piece down by striking the sheet on the top side. A good weld will bend through 90° without difficulty. Moreover a good weld will stand hammering flat after bending without breaking. If it break, over-heating is indicated.

FIG. 17.—***Lesson 3.*** *Practice with welding rod.* After practicing lesson 2, try the same thing, using a 1/16 in. steel welding rod. Hold the tip of the rod so it touches the puddle of molten metal as it melts. Never hold the rod so that it drips, drop by drop, into the weld. By fusing the welding rod into the weld as the welding proceeds, build up the surface of the sheet. Welds on light sheet metal are usually made just flush in this way for appearance. In making the corner weld, be sure to add enough metal from the welding rod so that a full corner is formed. This is a most important point in fabricating sheet metal products, such as gear guards, in which corner welds are used.

the "feel" of the blow pipe by simply melting the surface of a piece of sheet steel.

Take a piece of sheet steel about 6 in. square and place it flat on the welding table. Fit the blow pipe with a welding head or tip one size smaller than that recommended in the manufacturer's table for that thickness of metal. Put on goggles. Light the blow pipe and adjust the flame to neutral.

Hold the blow pipe with the head or tip inclined at an angle of about 60° to the surface of the sheet so that the flame points ahead along the line of weld.

With the tip of the inner cone of the flame about ⅛ in. away from the sheet, move the blow pipe along in a straight line across the sheet slowly enough just to melt the sheet without burning a hole clear through.

According to Chapman, flame should be close to inner tip as possible, but not in inner tip (cone).

Just before the edge of the sheet is reached, hold the blow pipe still until a hole is burned through the sheet. Observe how rapidly the steel melts and flows. Continue this practice until thoroughly familiar with the action of the flame on the metal.

Sheet welding procedure is illustrated in figs. 16 to 18 by courtesy of The Linde Air Products Co.

Welding Wire.—Welding wire should generally be similar in composition to the metal being welded, and wire should always be used in preference to strips cut from the sheet.

The wire should never be less than 1/16 in. diameter; in fact, for metals having a low melting point, such as aluminum, the welder can exercise more control over the molten puddle by using a rod as large as ⅛ in. diameter, for even 16 gauge sheet.

Ques. What rod should be used for thin aluminum sheets?

Ans. Aluminum wire containing silicon and having a lower melting temperature than pure aluminum.

Oxy-Acetylene Welding Fluxes.—Owing to the fact that all metals tend to oxidize when heated or melted and that the oxide of most non-ferrous metals melts at a higher temperature than the metal itself, a flux is necessary for non-ferrous metals in order to prevent oxide being trapped in the solidified weld metal.

Oxide of iron or steel melts at a lower temperature than the metal and is dissolved in the weld puddle so that a flux is not necessary.

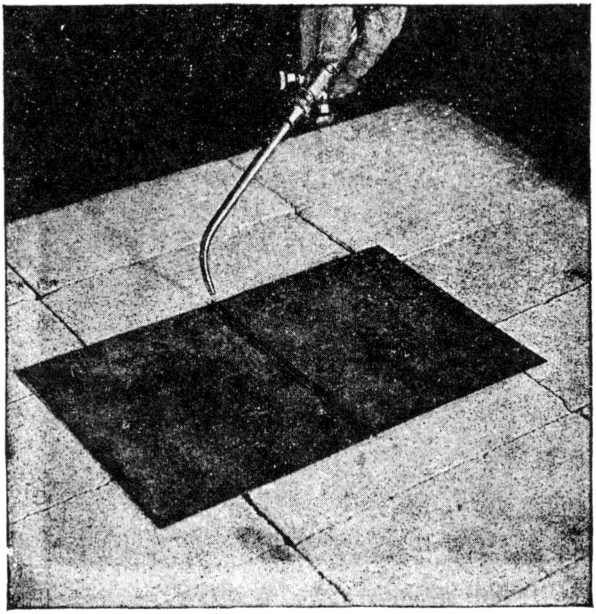

FIG. 18.—***Lesson 4.*** *Practice flange weld.* Take two pieces of 24 gauge sheet. To prepare the joint turn up a very thin edge along one side of two pieces of the sheet. The flange can be turned over the edge of a steel plate, or in any other way that is convenient. The height of the flange above the surface of the sheet should not be more than the thickness of the sheet; if higher, shear off the extra metal. Put the two flanged edges together so the flange stands up, and weld from one end to the other by melting down and fusing the flanges. No welding rod will be needed. It is a little more difficult to secure fusion to the under side in this kind of a weld because of so much more metal to melt down, but it can be done without burning holes through after acquiring sufficient skill. ***Testing.*** Such a weld should be tested in the same way as the previous type.

Ques. What precaution should be taken in using fluxes and why?

Ans. They should be used sparingly, partly because of the trouble of cleaning off burnt flux after welding, and partly because of the risk of including particles of flux in the weld metal.

Ques. What is the objectionable property of aluminum flux?

Ans. It is corrosive.

NOTE.—***In learning*** the welding rod technique, the tendency is to work too fast and melt the rod between the edges before these are properly melted through to the bottom of the sheet. A weld of this sort will break when hammered in the vise.

FIG. 19.—Ingenious swivel type jigs facilitate production welding of intricate assemblies.

FIG. 20.—Bronze welding in the manufacture of metal furniture.

When included in the solidified weld it sets up internal corrosion which cannot be stopped.

Ques. What should be done after making a weld with aluminum flux?

Ans. Wash off all traces of the flux from the front and back of the weld, before the weld has cooled, with hot water or a weak solution of nitric acid, finishing washing with clean water.

Ques. Should flux be applied by dipping the heated welding wire into the flux?

Ans. No.

This is because of the irregular and usually excessive amounts of flux that are supplied to the weld.

Ques. Should home made fluxes be used?

Ans. No.

Ques. How can welds in sheet steel, aluminum and copper be improved?

Ans. By light hammering or rolling after welding.

Steel and aluminum should be hammered cold, but copper hot. Hammering improves the weld strength and can be used to eliminate any buckling or distortion that may have taken place around the weld.

Only light hammer blows should of course be used, and the underside of the weld should be supported by an anvil plate or dolly.

CHAPTER 7

Off-set Pipe Connection

A simple lobster-back bend is one which is shown laying flat on the paper, board or floor. Such a bend is illustrated in fig. 1, in which the central axis AB, of the bend lies in the plane of the paper.

To develop one segment, the usual method of "unrolling" or developing by the parallel line method is then applied to obtain the pattern as shown in the diagram.

Angular Off-set Pipe Connection.—A pipe connection such as that shown in fig. 2 consists of three pipes, A, B and C, whose pattern development requires two projections for its proper solution.

As illustrated, the middle connecting piece A, joins two pipes B and C, whose center lines produced would pass at right angles to each other at a distance of three feet apart. Separately, the joints between A and B, and A and C, are simple elbows, but together they form complex angles with the middle part A and hence need auxiliary projections to place them correctly in relation to each other.

The solution of the problem is shown in fig. 3. In the first projection from the plan, the elbow between A and B is obtained so that it lies flat in the plane of the paper. Similarly, in the second projection from the elevation, the elbow between A and C is obtained so that it lies flat in the plane of the paper.

The development of the middle part A, is most easily obtained from the first projection as shown in fig. 3. Before the pattern can be "unrolled" the ellipse which represents the joint line of the elbow between A and C in the first projection must first be correctly drawn. The starting point in the determination of this ellipse is in the second projection from the elevation. To obtain this view, project the points M′ and O′ perpendicularly into the projection, and mark off the base line M°,N° in any convenient position parallel to M′,O′.

Next mark off the height N°,O° equal to the distance N,O in the plan. Also mark off O°,P° equal to the distance O,P in the plan. Join M°,O°.

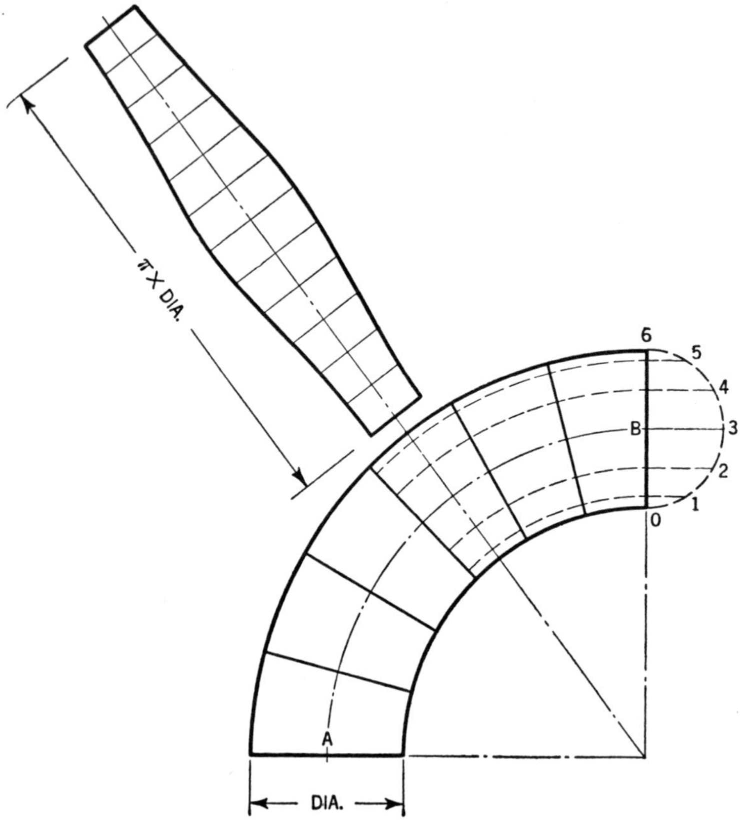

FIG. 1.—Pattern development of simple lobster-back bend.

Then the angle M°,O°,P° represents the true angle of the elbow, and the pipe may be drawn as shown, with the joint line passing through O° bisecting the angle M°,O°,P°. The usual procedure may now be followed, of drawing the semi-circle on the end of the pipe at P°, dividing it into six equal parts and projecting the points back perpendicularly to the end of the pipe.

These lines should be further produced right back to the circle around point O′ in the elevation, thus dividing the circle into similar parts and points to those on the semi-circle. The points obtained on the circle around O′ should now be projected vertically downwards into the plan to meet the pipe end at P. It will be observed that as the points around the circle do not fall vertically behind one another, there will be twelve lines projected downwards instead of the usual seven.

The next stage is to obtain the ellipse around the point O in the plan. To do this, take the distances in the second projection from the pipe end

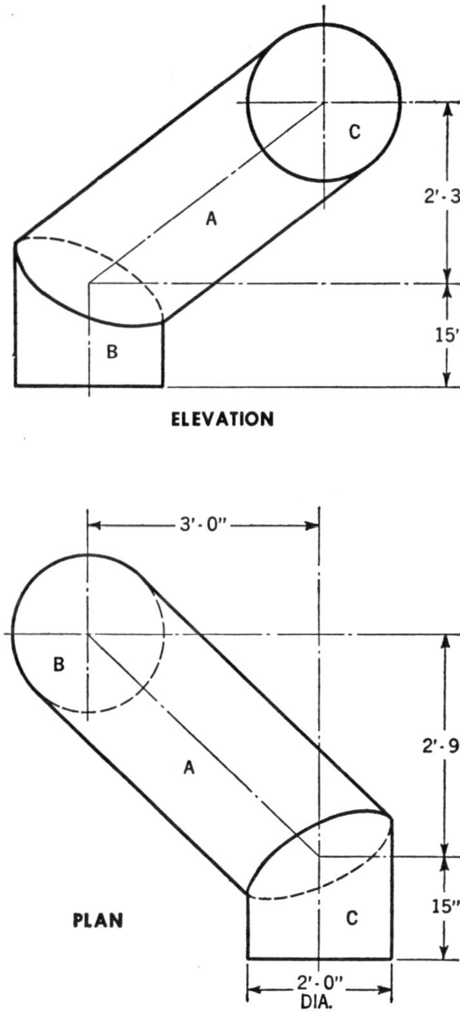

FIG. 2.—Typical connection between three pipes. For pattern development see fig. 3.

P° to the joint line through O°, and mark them upwards on the corresponding lines in the plan from the pipe end P.

For example, take the distance *a*°,*b*° from the second projection, follow the line of projection back to the circle in the elevation where it will be seen that two corresponding points, *b*′ and *b*′ are represented, follow these points downwards into the plan and mark the distances *a*,*b* upwards from the pipe end P. This process carefully applied to the other lines between the pipe end P° and the joint line through O° should produce sufficient points in the plan to draw the ellipse around the center point O.

The next stage in the development is to obtain the ellipse around the center point O″ in the first projection. To set out the projection, first project lines from the plan points M and O at right angles to the center line M,O and mark off a base line M″N″ in any convenient position parallel to M,O. Mark off N″,O″ equal to the height N′,O′ in the elevation, and

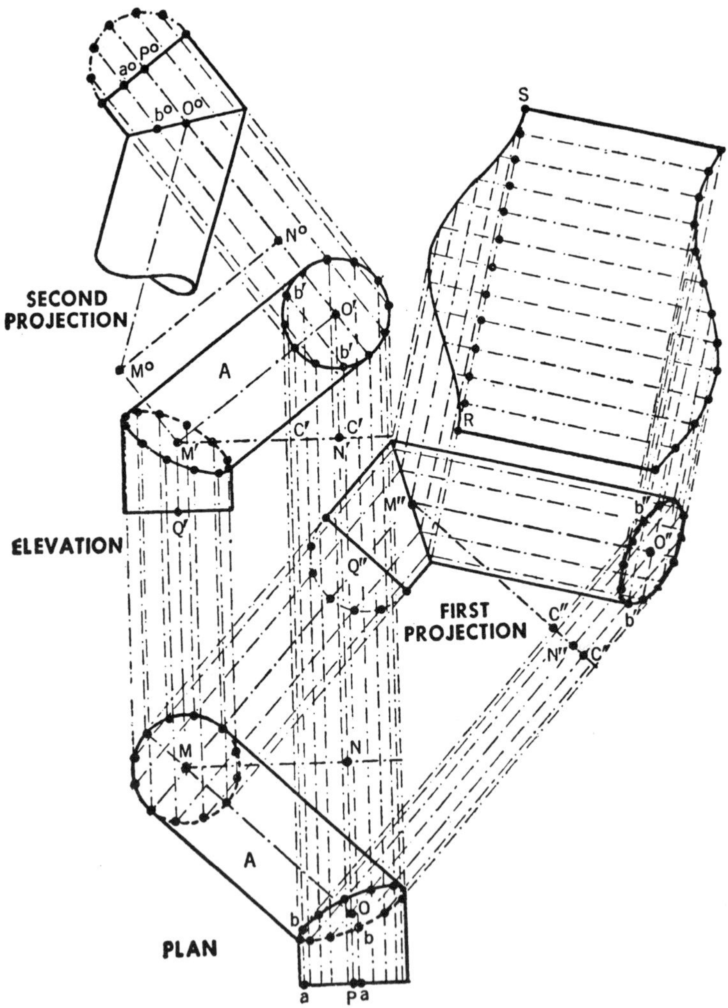

FIG. 3.—Solution to pipe connection problem shown in fig. 2.

join M″ and O″. Next mark off M″,Q″ equal to M′,Q′ in the elevation. Then the angle O″,M″,Q″ represents the true angle of the elbow at M″, and the pipe may be drawn as shown, with the joint line passing through M″, bisecting the angle O″,M″,Q″.

Now, to obtain the ellipse around the center point O″, project all the points on the ellipse O from the plan into the first projection. Then mark off the heights from the base line M″,N″, equal to the heights in the elevation from the base line M′,N′ to the corresponding points on the circle O′. For example, take the two points *b*, already located on the ellipse in the plan.

These points projected into the first projection give the two points *b″*,*b″*, at heights *c″*,*b″* and *c″b″* which correspond respectively to the heights *c*,*b′* and *c′*,*b′* in the elevation. Thus, this process repeated, of taking the heights from the base line M′,N′ to all the points on the circle O′, and marking these heights in the first projection from the base line M″,N″, should provide sufficient points for drawing the ellipse around the center point O″. This ellipse should be carefully drawn, as the accuracy of the pattern will depend largely on the care expended in plotting the ellipse.

The development or "unrolling" of the pattern may now be undertaken. First draw the semi-circle on the pipe end Q″, divide it into the usual six equal parts, project the joints perpendicularly back to the pipe end, and produce them to the joint line at M″. Next, project all the points on the joint line parallel to the center line M″,O″ to cut the ellipse around the center point O″. It is important here to note that the lines so drawn will *not* necessarily cut the ellipse at the points already located in plotting the ellipse. This does not matter however, provided that the ellipse has been carefully drawn.

Now project all the points on the joint line M″ into the pattern. Repeat this with the points on the ellipse where the corresponding parallel lines cut the ellipse. Next mark off twelve divisions along the girth line R,S equal to those on the semi-circle on the pipe end Q″, and draw parallel lines through them at right angles to R,S. It now remains only to plot the points for the curves at each end, and this process should be easily followed from the diagram.

Three-Way Junction of Tapered Segment Bends

In fig. 4 is shown a three-way junction, formed by the intersection of identical, tapered or lobster-backed bends. Although it is unlikely that a duct will be required to branch into three equi-angled directions, for the sake of expediency in pattern development, the transition piece is designed as shown, the three smaller ducts being left to follow their respective courses after the division.

This is one of those problems requiring a sound knowledge of projection, particularly conic sections. Further difficulty is added by the confusion of lines consequent upon the need to solve two geometrical problems on the one layout, that is, to ascertain the position and shape of the line of intersection, and the development of the segment patterns.

Layout.—Fig. 5 represents the layout. As mentioned previously, the three sections of the unit are identical in shape in order to minimize development problems, the chief advantages of similarity being that the line of intersection from a plan view is straight, and that the patterns for one bend will do for the other two. The first operation in constructing the layout is to draw the outline of one complete bend in elevation.

To do this the positions of the four extreme corners of the elevation are determined and an arc drawn from "*g*" on the base line through an angle of 90°, using a suitable radius compatible with the principles of good air-flow.

This arc is divided into six equal divisions—the number of segments in the bend. The center for drawing the arc is enclosed by the upper small circle (fig. 5). Thus the position and shape of the throat are fixed.

The back is drawn by first bisecting the length of the back and then locating on the bisecting line a suitable point to serve as a center, having regard for the fact that the efficiency of the bend requires the reduction in cross-sectional area to be constant throughout its length. The area of the smaller end is one-third that of the larger. The center for drawing the back arc is enclosed by the lower small circle, which by coincidence, falls on the edge of the plan view.

From the throat center, straight lines are drawn fan-wise, through the points of division in the throat arc till they intersect the back arc. These points of intersection, both in the back and throat are connected by straight lines to form the outline of the bend.

The segments are lined in completely and indicated by capital letters A to F, and to facilitate development, the edges of the segments are to be circular in shape.

Below the base of segment "A" the first and largest section, a semi-circle is drawn whose diameter is equal to the length of the bottom edge of the segment seen in elevation ("*a*" "*g*"). Another, smaller semi-circle is drawn on the top edge of segment A, then on segment B,C,D, etc. each succeeding semi-circle having a reduced diameter till segment F is reached. With the exception of the one in the plan view (segment A) these semi-circles are

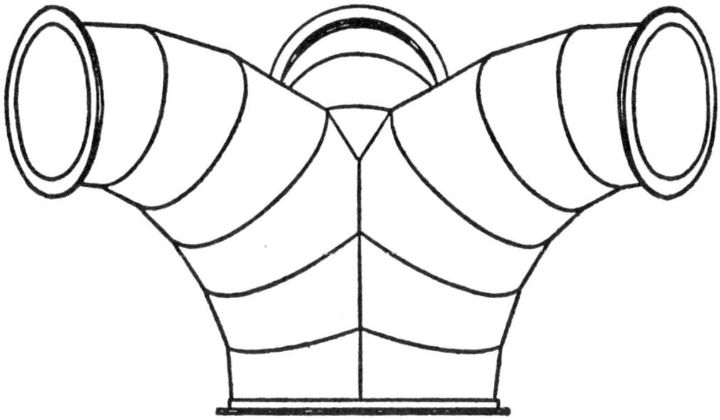

FIG. 4.—Typical three-way duct junction.

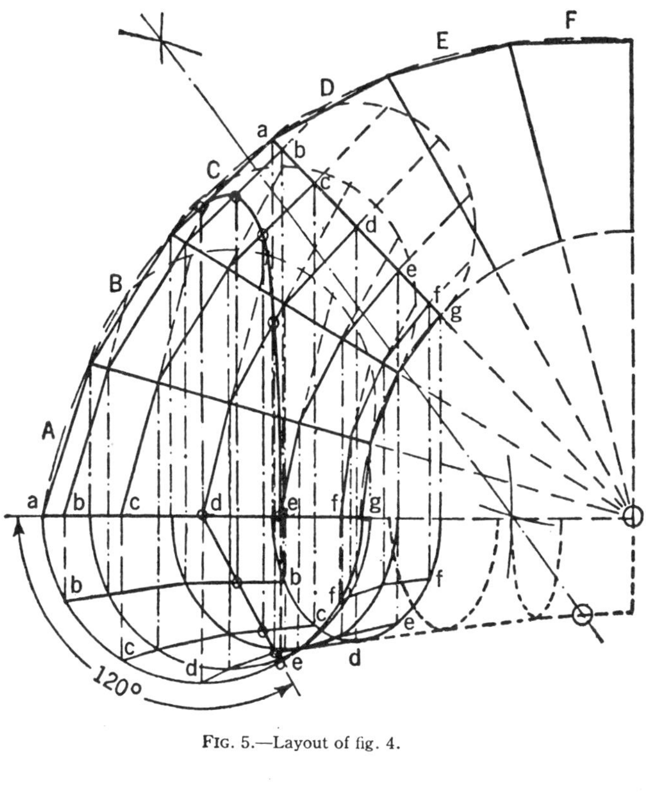

FIG. 5.—Layout of fig. 4.

drawn in broken lines in the elevation (fig. 5) (for reasons of clarity, main construction lines are shown solid, secondary construction lines broken and projection lines dash-dot). They are then divided into six, lines being projected from these points to their respective center lines, namely: the edges of the segments.

The points on the center-lines or segment edges, where these lines intersect are connected by straight lines, the points on one segment being joined to the corresponding points on the next, etc. These lines are drawn solid and lettered from "*a*" to "*g*". In order to avoid confusion, they will be assumed to be continuous throughout the length of segments A,B and C, as these three contain within them the line of intersection. This is the reason for lettering the extreme ends of the lines.

The half-plan view is constructed as shown by projecting vertically downwards all the points of intersection of lines "*a*", "*b*", "*c*", etc. with the segment edges.

To take a typical example, point "*a*" at the top of segment "C" is projected down to the base line, which fixes its position as it is on the center. The next point "*b*", is projected a distance past the base line sufficient to accommodate the length taken at right angles to "*b*" on segment "C" to the edge of the semi-circle (elevation), this length being stepped off from the base line along the extended projection line. The same operation is performed with the rest of the points, the result, in the plan view, being a half ellipse.

The lines "*a*", "*b*", "*c*", etc. in the elevation are reproduced in the half-plan view by connecting the development points in the ellipses which were projected from the elevation, segments A, B and C only being dealt with in this way.

As mentioned previously, the line of intersection from a plan view is straight, and as there are three bends branching away at equal angles to each other, the angle of intersection is 120°. Therefore, a line is drawn, as shown, at this angle from the center of the base (point "*d*") to the outer edge. It cuts across three segments.

The points where this line intersects lines "*a*", "*b*", "*c*", "*d*" and "*e*" in the half-plan are enclosed with small circles (fig. 5) and they are projected, vertically upwards until they intersect the corresponding lines in the elevation. These points also are enclosed in small circles and when joined by a curve give the shape of the line of intersection when seen in elevation.

It will be noted that the outer edge of the line of intersection falls on point "*e*". This is obvious as point "*e*" was obtained in the first place by the division of the base semi-circle into six, which gives "*e*" an angle of 120° with "*a*".

Pattern Development.—The large number of lines in fig. 5 may appear confusing, but if the projection lines be followed carefully, it will be seen that the principle of projection when dealing with single segments is fairly simple. It is the necessity of having to superimpose one segment over another to obtain

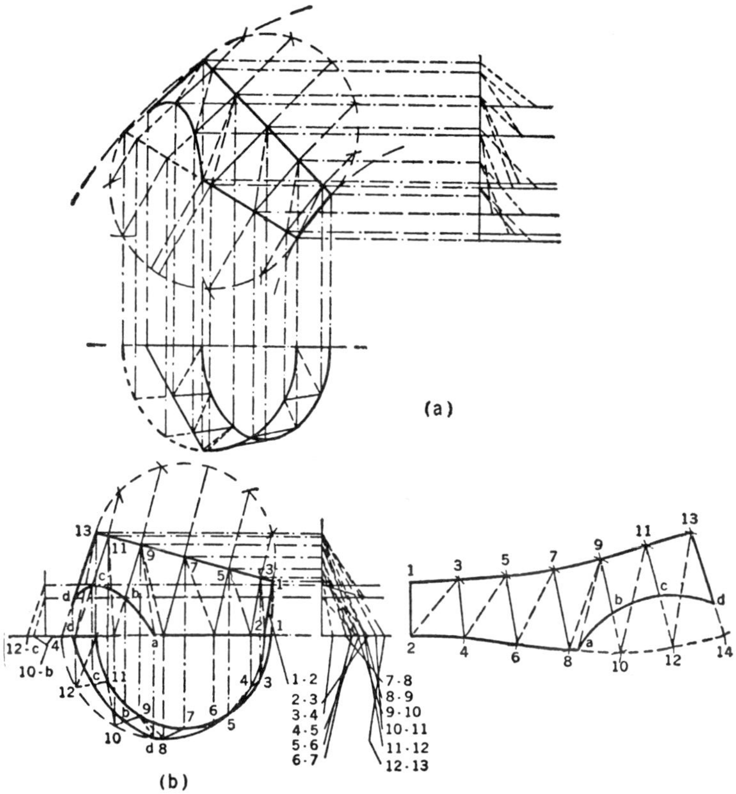

FIG. 6.—Pattern development.

the line of intersection which causes the difficulty. This multiplicity of lines is often unavoidable in surface development, and more errors are attributable to this fact than to lack of understanding of the geometrical principles involved.

In this case, for instance, it would be necessary in order to develop the segment patterns, to add still more lines to the layout in fig. 5. To obviate this further addition, it is preferable to lay each segment out separately. In addition, another important advantage to be gained from this arrangement can be readily appreciated by referring to fig. 6 (*a*), where a typical segment (C) is laid out in the position it would occupy in fig. 5.

In this slanting position the base lines for obtaining the true lengths of the development lines are constantly changing, whereas in (*b*) the same segment is laid out level, thus giving to the base, from a plan view, a circular shape. This means that all the true lengths are developed from a common level, the only addition being that the shape of the line of intersection in the elevation has to be projected to the half plan. Even this is offset by the reduced risk of error as it is now possible to draw the half-plan view directly below the elevation. This close proximity of the two views is desirable as it shortens the lengths of the projection lines.

The shape of the segment is taken directly from the main layout (fig. 5) and drawn in elevation (fig. 6 (*b*)), the half-plan view of the whole segment exclusive of the cut away position, being projected in a similar manner as described for the layout. The plan-view shape of the line of intersection is also found by projecting points "*a*", "*b*", "*c*" and "*d*" from the elevation to the plan, then joining them to form a curve.

The points in both views are numbered as shown, the cut-away portion being lettered. The numbers follow through consecutively from side to side, this system being extremely helpful, particularly when development lines become partially or wholly obscured as in the plan view (fig. 6 (*b*). The continuity of the numbers then becomes the only guide for obtaining the necessary true lengths.

The length 1-2 can be taken directly from the elevation to start the pattern; 1-3 is the length of one division in the semi-circle above the top of the elevation, while 2-4 is a similar division in the base semi-circle in the plan view. The diagonal development line (dotted) 2-3, is obtained by setting the plan of the line along the base of the true-lengths diagram and so finding

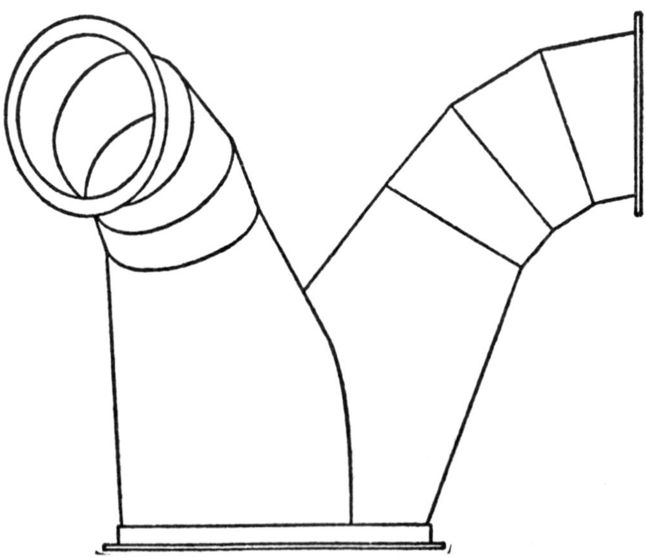

FIG. 7.—Alternative design of three-way duct junction shown in fig. 4.

the hypotenuse length of the triangle whose other side is the elevation height, "base" to 3. This is repeated until the whole of the pattern has been developed, assuming there is no cut-away. This cut-away section can of course, be developed as an integral part of the pattern, but in this case it is simpler to develop the whole half segment and then step off along the main development lines (solid) the points at which the cut-away curve intersects.

Point "*a*" of the cut-away is found by taking the length "8-*a*" along the circumference of the base and stepping it off along the bottom curve of the pattern; "*b*" is obtained by using "*b*"-10 (plan) as a plan length, and "base"-"*b*" as an elevation-height (elevation). The hypotenuse-length is the true length of "*b*"-10 and can be stepped off along the development line in the pattern. Similarly, point "*c*" is found, while "*d*" can be taken directly from the elevation.

Alternate Design.—As with many development problems, design can eliminate much of the work and fig. 7 illustrates an alternative shape.

In this instance a transition divides the main duct into three, both with respect to direction and volume, and is formed by the juncture of three identical conic sections. The change in direction of air flow is then completed by means of ordinary segment bends which adjoin the smaller ends of the unit.

It will be readily appreciated that this design requires only two patterns—one for each of the conic sections and another for a segment.

Pattern for Fume Hood

The fume hood illustrated in fig. 8 supplies a typical example on the elusive character of pattern drafting. The point of difficulty here occurs on the sides or cheeks of the hood as at MNOP in the left hand hood in the side elevation.

It will be observed that there are four straight lines, MN, NO, OP and PM enclosing the quadrilateral MNOP. The point is, however, that the metal inside MNOP is not flat or in one plane. Nor can the hood be made with this portion flat although from the figure it might appear it could be. There must be a bend or kink across one of its diagonals as illustrated at A or B, fig. 9.

A bend across either diagonal will enable the metal to assume the correct form as shown in fig. 8, but with the difference that a bend across the longer diagonal at A, will be inwards or "knuckle in", whereas a bend across the shorter diagonal at B, will be outwards or "knuckle out".

It is now readily apparent that the patterns developed to these two alternatives will be appreciably different, although they may both be worked up to meet the conditions shown in the drawing. The two patterns are shown in fig. 10, and it will be observed that the angle 3″7″8″ in the figure at (*a*) is not a right angle, whereas in the corresponding patterns at (*b*) the corresponding angle 3″7″8″ is 90°. Also, the straight cross distance between points 3″ and 8″ is greater in (*a*) than in (*b*). In addition the difference in the angle at 3″7″8″ also affects to some extent the shape of the curve between points 8″ and 13″.

These differences in the two patterns emphasize the need for care in the process of dividing up the surface for development by triangulation, and the importance of subsequently shaping the article in accordance with the method of surface division.

In dividing up the surface of this hood for triangulation, beginning at the center of the back and numbering one half only, the first portion from 1 to 6 is straightforward tallboy triangulation. Then comes the quadrilateral 6,8,7,3. Perhaps the better choice of diagonal would be from 3 to 8, since this, as the bend is "knuckle outwards" would give the hood a little greater capacity than the alternative diagonal. In this construction the triangle

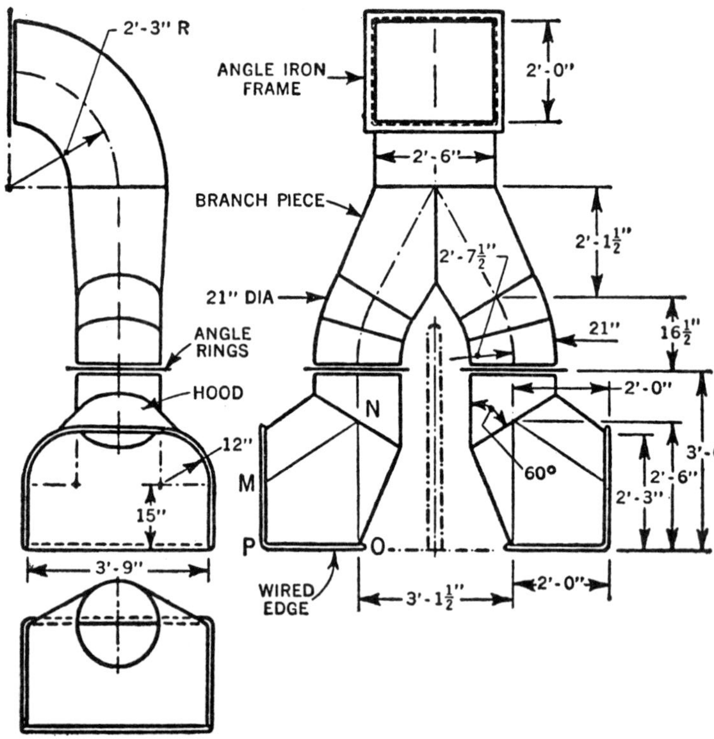

Fig. 8.—Illustrating typical fume hoods and extraction ducts.

3,7,8, falls in a vertical plane. The next section from point 8 to point 14, is simple triangulation between the two curved edges comprising points 8,9,11,13 on the front of the hood, and points 6,10,12,14 on the joint edge at the top. Finally, the triangle 13,14,15 forms a flat piece on the front of the hood.

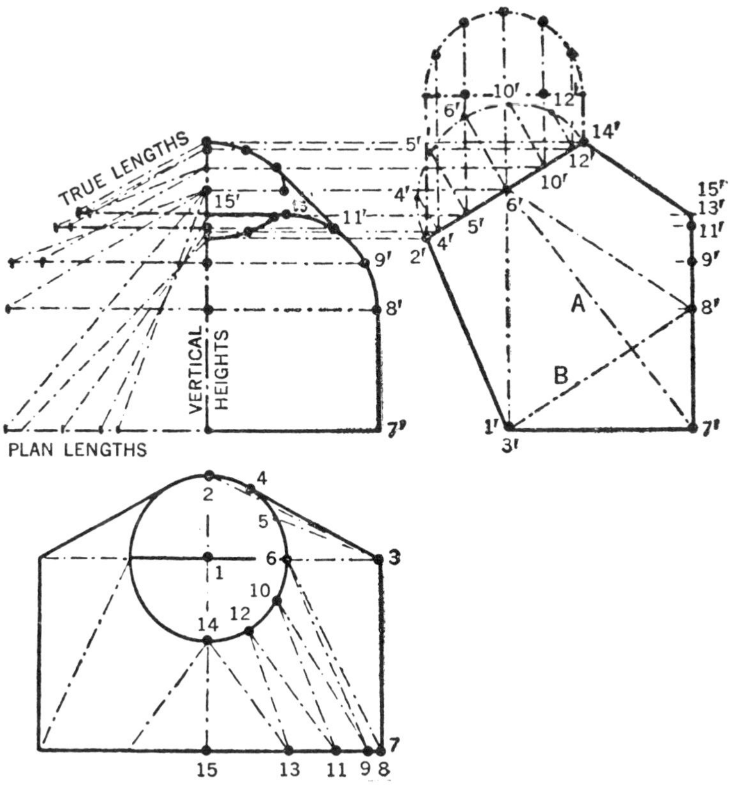

FIG. 9.—Hood development.

Prior to developing the pattern, first erect a vertical height line, and project all the points in the elevation to it. In the illustration at Fig. 9, only one half of the front elevation is shown, and the centre line of this view is used as the vertical height line. Next, since the true shape of the joint line between points 2 and 14 is elliptical, lay out half of the true ellipse as shown in the side elevation on the axis 2′14′. The spacings around this elliptical curve will then give the true distances between the points 2′,4′,5′,6′,10′,12′, and 14′, which will be required for the pattern.

To develop the pattern first take the true distance 1′,2′, direct from the side elevation, and mark off 1″,2″, in the pattern in any convenient position. Next take the true distance 1,3, direct from the plan and from point 1′ in

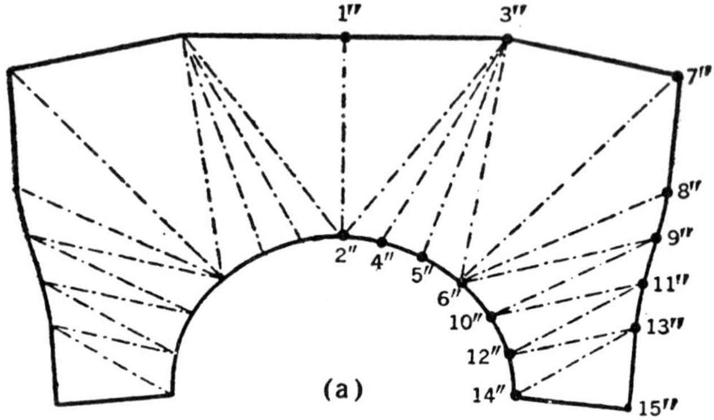

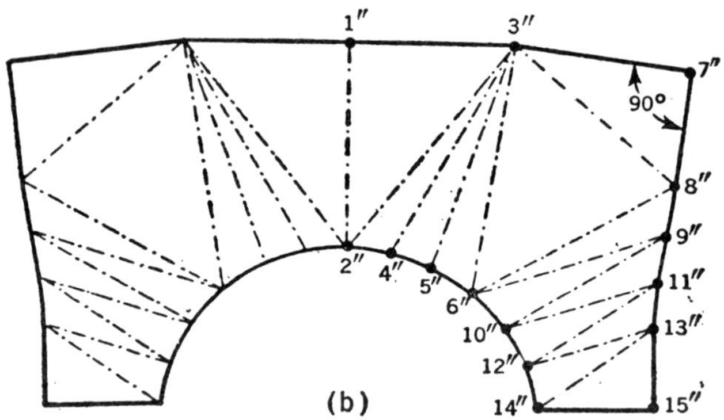

FIG. 10.—Pattern for hoods shown in fig. 8.

the pattern swing an arc through point 3″. Now take the plan length 2,3 and mark it off along the base line at right angles to the vertical height; take the true length diagonal up to the point level with 2′ and from point 1″ in the pattern describe an arc cutting the previous arc in point 3′. This completes the first triangle.

For the second triangle take the plan length 3,4 and mark it off along the base line at right angles to the vertical height; take the true length diagonal up to the point level with 4′, and from point 3″ in the pattern swing an arc through point 4″. Now take the true distance corresponding to 2′4′ round the ellipse in the side elevation, and from point 2″ in the pattern describe an arc cutting the previous arc in point 4″.

For the third triangle take the plan length 3,5 and mark it off at right angles to the vertical height; take the true length diagonal up to the point level with 5′, and from point 3″ in the pattern swing an arc through point 5″. Now take the true distance corresponding to 4′5′ round the curve of the ellipse in the side elevation, and from 4″ in the pattern describe an arc cutting the previous arc in 5″.

For the next triangle take the plan length 3,6 and mark it off at right angles to the vertical height; take the true length diagonal up to the point level with 6′ and from point 3″ in the pattern swing an arc through point 6″. Now take the true distance corresponding to 5′6′ round the curve of the ellipse in the side elevation, and from 5″ in the pattern describe an arc cutting the previous arc in point 6″.

The next triangle, 3″,6″,8″ is obtained by taking the plan length 6,8,—it will be noted that in the plan the same point represents both 7 and 8, and marking this distance 6,8, off at right angles to the vertical height which is not taken from the bottom base line, but from a new base line level with point 8′.

Take the true length diagonal up to the point level with 6′ and from point 6″ in the pattern swing an arc through point 8″. Now, it will be observed that in the elevation the distance between 3′ and 8′ represents its true length. Therefore, take the true length 3′8′, direct from the elevation, and from point 3″ in the pattern describe an arc cutting the previous arc in 8″.

Since the triangle 3′,7′,8′ is in a vertical plane, the view in the side elevation presents its true size. Therefore, to add this triangle to the pattern, take the true distance 3′7′, direct from the side elevation, and from point 3″ in the pattern swing an arc through point 7″. Then take the true distance 8′,7′, direct from the side elevation and from point 8″ in the pattern describe an arc cutting the previous arc in point 7″.

For the next triangle take the plan length 6,9, and mark this off at right angles to the vertical height, but again it will be noted that its base line must be level with point 9′. Take the true length diagonal up to the point level with 6′ and from point 6″ in the pattern swing an arc through point 9″. Now take the true distance 8′,9′ direct from the front elevation and from point 8″ in the pattern describe an arc cutting the previous arc in point 9″.

For the next triangle take the plan length 9,10, and mark this off at right angles to the vertical height along the base line level with point 9′. Take the true length diagonal up to the point level with 10′ and from point 9″ in the pattern swing an arc through point 10″. Now take the true spacing 6′,10′ from the curve of the ellipse in the side elevation, and from point 6″ in the pattern describe an arc cutting the previous arc in point 10″.

From this point the remainder of the pattern should be readily followed from the directions previously given. The chief points to observe are that in triangulating the zigzag diagonals 10,11; 11,12; 12,13; 13,14; and 14,15, these should be marked off at right angles to the vertical heights along the appropriate base lines. The true spacings for the top edge, 10′,12′, and 12′,14′ should be taken from the curve of the ellipse in the side elevation, and the true spacings for the bottom edge 9′,11′; 11′,13′; and 13′,15′, should be taken direct from the front elevation.

The Branch Piece.—Referring now to the second problem, the branch piece shown on the drawing in fig. 8, this also may be developed in a number of different ways, resulting in a series of different patterns, all of which could be made to suit the requirements of the drawing.

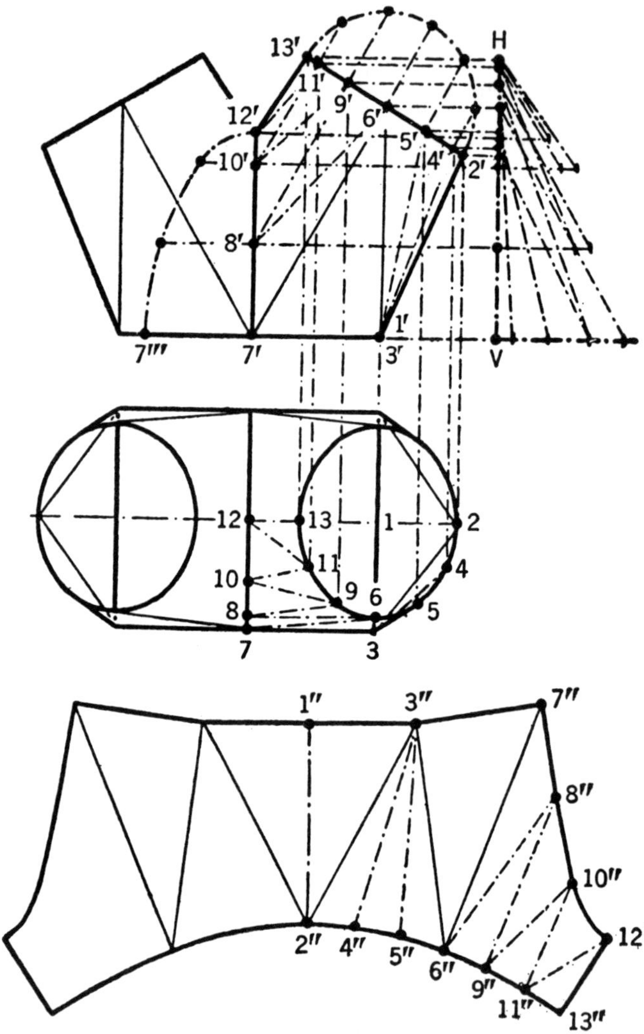

FIG. 11.—Pattern development of branch piece illustrated in fig. 8.

According to the details given, the branch piece is composed of two limbs which together transform from a rectangle at one end to two circles at the other. Various geometrical constructions could be applied to meet the conditions given, some of which are illustrated in fig. 12. The differences lie on the inner part of the branch piece which connects the semicircle at the top to the joint line between the limbs. The form of the joint could be almost any reasonable shape, which could be transformed or connected to the upper half of the circle at the top of the limb.

Perhaps the best construction is that shown at (*d*) fig. 12, which is selected for development in fig. 11. The two limbs which form the branch piece are similar so that the development for one serves for both. The elevation and plan of the branch piece are shown in fig. 11, in which one limb is divided up for triangulation.

It will be noted, perhaps more easily in the elevation, that the part numbers 1′ to 7′ constitutes one half of an ordinary tallboy, transformed from rectangle to circle. The other part, between the two lines 6′,7′ and 12′,13′, transforms from a semi-circle at the top 6′,13′ to a semi-ellipse at the joint line, 7′,12′. Half of the semi-ellipse at the joint line is shown swung back through 90° to present its true shape as seen by the curve between 7‴ and 12′.

Prior to developing the pattern erect a vertical height line V,H. and project all the points in the elevation horizontally to V,H. The lines from 8′,10′ and 12′ should be extended beyond V,H. to serve as base lines in the process of triangulation.

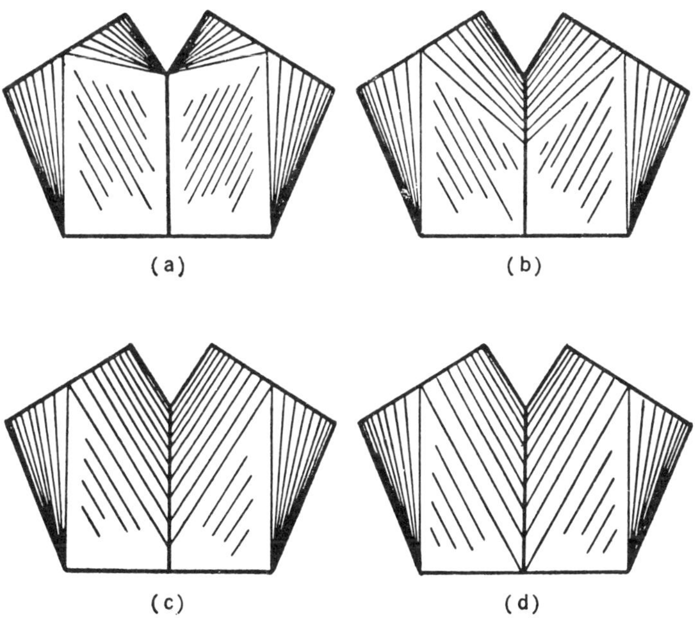

FIG. 12.—Various possible geometrical construction for extraction ducts.

To develop the pattern, first take the true distance 1′,2′ direct from the elevation, and in any convenient position mark off 1″,2″ in the pattern. Next take the plan length 2,3 and mark this distance off along the bottom base line level with 3′. Take the true length diagonal up to the point level with 2′, and from point 2″ in the pattern swing an arc through point 3″. Now take the true distance 1,3, direct from the plan and from point 1″ in the pattern describe an arc cutting the previous arc in point 3″. This completed the first triangle.

For the second triangle, take the plan length 3,4, and mark it off along the bottom base line at right angles to the vertical height. Take the true length diagonal up to the point level with 4′ and from point 3″ in the pattern swing an arc through point 4″. Now take one of the equal spacings from the circumference of the semi-circle in the elevation and from point 2″ in the pattern describe an arc cutting the previous arc cutting the previous arc in point 4″.

For the next triangle take the plan length 3,5 and mark it off along the bottom base line at right angles to the vertical height. Take the true length diagonal up to the point level with 5′ and from point 3″ in the pattern swing an arc through point 5″.

Now take another of the equal spacings from the circumference of the semi-circle in the elevation and from point 4″ in the pattern describe an arc cutting the previous arc in point 5″.

For the next triangle take the plan length 3,6, and mark it off along the bottom base line at right angles to the vertical height. Take the true length diagonal up to the point level with 6′ and from point 3″ in the pattern swing an arc through point 6″. Now take another of the equal spacings from the circumference of the semi-circle in the elevation and from point 5″ in the pattern describe an arc cutting the previous arc in point 6″.

For the next triangle take the plan length 6,7, and mark it off along the bottom base line at right angles to the vertical height. Take the true length diagonal up to the point level with 6′ and from point 6″ in the pattern swing an arc through point 7″. Now take the true distance 3,7 direct from the plan and from point 3″ in the pattern describe an arc cutting the previous arc in point 7″.

Next take the plan length 6,8, and mark it off at right angles to the vertical height along the base line which, this time, is level with point 8′. Take the true length diagonal up to the point level with 6′ and from point 6″ in the pattern swing an arc through point 8″. Now take the first true spacing up the curve of the semi-ellipse in the elevation corresponding to 7′8′, that is, the first spacing up from point 7″′, and from point 7″ in the pattern describe an arc cutting the previous arc in point 8″.

Next take the plan length 8,9 and mark it off along the base line level with point 8′. Take the true length diagonal up to the point level with 9′ and from point 8″ in the pattern swing an arc through point 9″. Now take another of the equal spacings from the semi-circle in the elevation and from point 6″ in the pattern describe an arc cutting the previous arc in point 9″.

Next take the plan length 9,10 and mark it off at right angles to the vertical height along the base line level with point 10′. Take the true length diagonal up to the point level with 9′ and from point 9″ in the pattern swing an arc through point 10″. Now take the second true spacing up the curve of the semi-ellipse in the elevation corresponding to 8′,10′, and from point 8″ in the pattern describe an arc cutting the previous arc in point 10″.

The remaining three triangles needed to complete the pattern should readily be obtained by following the directions for the last three triangles, but applying them to the appropriate remaining points.

Screw Conveyor Chute Patterns

Delivery chutes of various forms are used in numerous industrial machinery and processes. In fig. 13 is shown a typical arrangement of a screw conveyor feed to a grading machine, the conveyor having three chutes, namely a *feed chute*, a *delivery chute* and a *bypass chute*.

The methods for developing these patterns differ somewhat from the usual simple type, since in two cases auxiliary projections are used to facilitate the development.

The Delivery Chute.—With reference to the middle transforming piece of the delivery chute, fig. 14, it will be observed that if the ordinary plan were used in conjunction with the elevation for the purpose of developing the pattern, every line used in triangulation for a true length would have a different vertical height.

Since the top and bottom joint edges of the transforming piece are in parallel planes, the development of the pattern will be much simplified by using a projected plan at right angles to the bottom joint line.

By turning the drawing around so that the projected plan occupies the position of an ordinary plan, and the bottom joint line in the elevation becomes a horizontal base line, then it will be seen that only one vertical height will be needed for the purpose of triangulation. The top joint line in the projected plan becomes an ellipse, and represents its true shape.

Prior to developing the pattern, erect a vertical height line VH, at right angles to the base line as shown in the illustration at fig. 14. Divide the plan into triangles and number the points from 1 to 17, arranging for the joint to occupy the position from 1 to 2.

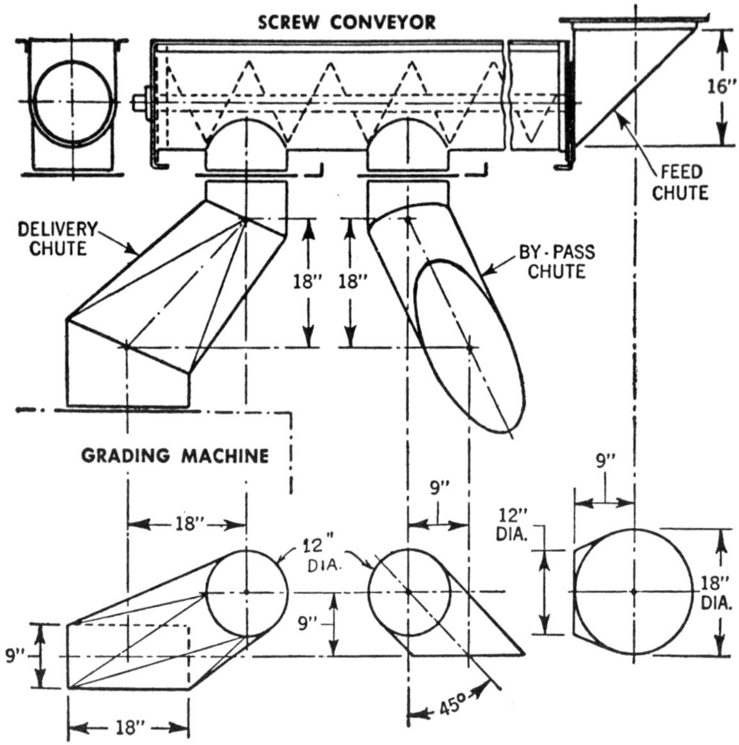

FIG. 13.—Screw conveyor arrangement.

To develop the pattern take the plan length 1,2 and mark it off from point V along the base line at right angles to the vertical height. Take the true length diagonal up to point H, and in any suitable position in the pattern mark off 1′,2′ as in fig. 15. Next take the plane length 2,3 and from V, mark it off along the base line at right angles to VH. Take the true length diagonal up to point H, and from point 2′ in the pattern swing an arc through point 3′. Now take the true length 1,3 direct from the plan, and from point 1′ in the pattern describe an arc cutting the previous arc in point 3′. This completes the first triangle.

For the second triangle take the plan length 3,4 and mark it off from V along the base line at right angles to V,H. Take the true length diagonal, and from point 3′ in the pattern swing an arc through point 4′. Next take the true distance 2,4 direct from the plan and from point 2′ in the pattern describe an arc cutting the previous arc in point 4′.

For the third triangle take the plan length 3,5 and mark it off from V along the base line. Take the true length diagonal, and from point 3′ in the pattern swing an arc through point 5′. Next take the true distance

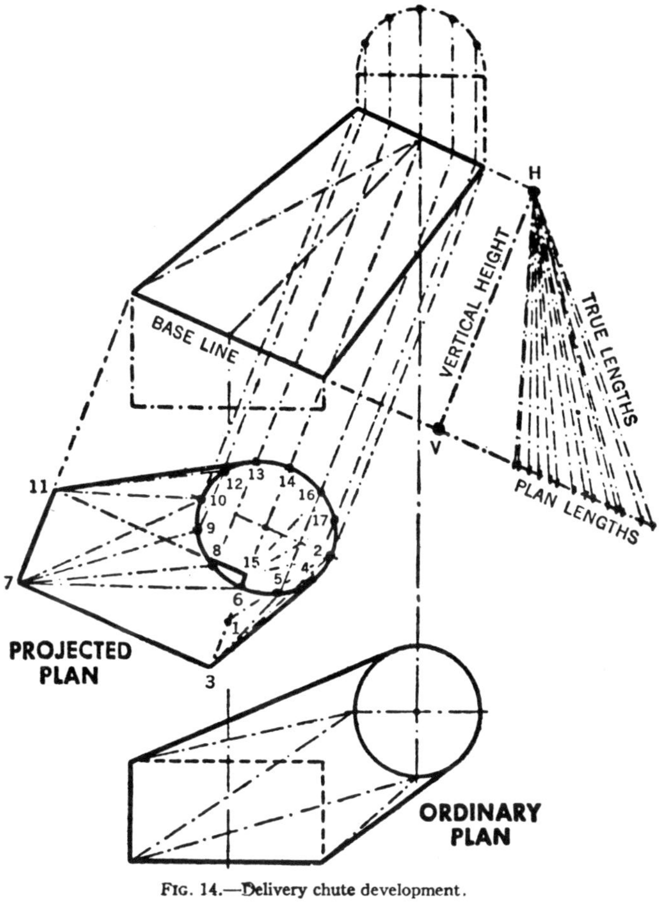

FIG. 14.—Delivery chute development.

4,5 direct from the plan and from point 4′ in the pattern describe an arc cutting the previous arc in point 5′.

For the next triangle take the plane length 3,6 and mark it off from V along the base line. Take the true length diagonal and from point 3′ in the pattern swing an arc through point 6′. Next take the true distance 5,6 from the plan and from point 5′ in the pattern describe an arc cutting the previous arc in point 6′.

The next triangle is reverse in position. Take the plan length 6,7 and mark it off from V along the base line. Take the true length diagonal and from point 6′ in the pattern swing an arc through 7′. Next take the true distance 3,7 direct from the plan, and from point 3′ in the pattern describe an arc cutting the previous arc in point 7′.

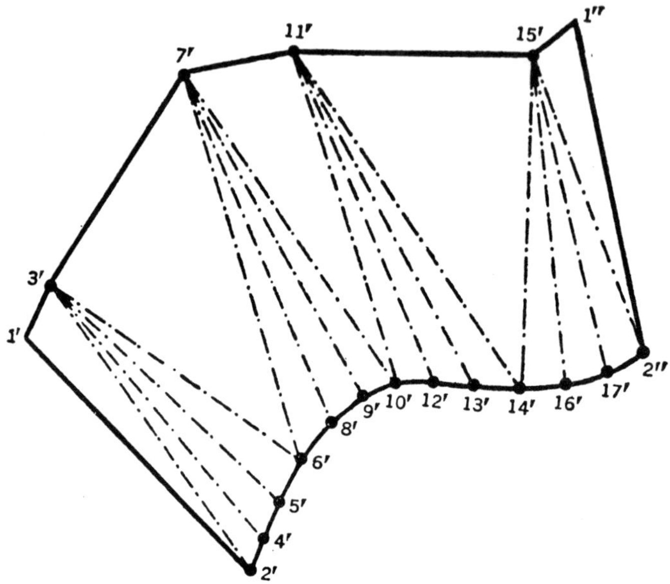

FIG. 15.—Pattern for delivery chute.

For the next triangle take the plan length 7,8 and mark it off from V along the base line. Take the true length diagonal and from point 7′ in the pattern swing an arc through point 8′. Next take the true distance 6,8 direct from the plan and from point 6′ in the pattern describe an arc cutting the previous arc in point 8′.

The remainder of the pattern should be readily followed from this point. It will be noted that all the spacings around the ellipse in the plan are already true distances, as are also the spacings around the rectangle 3,7,11,15 and as such do not need triangulation. On the other hand, all the lines passing from the rectangle to the ellipse, or vice versa, must be triangulated against the vertical height in order to obtain their true lengths.

The Bypass Chute.—In the case of the bypass chute the auxiliary projection takes the form of a new elevation obtained from the plan, fig. 16, and is projected at right angles to the plan centerline A′B′ of the chute. In the auxiliary projection, the height A″C″ is equal to the height A,C in the elevation, and the horizontal distance A″B″ is equal to the distance A′B′ in the plan. In this projection the angle D″C″B″ is the true angle between the centerlines of the chute, and the joint line *x,y* bisects the angle D″C″B″.

To obtain the ellipse around the center point B″ in the projection, first divide the circle in the plan into 12 equal parts as shown in the figure, and from them draw lines parallel to A′B′ to meet the oblique edge at B′.

From the points around the circle, project lines into the auxiliary view to meet the joint line *x,y*. From the points on the joint line *x,y*, project lines downwards parallel to the centerline C″B″ to meet the series of lines projected from the points on the edge at B′. Where the last two sets of lines meet, points will be obtained through which to draw the ellipse as shown in the figure.

To develop the pattern, project lines at right angles to C″,B″, from all the points on *x,y* and also from the points on the ellipse around B″. Since the design of the chute is based on right cylinder construction, the length of the girth line is spaced straight along one of the projected lines. In the pattern at fig. 16, the spacing of the girth line is shown as from 1′ to 1″, and the spacings are equal to these around the circle in the plane. A series of lines should now be drawn parallel to C″B″, through all the points along the girth line 1′1″. Where these parallel lines meet those projected from the joint line *x,y* and the ellipse around B″, sufficient points will be obtained for drawing in the curves in the pattern.

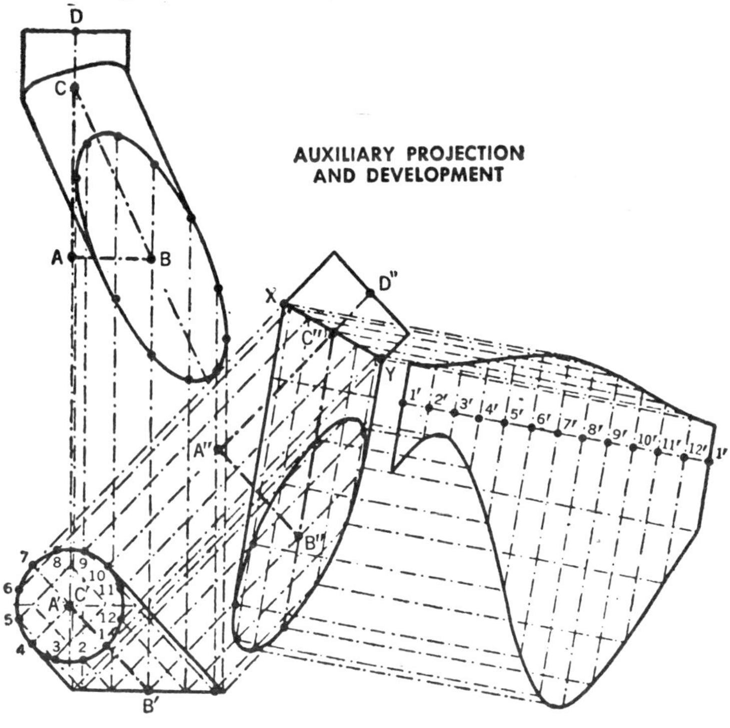

FIG. 16.—Auxiliary projection and development for by-pass chute.

The pattern for the short vertical part of the chute will be similar to the upper part of the pattern above the line 1′1″. as shown in the figure.

The Feed Chute.—The chute at the feed end of the conveyor is a transforming piece from a circular top to a circular inlet to the conveyor. The development of the pattern is shown in fig.

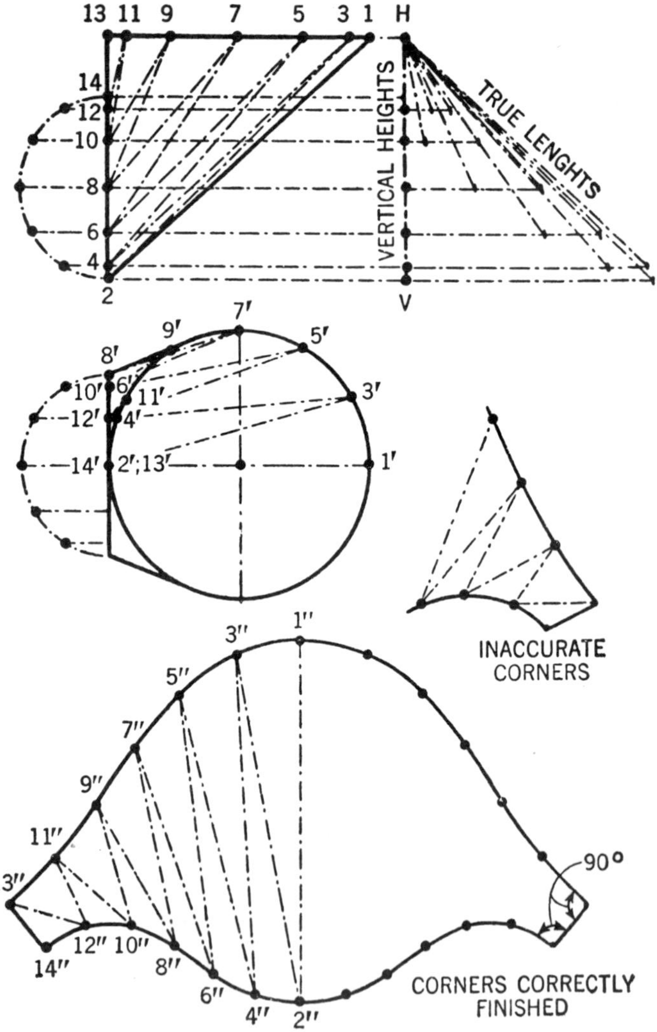

FIG. 17.—Pattern development for feed chute.

17. No auxiliary projection is needed in this case, but a point of special interest in the development arises from the fact that as the edge of the circular top lies vertically over the plane of the inlet circle, one single point in the plan represents three points 2, 13 and 14 in elevation. Also points 4 and 12, and points 6 and 10 fall one below the other and are represented by single points in the plan. This needs to be borne in mind when triangulating for the pattern.

Prior to developing the pattern, draw a semi-circle on the diameter of the inlet in the plan and the elevation. Divide the semi-circle into six equal parts and project the points perpendicularly back to the diameter. Number the points on the diameter in the elevation beginning at the bottom 2,4,6,8,10,12 and 14. Number the same points in plan as shown in the illustration, remembering that in this case 2,4 and 6 fall vertically below points 14,12 and 10 respectively. Number the points on the top circle from 1 to 13 as shown. In this case point 13 falls on the same points in the plan as points 2 and 14.

Erect a vertical height line V,H and project all the points in the elevation to it, and beyond, as shown in fig. 17.

To develop the pattern, take the true distance 1,2 direct from the elevation and mark it off in any convenient position as at 1″2″ in the pattern. Next take the plane length 2′3′ mark it off along the bottom base line level with point 2. Take the true length diagonal up to the point H and from point 2″ in the pattern swing an arc through point 3″. Now take the true spacings 1′3′ direct from the plan and from point 1″ in the pattern, describe an arc cutting the previous arc in point 3″. This completes the first triangle.

For the second triangle take the plan length 3′4′ and mark it off at right angles to V,H, but this time along the base line level with point 4. Take the true length diagonal up to point H, and from point 3″ in the pattern swing an arc through point 4″. Next, for the true spacing between points 2 and 4, take one of the divisions on the inlet semi-circle and from point 2″ in the pattern describe an arc cutting the previous arc in point 4″.

The remainder of the pattern is straight forward triangulation, and the directions for the first two triangles may be repeated right through to the end, except that the appropriate numbers should be used for each triangle as the pattern progresses.

CHAPTER 8

Practical Drawing

By definition the term mechanical drawing means ***drawing executed mechanically by aid of drawing instruments as distinguished from that executed by the unguided hand.*** Accordingly the first thing to consider is the drawing instruments, and secondly, how to use them.

Drawing Instruments.—A good draftsman should have good instruments; in fact the best are none too good and are easily rendered unfit for use unless they are properly handled. The advice given by some instructors for beginners to buy a cheap set of instruments for use until he finds out if he is gifted in the art of drafting, is rather questionable. If an experienced draftsman cannot do good work with poor instruments, then how can a beginner be expected to accomplish anything, or determine if he has any talent for drawing?

There are two general classes of drawing instruments—those with circular cross section friction joints of the Riefler pattern, and those of angular cross section with set screw joints. The author very strongly recommends the purchase of the circular pattern instruments, as they are superior in every way to the other type.

The following is a list comprising everything needed for general drafting work:

1 drawing board
1 set of instruments
1 tee square
2 triangles, 30° and 45°
1 drawing scales
2 pencils 3H and 6H

1 bottle of drawing ink
1 box of thumb tacks (small)
1 ink and pencil eraser
1 sponge eraser
1 pen holder and lettering pens
1 irregular curve
1 protractor

In some cases involving enlarging or reducing the size of drawings proportional dividers are necessary.

Drawing board.—The size of the board should be about 2 ins. longer and 2 ins. wider than the size of paper to be used. The

FIG. 1,480.—Dietzgen white pine drawing board with dovetailed hardwood cleats or "ledges." This construction permits expansion and contraction. The dovetail grooves are sunk in ¼ the thickness of the board, thus securing a firm grip on the narrow wooden strips. *Standard sizes:* 31 × 43; 37 × 55; 43 × 61; 49 × 73; 49 × 85 ins.

board should be made of well seasoned, straight grained pine, free from all knots; the grain should run lengthwise of the board.

The edges of the board should be square to each other and perfectly smooth in order to provide a good working edge for the head of the tee square to slide against.

A pair of hard wood cleats is screwed or dove tailed to the back of the board. The board should be about three-quarter inch in thickness. The cleats, fitted at the back of the board, at right angles to its longest side, may be about two inches wide and one inch thick. Such cleats will keep the board from warping through changes of temperature and moisture.

Fig. 1,480 shows a board of this type bottom side up. Another method of preventing warping is by two transverse end pieces into which the several pieces forming the board proper are secured by tongue and groove

FIG. 1,481.—Dietzgen white pine drawing board with tongue and groove end ledges. This board has two drawing surfaces. *Standard sizes:* 12 × 7; 16 × 22; 20 : 24½; 20 × 26; 23 × 31; 31 × 43 ins. A good size for the beginner and for ordinary use is the 16 × 22 ins.

joints as shown in fig. 1,481. The standard dimensions given in figs. 1,480 and 1,481 will indicate the ordinary sizes as regularly manufactured.

The paper is fastened to the board usually and most conveniently by *thumb tacks*. Under no circumstances should the large size tacks be used—get the smallest and thus increase the life of the board, as no board can be expected to remain in con-

FIG. 1,482.—Dietzgen Eureka adjustable drawing table. Hardwood frame, iron legs, pine wood board having hardwood ledges. Height adjustment 32 to 40 ins. The wheel clamp at right locks table when tilled to any desired angle.

dition if jumbo tacks or railroad spikes be used to fasten down the paper.

The board is sometimes mounted on legs and arranged to fold up when not in use, such device being called a drawing table. There are numerous kinds of these tables on the market. Most of these are too rickety to be satisfactory. Fig. 1,482 shows a rigid design of board having adjustments for height and inclination of board.

"Set" of Instruments.—Drawing instruments usually come in sets, that is, several instruments in a case. For beginners, and for general use, a set containing the following instruments is all that is necessary.

1 compass
1 hair spring dividers
3 spring bows (pencil, pen and points)
1 extension bar
1 ruling pen

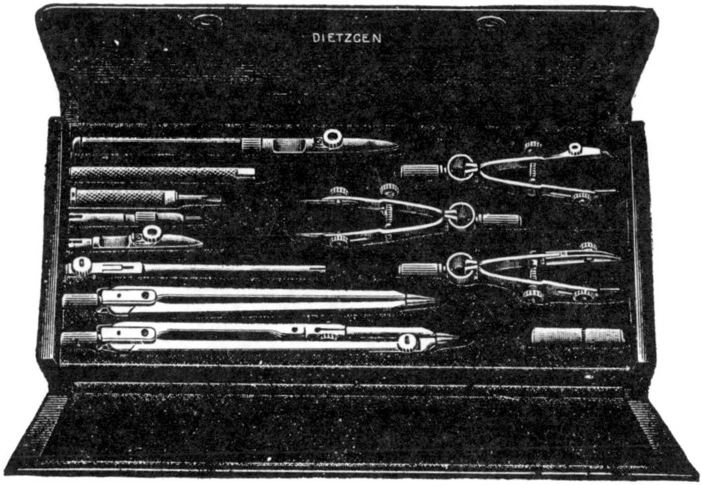

FIG. 1,483.—Dietzgen Riefler pattern or cylindrical set of drawing instruments, ***comprising:*** 5½ in. compass with detachable needle and pencil points; 5 in. hair spring dividers; 5½ and 4½ in. ruling pens; extension rod; 3 spring bows, pencil, pen, and point; box of leads; key. The extra or small size ruling pen is not necessary but can be used to advantage in some instances.

Fig. 1,483, shows a set comprising the above instruments of the Riefler pattern with exception that set contains an extra ruling pen.

Compasses.—This instrument is for describing arcs or circles

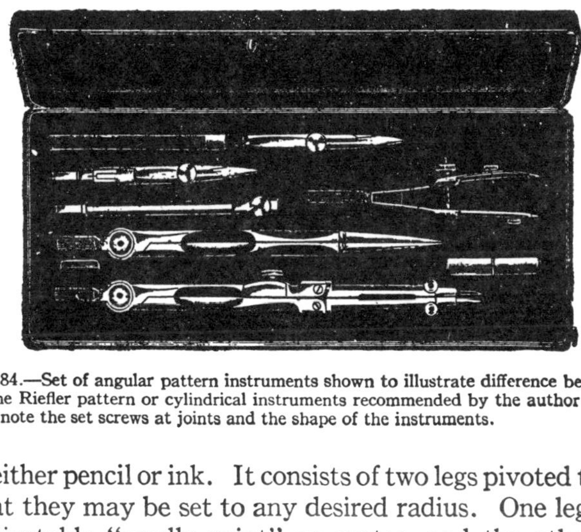

FIG. 1,484.—Set of angular pattern instruments shown to illustrate difference between these and the Riefler pattern or cylindrical instruments recommended by the author. In above figure note the set screws at joints and the shape of the instruments.

with either pencil or ink. It consists of two legs pivoted together so that they may be set to any desired radius. One leg carries an adjustable "needle point" or center, and the other has a joint in which may be secured the pencil or pen arms. Each leg has a pivoted joint permitting adjustment of the ends, so that the end arms which carry the center needle point and pen or pencil may be adjusted perpendicular to the paper for various radii. Figs. 1,485 to 1,487 show six inch compasses with pencil and pen arms and extension bar. The important requirement of good compasses is that the legs may be moved to any radius without any spring back; cheap instruments always spring back making it difficult to set them with precision. Figs. 1,488 to 1,493 show some construction details of compasses and dividers.

Hair Spring Dividers.—Compasses and dividers are very much alike but each has its special use.

Dividers consist of two legs pivoted at one end and provided with sharp needle points at the other, for use in spacing off distances, as shown in fig. 1,494. For precision, they are fitted with a "hair spring" device, consisting of an adjustable screw controlled by a steel spring in one leg, as shown in fig. 1,495. In operation the legs are set to the approximate desired position and brought to the exact position by turning the adjusting screw.

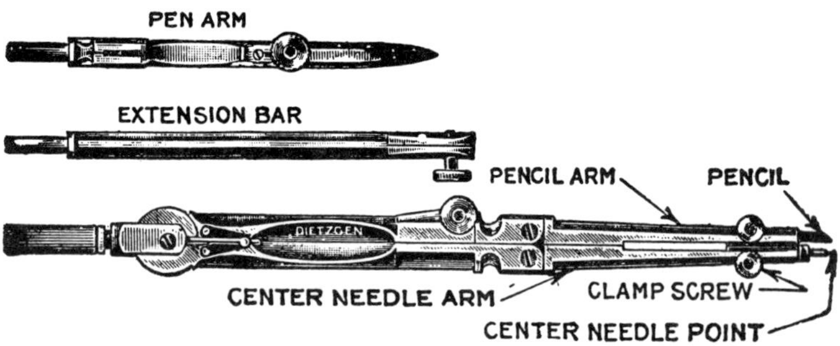

FIG. 1,485 to 1,487.—Compasses with pencil and pen arms and extension bar. Fig. 1,485, pen arm; fig. 1,486, extension bar; fig. 1,487 compasses with pencil arm clamped to leg.

Spring Bows.—These small compasses and dividers made with two spring legs whose distance apart is regulated by a small through bolt and thumb screw. They usually come in sets of three: pen bow, pencil bow and dividers, as shown in figs.

1,496 to 1,498. Fig. 1,497 is the pencil bow shown without the lead. Spring bows are used for describing circles of small diameter, and for minute spacing.

Extension Bar.—In order to extend the range of compasses, a lengthening or extension bar, as shown in fig. 1,486, is generally provided which greatly increases the diameters of circles which may be described.

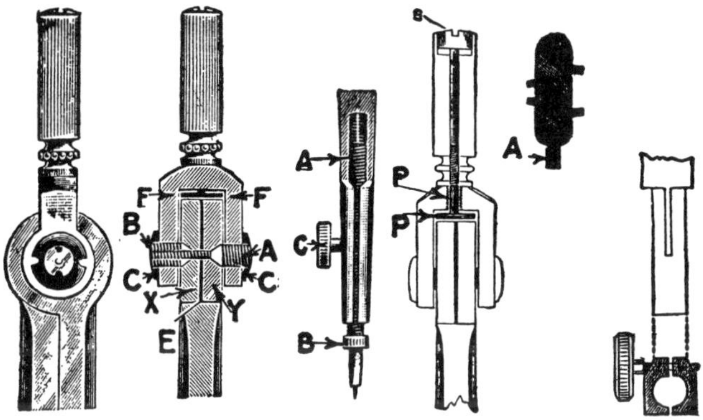

FIG. 1,488 and 1,489.—Main pivot as constructed for compasses and dividers. FF, on the pivot forks. The bolt AB, goes entirely through the legs and bolts the forks together. The conical parts of A and B, form the pivot joints which are securely held by lock nuts CC. E, is a steel disc which acts as an anti-friction bearing for the heads of the legs X, and Y. To apply tension in adjusting, loosen only one of the lock nuts CC.

FIGS. 1,490.—Screw thread needle point. *In construction,* the portion A, is threaded to the extremity of the arm. The portion B, is knurled to be more easily turned with the fingers. The thumb screw C, clamps the needle point rigidly.

FIGS. 1,491 and 1,492.—Clamping device. *In operation* by a turn of the key A, the screw S, is pressed down on pin P, which is fastened to the small plate P. The plate P, rests on the top of the legs of the compasses or dividers and when pressed down by turning the screw S, holds the legs firmly in the desired position. The device is useful when spacing or using the same opening of compasses or dividers repeatedly.

FIG. 1,493.—Shank and clamp socket. In the round form, the feathered shanks fit into side clamping spring sockets. By this construction the interchangeable parts of the compasses are firmly locked twice. First, by the steel feather of the shank, and secondly, by the clamping sockets being drawn together with the screw.

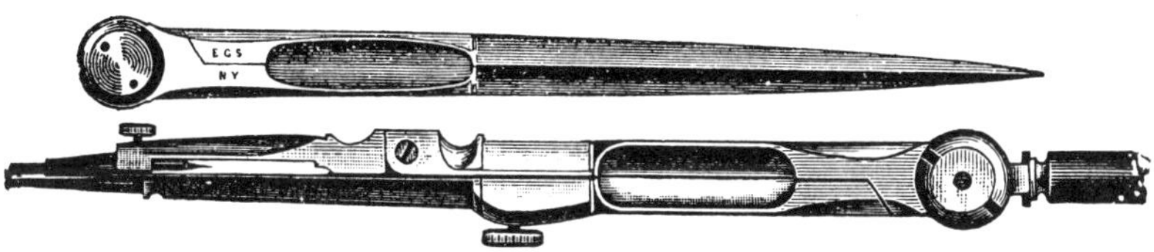

FIGS. 1,494 and 1,495.—Plain and hair spring dividers. Note the sharp pointed legs, and the adjusting screw of the hair spring in.

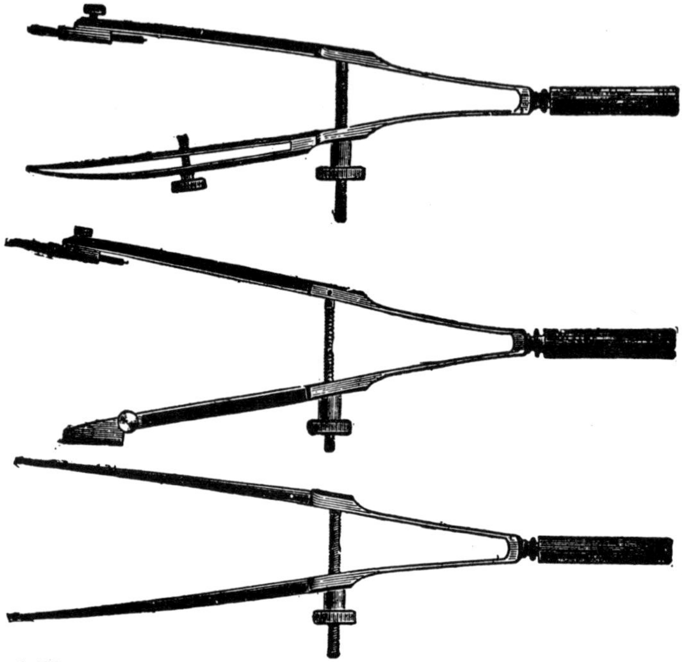

FIGS. 1,496 to 1,498.—Spring bows; fig. 1,496. bow pen; fig. 1,497, bow pencil (shown without lead); fig. 1,498, bow dividers.

Drawing or Ruling Pens.—A special pen is used for drawing lines, as shown in figs. 1,499 and 1,500. The points are made of two steel blades which open and close as required for thickness of lines by a regulating screw.

A good drawing pen should be made of properly tempered steel, neither too soft nor hardened to brittleness. The nibs should be accurately set, both of the same length, and both equally firm when in contact with the drawing paper. The points should be so shaped that they are fine enough to admit of absolute control of the contact of the pen in starting and ending lines, but otherwise as broad and rounded as possible, in order to hold a convenient quantity of ink without dropping it. The lower (under) blade should be sufficiently firm to prevent the closing of the blades of the pen, when using the pen against a straight edge.

The spring of the pen, which separates the two blades, should be strong enough to hold the upper blade in its position, but not so strong that it would interfere with easy adjustment by the thumb screw. The thread of the thumb screw must be deeply and evenly cut so as not to strip. Figs. 1,499 and 1,500 are side and end views of an ordinary ruling pen.

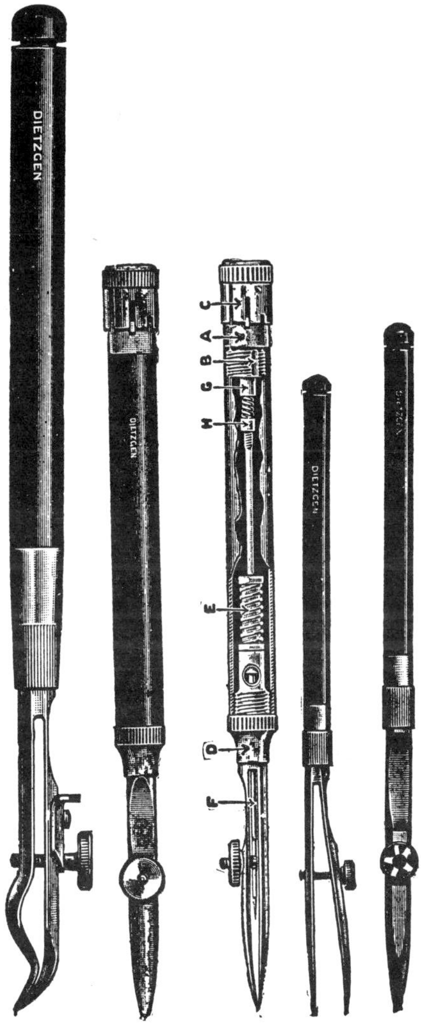

FIG. 1,499. — Dietzgen reservoir ruling pen.

FIGS. 1,500 and 1,501.—Dietzgen fountain ruling pen. In fig. 1,501, A is the metal top; B, the air escape; C, the cap; D, the pen socket; E, the plunger; F, the ink tube; G, the packing nut; H, the cap tension. The barrel is filled by unscrewing the metal top and dropping the ink into the barrel by means of an ink dropper which is furnished with each instrument. The ink is conveyed to the pen point by engaging the stud on the metal top in the longest slot in the cap and pressing the cap gently with the thumb or the forefinger. The intermediate slot is used when less ink is desired, and when the pen is not in use the stud is engaged in the smallest slot, thus preventing the cap being pressed down accidentally.

FIGS. 1,502 and 1,503. — Side views of ordinary ruling pen spring blade, polished ebony handle.

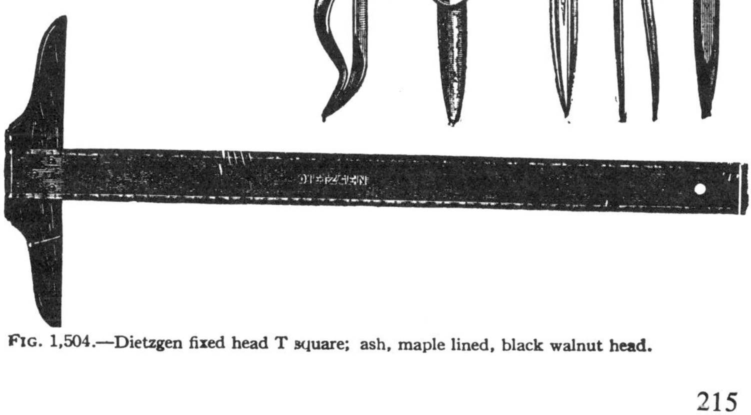

FIG. 1,504.—Dietzgen fixed head T square; ash, maple lined, black walnut head.

Tee Square.—This instrument is used for drawing lines parallel to the lower edge of the board, and consists of two parts, the head and the blade, these being fastened at 90° to each other

FIG. 1,505.—Dietzgen movable head T square; ash, maple lined blade, black walnut. This type of square has a fixed head on one side and movable head on the other. The movable head is pivoted so that it may be shifted to any angle with the blade, a clamp screw being provided to lock the head in any position.

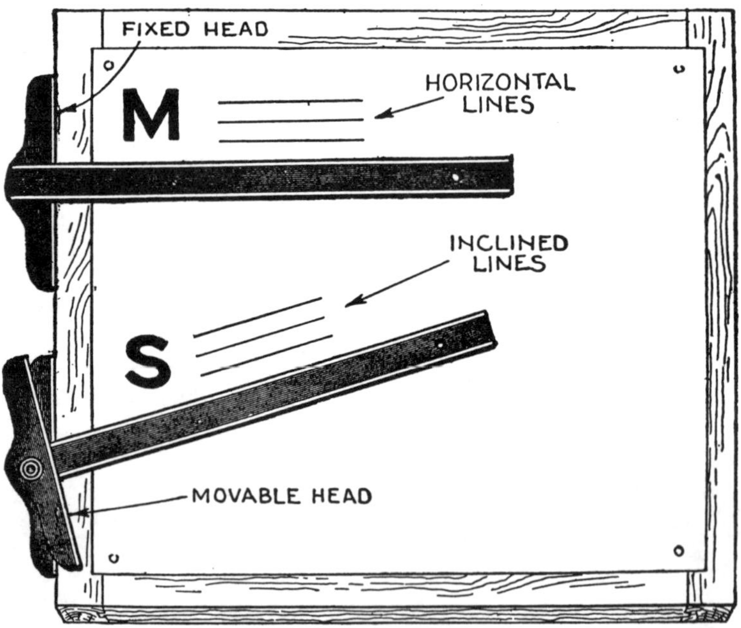

FIG. 1,506.—Fixed and movable head T squares in position on board showing horizontal lines that may be drawn with the fixed head square **M**, and inclined lines with movable head square at **S**.

as shown in fig. 1,504. This is the fixed head type of square. The square is sometimes fitted with a movable head, as shown in fig. 1,505, permitting drawing lines inclined to the edge of the board. Fig. 1,506 shows the two types of squares in position on the board.

Triangles.—For drawing other than horizontal lines, "triangles" are generally used. It is inadvisable to buy cheap wooden triangles, as they soon warp out of true; get only the best made

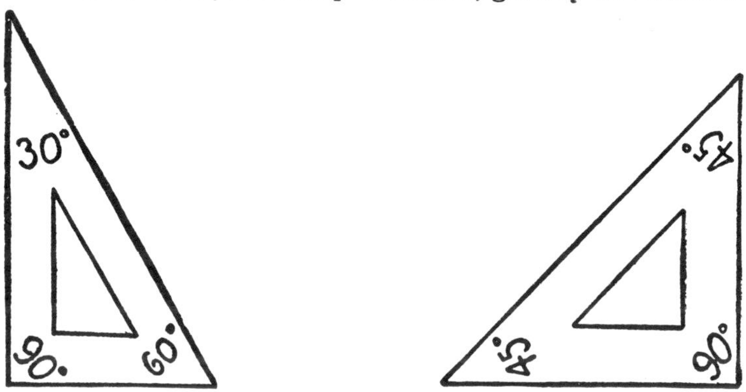

FIGS. 1,507 and 1,508.—45° and 30° triangles.

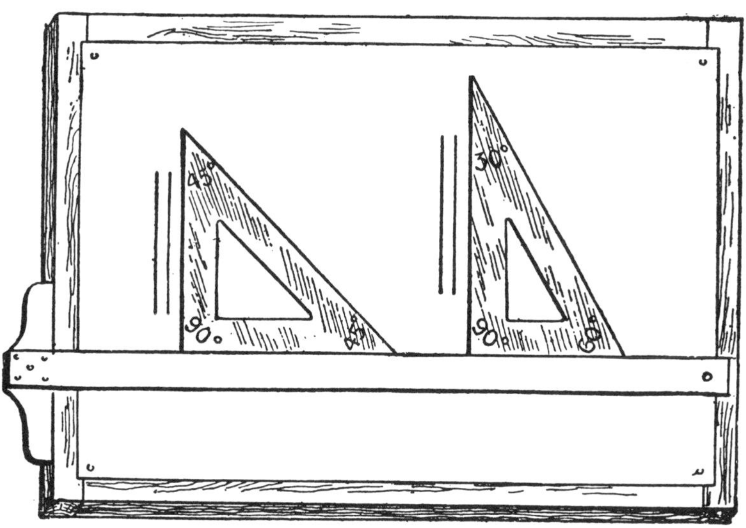

FIG. 1,509.—45° and 30° triangles in position on T square.

of transparent ambro. Two triangles will be ordinarily required—the 45° and the 30°, as shown in figs. 1,507 and 1,508. The first one has two sides at right angles and the third at 45°; the second two sides at right angles and the third making a 30° angle with one side and 60° with the other. By placing these triangles on the T square as in fig. 1,509, vertical, or inclined lines may be drawn.

Rule and Scales.—The rule is used for measuring and comparing dimensions the same as the ordinary carpenter's 2 ft. folding rule. For the drawing board however it is an instrument

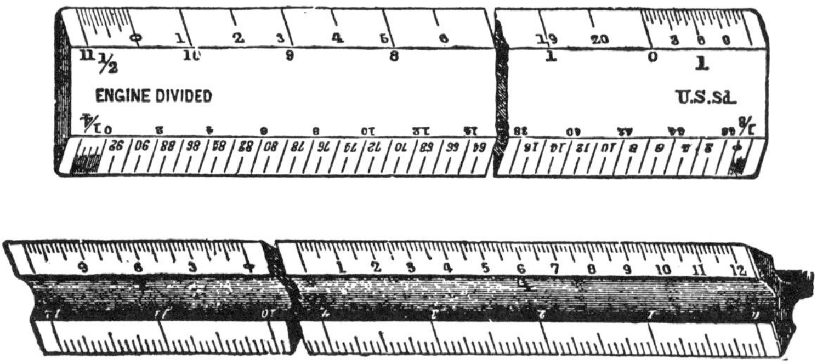

FIGS. 1,510 and 1,511.—Flat and triangular draughtsmans' box wood scales. ***An explanation*** of the 1 in. and ½ in. scales will suffice for all. Where it is used as a scale of 1″ to one foot, each large space, as from 0 to 12 or 0 to 1, represents a foot, and is a foot at that scale. There being 12″ in one foot, the twelve long divisions at the left represent inches; each inch is divided into two equal parts, so from 0 to one division at the left of 9 is 9½″ and so on. The 1″ and ½″ scales being at opposite ends of the same edge, it is obvious that one foot on the 1″ scale is equal to two feet on the ½″ scale, and conversely, one foot on the ½″ scale is equal to six inches on the 1″ scale; and 1″ being equal to one foot, the total feet in length of scale will be 12; at ½″ to 1 foot the total feet will be 24.

of greater precision being usually 10 or 12 ins. long and divided into 32nds. The face is beveled to an edge so that the division lines will lie very close to the drawing paper, thus permitting distances to be marked off accurately.

When drawings are made the same size as the object that is "full size," a rule of the above description answers the purpose, however, when drawings are to be made smaller or larger than the actual size of the object to be drawn, *scales* are employed. For architectural drawing the various scales are divided into feet and inches with sub-divisions. The most convenient forms are the usual flat or triangular box wood scales as shown in figs. 1,510 and 1,511.

The triangular scale is the one generally used as it contains six different scales as shown. The usual scales are:

3 ins. = 1 ft.	¾ in. = 1 ft.
1½ ins. = 1 ft.	⅜ in. = 1 ft.
1 in. = 1 ft.	¼ in. = 1 ft.

FIG. 1,512.—Dietzgen's India ink.

the scales being usually designated by the length of the foot division as for instance the 1½ or ¾ in. scale. On each scale, as can be seen, the first foot is divided into inches, and where the scale is large enough, into fractions of an inch.

Drawing Ink.—India ink, and not ordinary ink is used. It can be obtained either in the dry (stick) or liquid form Although the dry ink is considered the best, it is not generally used because of the time and skill required in mixing. The liquid India ink comes in small bottles having a quill attached to the cork, by means of which the pen is easily filled. Fig. 1,512 shows the type bottle used, and a brand that can be recommended. It can be had waterproof.

Pencils.—Drawings are generally made "*in pencil*" and then "*inked in.*" These are made in various degrees of "hardness," the hardest being designated 9H. The choice as to hardness depends upon the kind of drawing and precision. For ordinary

FIGS. 1,513 to 1,517.—Chinese or India inks. Fig. 1,513 oval black; figs. 1,514 and 1,515, square black; fig. 1,516, oblong gilt; fig. 1,517 oblong black.

work, as in laying out house frames on a large scale, a 3 or 4H would do. However, in drawing a roof stress diagram on, say,

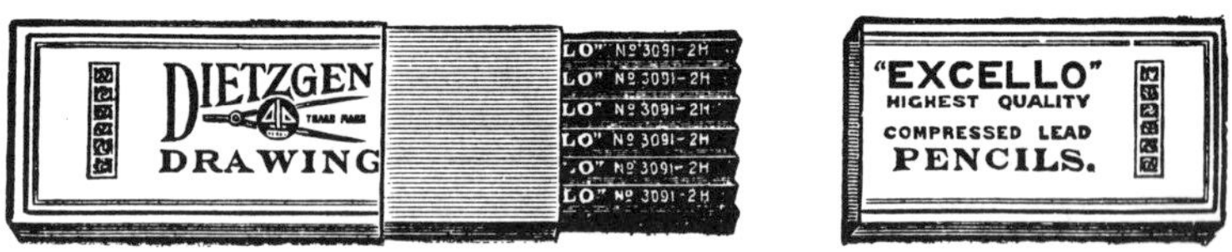

FIG. 1,518.—Dixon's "Excello" drawing pencil. Note that the four H hardness is expressed "HHHH." On some pencils it would be expressed as "4H."

a scale of 1″ = 2,000 lbs., a pencil no softer than 6H should be used, because a sharp point could not be maintained with a soft pencil, and precision could not be expected with a pencil having *"an acre of lead" on its point.*

Pencils are generally sharpened to a conical point, as in fig. 1,519, but some sharpen them so the lead is wedge-shape as in figs. 1,520 and 1,521.

FIG. 1,519.—Pencil with conical point.

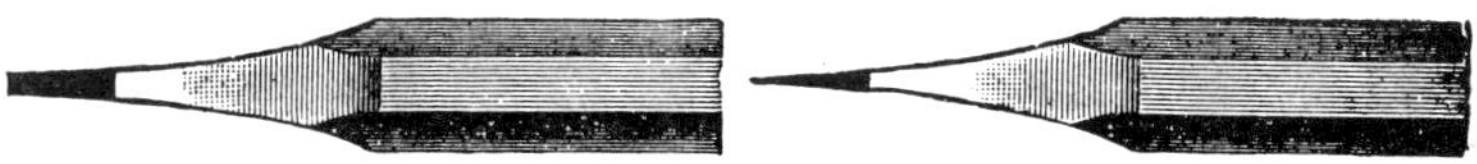

FIGS. 1,520 and 1,521.—End and side views of drawing pencil with lead sharpened "wedge shape."

Thumb Tacks.—As stated elsewhere, don't expect to get long service from a drawing board if "spikes" be used for holding down the paper. Thumb tacks are so called because they have a large head so that they may be easily pressed into the board by the thumb. Use the smallest tack that will hold the paper on the board; this will depend somewhat upon the size of the paper, its weight, backing, etc.

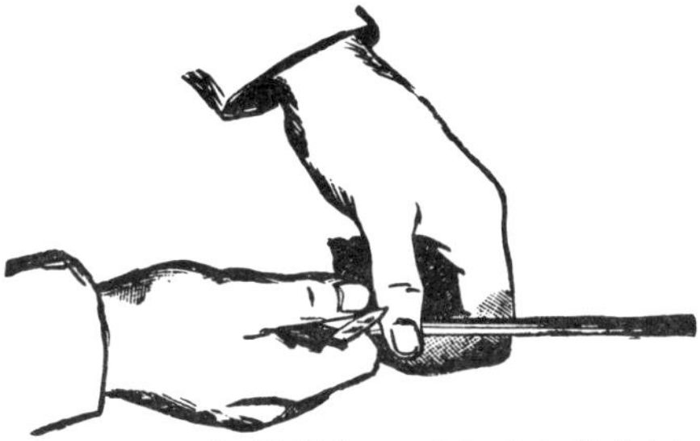

FIG. 1,522.—How to sharpen a pencil. Hold the pencil firmly in the left hand, as in the drawing, allowing about an inch to project beyond the fingers, and turn it gradually as the knife removes the wood. The knife should be held so that the blade alone projects beyond the fingers, and the part of it nearest the handle used for cutting. The pencil should be placed against tne inside of the thumb of the right hand, as shown, and the wood removed by slight shaving. The lead should not be cut at the same time as the wood, but rested on the thumb and pared gently afterwards; by attention to these directions the pencil will be economized.

FIGS. 1,523 to 1,525.—Steel stamped thumb tacks. They come in cardboard boxes of 100.

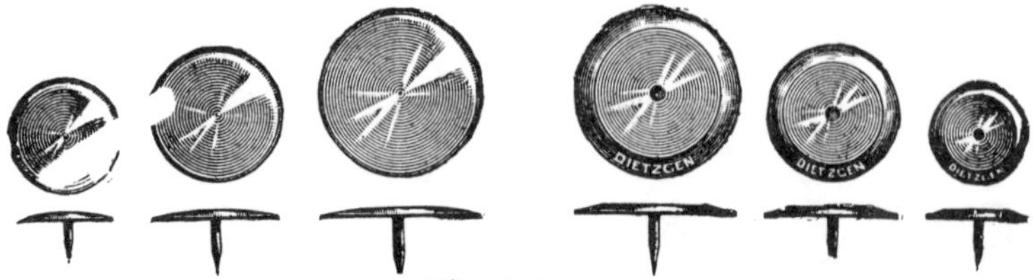

FIGS. 1,526 to 1,537.—Gem union (solid head) thumb tacks.

The cheap steel stamped thumb tacks as shown in figs. 1,523 to 1,525 are preferable to the more expensive solid tacks shown in figs. 1,526 to 1,537. For large tacks they are conveniently removed with a tack lifter as in fig. 1,538, but this is not necessary with the small tacks as they are easily removed by aid of the finger nail.

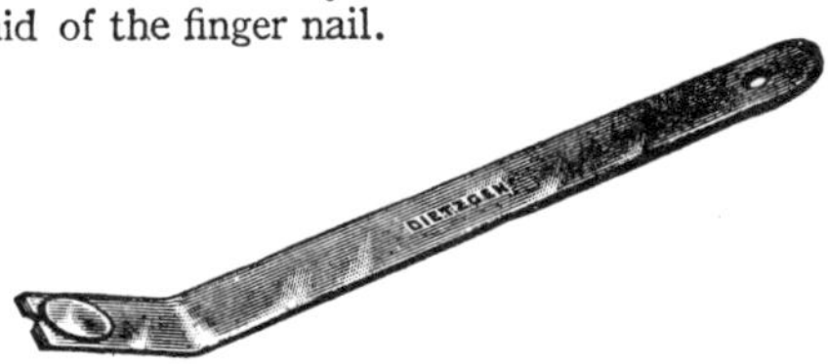

FIG. 1,538.—Tack lifter. Made of metal, nickel-plated. Very convenient for pushing in or for extracting tacks from drawing boards, without injuring the points. The handle can be used as a paper cutter, and is also serviceable for pressing down the edges when stretching paper or for removing sheets which have been gummed to the board.

FIG. 1,539.—Ordinary form of pencil eraser.

FIG. 1,540.—Circular form of rubber ink eraser.

Erasers.—There should be three kinds of erasers included in the drawing outfit:

1. Pencil
2. Ink
3. Sponge

For erasing any portion of a line in pencil, a piece of prepared white vulcanized rubber is the best, small in size and of rectangular shape, as in fig. 1,539.

FIG. 1,541.—Steel ink eraser.

An ink eraser is made of a composition of rubber and ground glass, and it should be used as sparingly as possible on drawings, as it roughens the paper and removes the gloss from its surface. Fig. 1,540 shows an ink eraser. Steel ink erasers, as shown in fig. 1,541, are useful in removing defects, overrun lines, joint of lines if swollen, etc.; they have a fine point and can be used to advantage with a little practice; they are used with a scratching, not a cutting, motion.

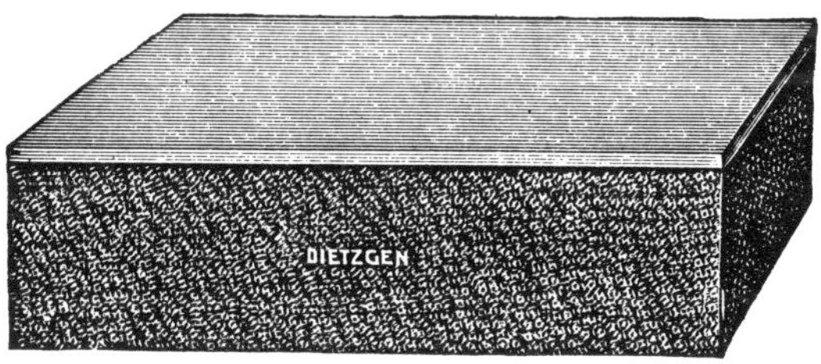

FIG. 1,542.—Sponge rubber, made of soft cellular rubber. Ordinary sizes 1 × 1 × 1 to 4 × 2 × 1 ins.

After a drawing has been made in pencil and inked in it is in a more or less soiled condition. By means of a so called "sponge" rubber as shown in fig. 1,542, which is very soft, friable and entirely free of grit, the drawing may be cleaned of dirt and projecting pencil lines without disturbing the inked lines or marring the surface of the paper.

Lettering Pens.—Nearly all lettering is executed by a pen similar to a common writing pen, but having a fine or wide point. The width of the point depends upon the desired thickness of the letters. The pen must at all times be kept clean as otherwise no clean cut line can be obtained. Figs. 1,543 to 1,550 show usual styles of pen used for lettering.

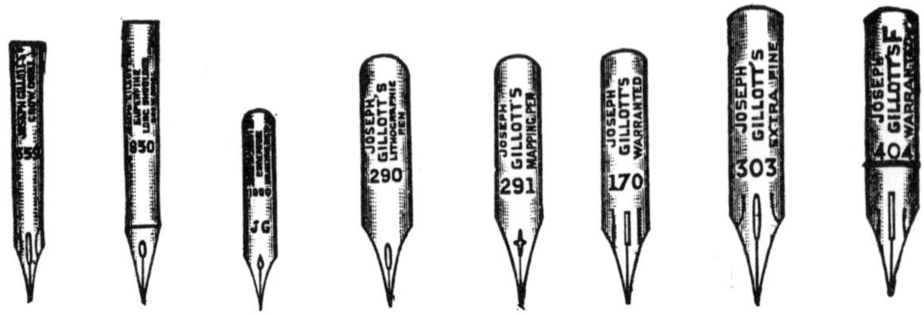

FIGS. 1,543 to 1,550.—Various lettering pens.

FIGS. 1,551 to 1,554.—Various irregular curves (sometimes called sweeps). They are useful when elliptical or parabolic curves are to be described.

Irregular Curves.—For describing curves other than circles, special cut forms called irregular curves are used to guide the pen. These may be obtained in great variety.

A set of two or three will be found frequently useful; several forms of these curves are shown in figs. 1,551 to 1,554.

Protractor.—This instrument is *for laying off or measuring angles*. Fig. 1,555 shows the ordinary form of protractor.

Its outer edge, as shown in the illustration, is a semi-circle with center at **O**, and for convenience is divided into 180 equal

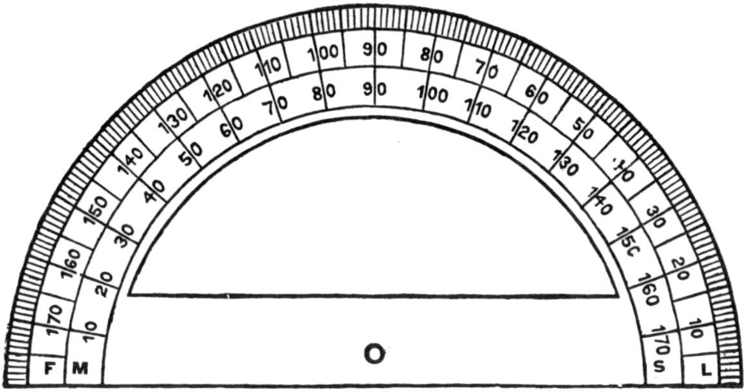

FIG. 1,555.—Protractor for laying out or measuring angles.

parts or degrees from **M** to **S**, and in reverse direction from **L** to **F**.

Protractors are often made of metal in which case the central part is cut away to allow the drawing under it to be seen.

A fine precision protractor is shown in fig. 1,556.

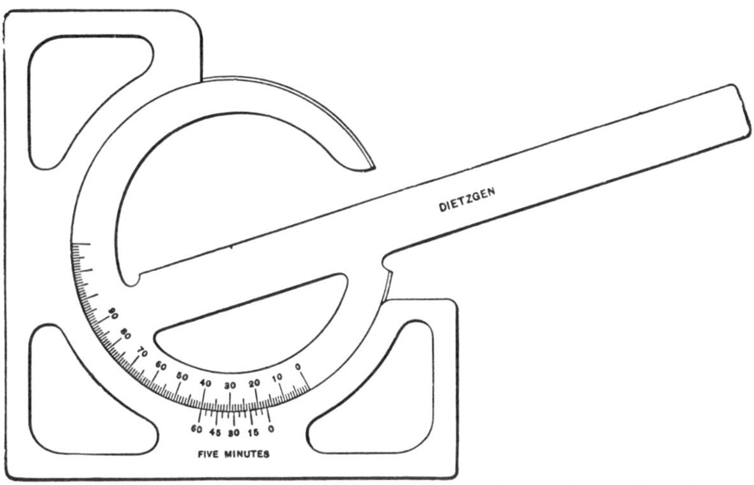

FIG. 1,556.—Dietzgen steel protractor. Blade 8½ ins. long and graduations reading to degrees, with vernier reading to 5 minutes. There are no projections on either face, and therefore it can be used on either edge of the blade or with either side up. This is an advantage when dividing circles, transferring angles, drawing oblique lines at right angles to each other, or laying off given angles each side of a vertical or a horizontal line without changing the setting.

How to Draw

Preparing for Work.—The paper is first secured to the drawing board by means of *thumb tacks*, one at each corner of the sheet. It should be stretched flat and smooth; to obtain this result proceed as follows: press a thumb tack through one of the corners about ½ inch or ¼ inch from the edge. Place the tee square in position as in drawing a horizontal line, and straighten the paper so that its upper edge will be parallel to the edge of the tee square blade. Pull the corner diagonally opposite that in which the thumb tack was placed, so as to stretch the paper slightly and push in another thumb tack. Proceed in the same manner for the remaining two corners.

Another method consists in stretching the paper while it is damp. For stretching the paper in this way moisten the whole sheet on the under side, with the exception of a margin all around the sheet, of about half an inch and paste the dry border to the drawing board. To do this properly requires a certain amount of skill. The paper thus stretched gives undoubtedly a smoother surface than can be obtained when using thumb tacks, but there are objections to this process as the paper stretched in this way is under a certain strain and may have some effect on the various dimensions of the drawing when cut off the board.

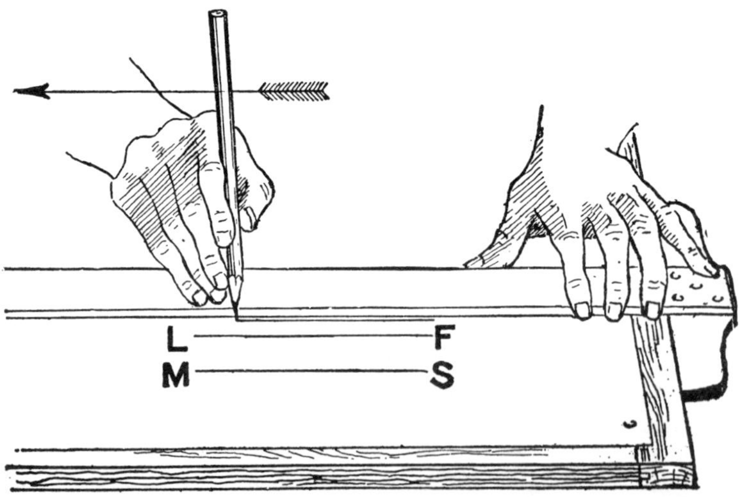

FIG. 1,557.—Parallel horizontal lines. These lines as MS, LF, are drawn by moving the pencil in the direction of the arrow, guided by the edge of the T square.

Straight Lines.—To draw a straight line use is made of the T square, or triangle, or both, depending upon the direction of the line. Horizontal lines are drawn by aid of the T square as in fig. 1,557 and sometimes vertical lines by applying the head of the square to the lower or horizontal edge of the board, as in fig. 1,558.

The usual method of drawing vertical lines is by aid of both the T square and one of the triangles as shown in fig. 1,561. Here one of the "legs" of the triangle is used to guide the pencil. By using the hypothenuse of the 45° triangle, oblique parallel lines may be drawn as in fig. 1,560, and by using the hypothenuse of the 30° triangle, oblique lines may be drawn at 30° or 60° as in fig. 1,559 and 1,562.

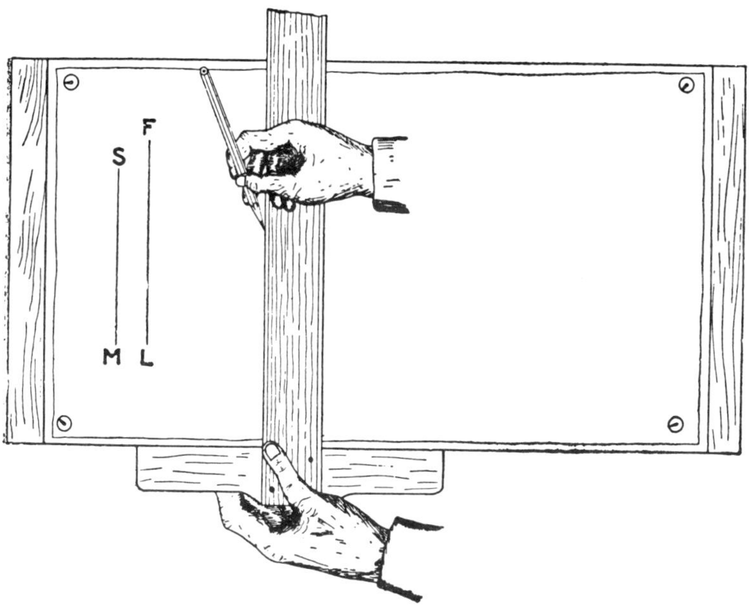

FIG. 1,558.—Parallel vertical lines with T square only. These lines as MS, LF, are drawn similarly as in fig. 1,557, but with the head of the T square in contact with the lower edge of the board.

By a combination of both triangles as in fig. 1,563, various other angles, such as 15°, 75°, 135°, may be obtained. Sometimes it is desired to draw a line parallel to another line which is not inclined at any of the angles obtained with the triangles. This is done by placing the edge of one triangle parallel with the given line and sliding it on the other triangle as in fig. 1,564.

In drawing a line it is important that the pencil be held correctly. Figs. 1,565 to 1,567 show the wrong ways and correct way to hold the pen.

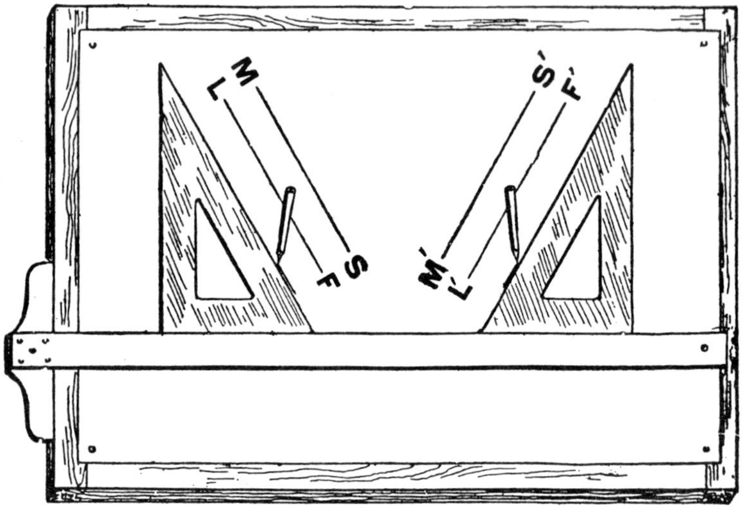

FIG. 1,559.—Parallel 30° oblique lines with T square and 30° triangle, MS, LF, right hand lines M'S', L'F', left hand lines.

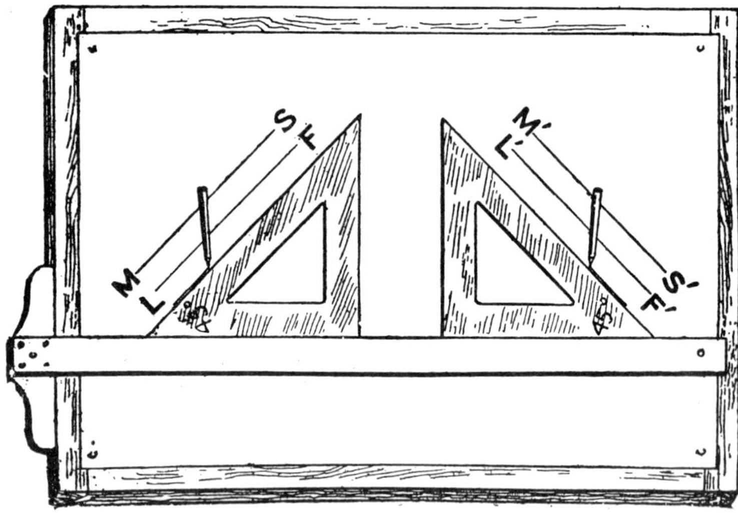

FIG. 1,560.—Parallel 45° oblique lines with T square and triangle MS, LF, right hand lines M'S', L'F', left hand lines.

It should not be inclined laterally, but in drawing a line (with either pencil or pen) it should be held with its axis in a plane perpendicular to the plane of the paper, slightly inclined in the direction in which it is being moved.

If held as in fig. 1,565, the inclination is likely to vary resulting in a wavy line; if held as in fig. 1,566, a reference point R, through which the line is to be drawn, may not be visible or only partially visible. When held for a Note here the right and wrong way to arrange the triangles. Always so place the triangles that the desired point may be reached without too much shifting.

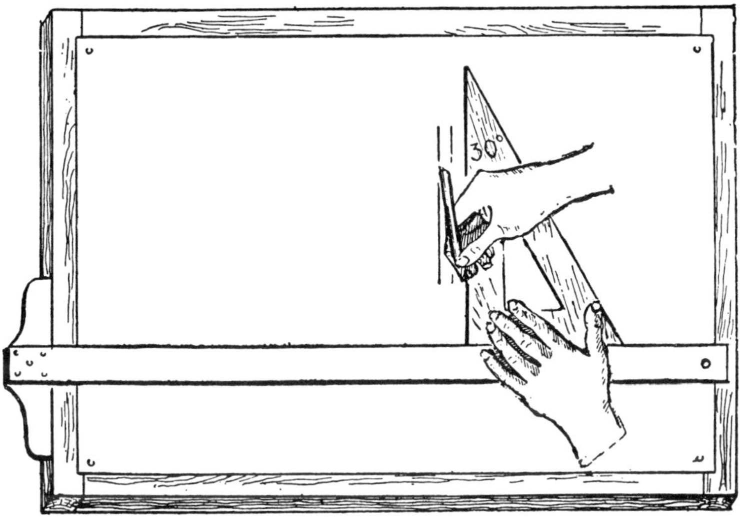

FIG. 1,561.—Parallel vertical lines with T square and triangle. The triangle in contact with the T square is shifted to any position at which it is desired to draw a vertical line.

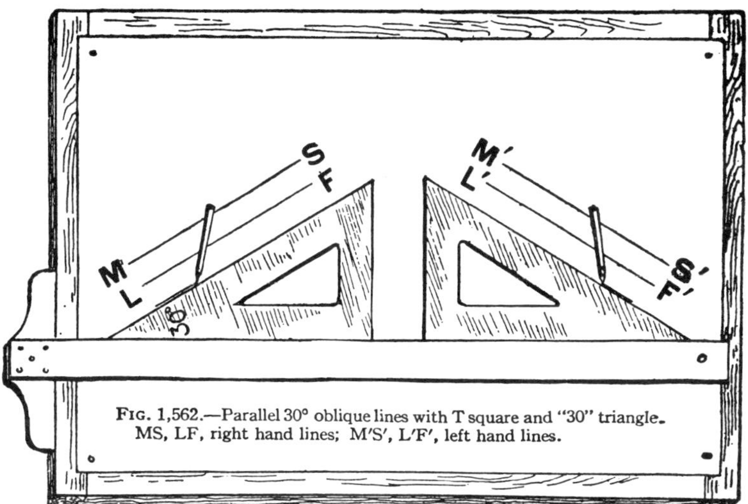

FIG. 1,562.—Parallel 30° oblique lines with T square and "30" triangle. MS, LF, right hand lines; M′S′, L′F′, left hand lines.

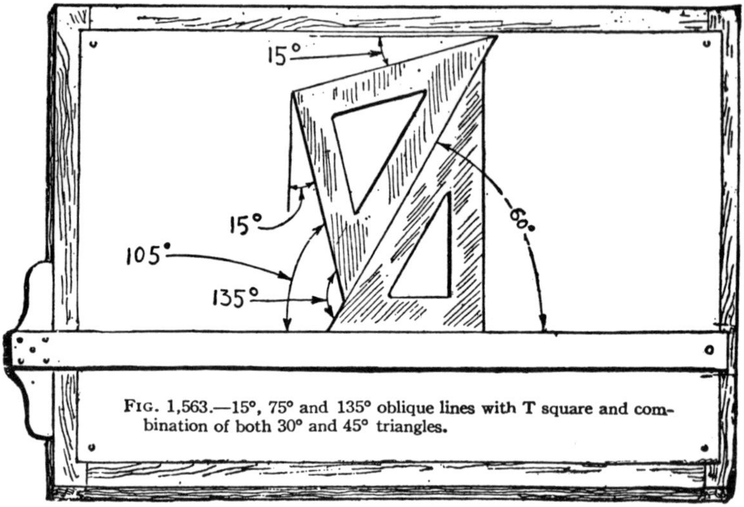

FIG. 1,563.—15°, 75° and 135° oblique lines with T square and combination of both 30° and 45° triangles.

perpendicular plane, as in fig. 1,567, the line comes very close to the lower edge, the reference point R, can be plainly seen, and there is least chance of drawing a wavy line.

In drawing lines with a pencil sharpened to a conical point the pencil should be given a slight twisting motion while the line is being drawn, as this tends to keep the point sharp.

Arcs and Circles.—These are "described" with the compass. The compass and proper positions for its use are shown in fig. 1,568. Both points should be nearly perpendicular to the paper—slightly inclined in the direction of movement.

The starting position should be such that the entire movement can be made in one continuous sweep by grasping the little handle at the pivot end by the thumb and forefinger, obtaining a twisting motion by moving the thumb forward without stopping to shift the hold on the compass.

Never hold the compass by the legs, even when the lengthening bar is used, as to do so will tend to move the legs to a different radius, as shown in fig. 1,569.

For very small circles a smaller compass called the bow compass is used; it is more convenient and having screw adjustment can be set with greater precision than the large compass. Particular attention is called to the result obtained by inclining the center point of a compass in describing circles as shown at S, fig. 1,570. Since these centers must be again used in inking in a drawing, accurate work cannot be done if the center indentations are spoiled as at S, by wrong use of the compass.

Spacing.—To accurately divide a given distance into several equal parts hair spring dividers are used. If an exact length is to be laid off with the dividers, a large multiple of that length should be first laid off with the scale on a right line and then exactly

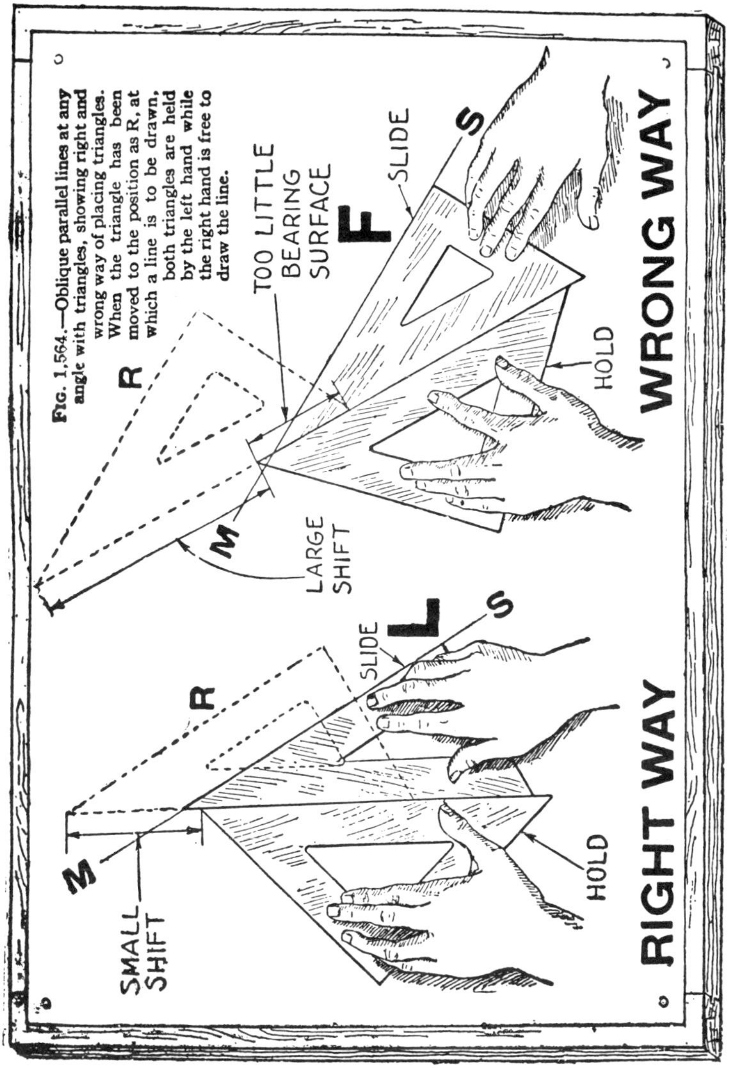

Fig. 1,564.—Oblique parallel lines at any angle with triangles, showing right and wrong way of placing triangles. When the triangle has been moved to the position as R, at which a line is to be drawn, both triangles are held by the left hand while the right hand is free to draw the line.

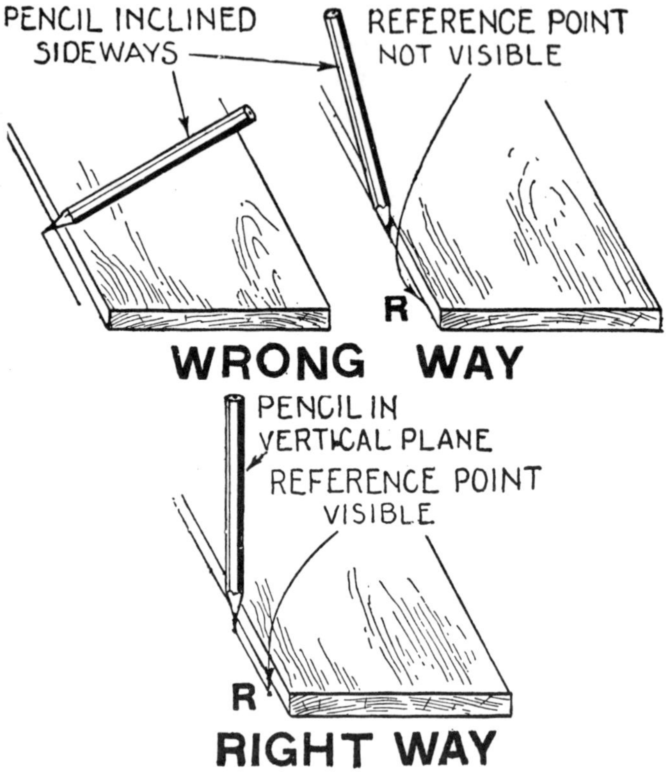

FIGS. 1,565 to 1,567.—Wrong and right ways of using the pencil in drawing lines.

sub-divided into the desired exact length by the dividers. This involves several trials. Set the dividers as near as can be to the desired length. Then test by spacing with the dividers along the line. The setting of the dividers after each trial is adjusted by turning slightly the hair spring nut until the correct length is obtained. For very fine divisions the bow dividers are more conveniently used. For precision the divisions L,A,R,F, should be marked off with a "pricker point."

Hints on Penciling.—The pencil drawing should look as nearly like the ink drawing as possible. A good draughtsman leaves his work in such a state that any competent person can without difficulty ink in what he has drawn.

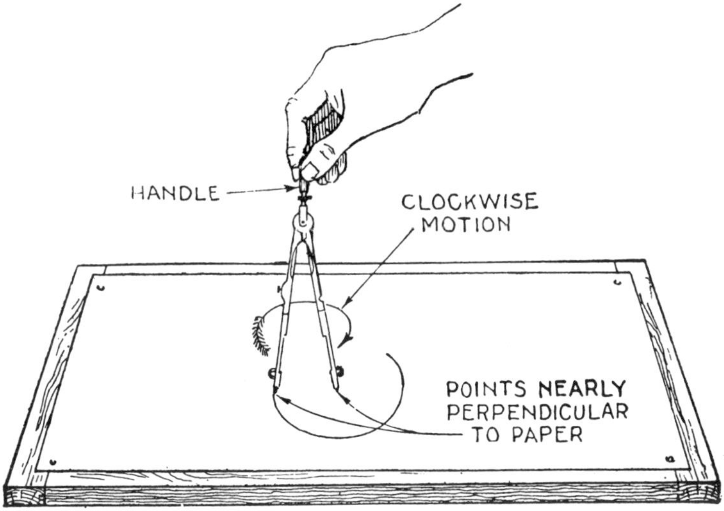

FIG 1,568.—Correct use of compass. Note method of holding compass grasping only the handle; also, upright positions of points, and clockwise motion.

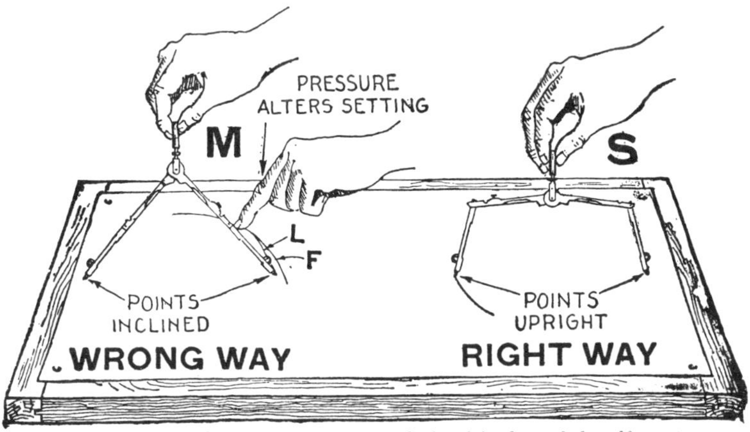

FIG. 1,569.—Wrong and right way to use compasses in describing large circles. Always have the points, especially the center point upright, otherwise the center indentation in paper will be enlarged and untrue. If both hands be used as at M, the setting may be altered by pressure on the leg, and part of the circle, as at L, will vary from the true part F.

The pencil should always be *drawn, not pushed*. Lines are generally drawn from left to right and from the bottom to the top or upwards. Pencil lines should not be any longer than the proposed ink lines. By keeping a drawing in a neat, clean condition when penciling, the use of the rubber upon the finished inked drawing will be greatly diminished.

Inking.—A drawing should be *inked in* only after the penciling is entirely completed. Always begin at the top of the paper, first inking in all small circles and curves, then the larger circles and curves, next all horizontal lines, commencing again at the top of the drawing and working downward. Then ink in all vertical lines, starting on the left and moving toward the right; finally draw all oblique lines.

Irregular curves, small circles and arcs are inked in first,

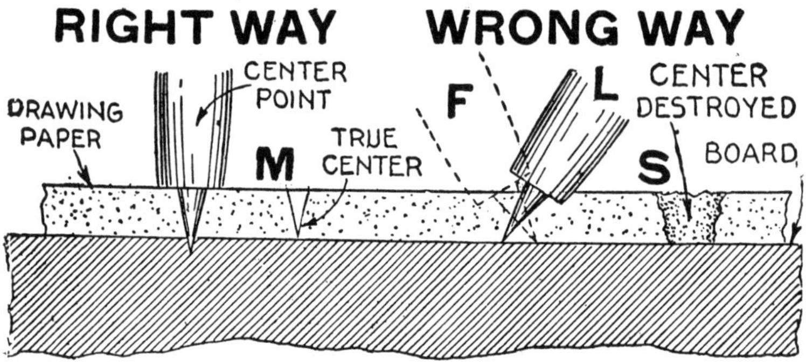

FIG. 1,570.—Right and wrong position of the compass center point (greatly enlarged for clearness). If the point be perpendicular to the paper in describing a circle as shown at the left, a clean cut indentation will be made in the paper as shown at M, with point removed, being in proper condition to use as a center. Again if the point be inclined in describing a circle the center point will rotate as indicated at L and F, the end of the point enlarging and tearing the indentation so that when the point is removed the indentation will be the condition shown at S, totally unfit to be again used as a center. The center point of a compass should be provided with a shoulder, as shown, instead of being simply conical—this limits the depth of indentation.

because it is easier to draw a straight line up to a curve than it is to take a curve up to a straight line.

In inking in lines the "ruling or lining" pen is used against the edge of the T square and triangles in making ink lines. The tool has two separate blades, or jaws, one of which is equipped with a spring to spread them apart. It is brought close to the other by a thumb screw or allowed to come away by turning the thumb screw. In the better class of drawing pens a hinge is wrought on one blade at the base, thus permitting the blade to open to a right angle, so that it can be cleaned of any ink which may cake or adhere to.

The drawing pen is applied by holding it perpendicular between the first and second fingers and thumb of the right hand, keeping the smooth blade close against the T square or triangle as required.

The ink is inserted between the jaws with a common writing pen or pen cork, which is now provided with every bottle of drafting ink sold, and any ink which remains outside must be cleaned off with a soft rag or piece of soft blotting paper. Similar directions must be followed when using the pen leg and point of compasses.

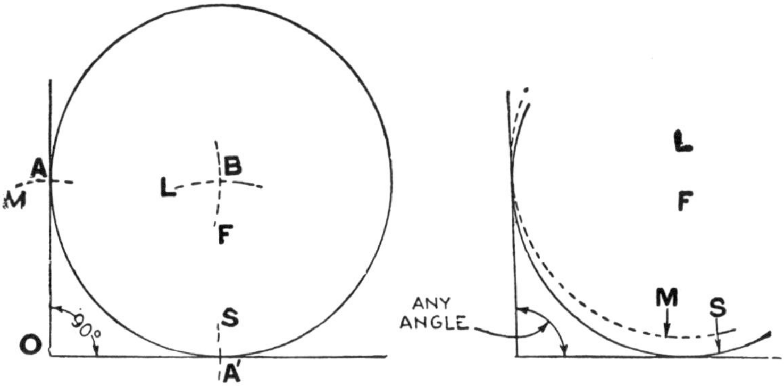

FIGS. 1,571 and 1,572.—Two methods of describing a circle of given radius tangent to two lines meeting at a point. In fig. 1,571 let OA be the given radius. With O as center, describe arcs M, and S. With points A, A', where these arcs cut the lines describe arcs intersecting at B, then will a circle described with B, as center, and radius OA, be tangent to both lines. ***In the second method,*** set compass to given radius and describe a trial arc M, from a trial center L, selected with pencil resting on one of the lines. If trial arc be not tangent to the other line, bring compass back to initial position with pencil point resting on line, lift center point, by inclining compass a little and select another trial center as F. Continue the process until a center is obtained about which a circle described will be tangent to both lines. With a little practice center can be quickly located by this method. The first method, fig. 1,571, holds when the lines are at 90° to each other, but the second method applies to any inclination.

The thicknesses of the lines are determined by the distance between the points of the blades, operated by the regulating screw. By separating or bringing together the blades a thick or thin line may be drawn, as desired.

Arcs, or circles, are inked in by removing the pencil end and inserting the pen end into the compass leg, or by using the bow compass.

Drawing to Scale.—The meaning of this is, that the drawing when done bears a definite proportion to the full size of the particular part, or, in other words, is precisely the same as it would appear if viewed through a diminishing glass.

When it is required to make a drawing to a reduced scale, that is, of a smaller size than the actual size of the object, say, for instance, ½ full size, every dimension of the object in the drawing must be one-half the actual size; in this case one inch on the object would be represented by ½ inch. Such a reduced drawing could be made with an ordinary rule. This, however, would require every size of the object to be divided by the proportion of the scale, which would entail a very great loss of time

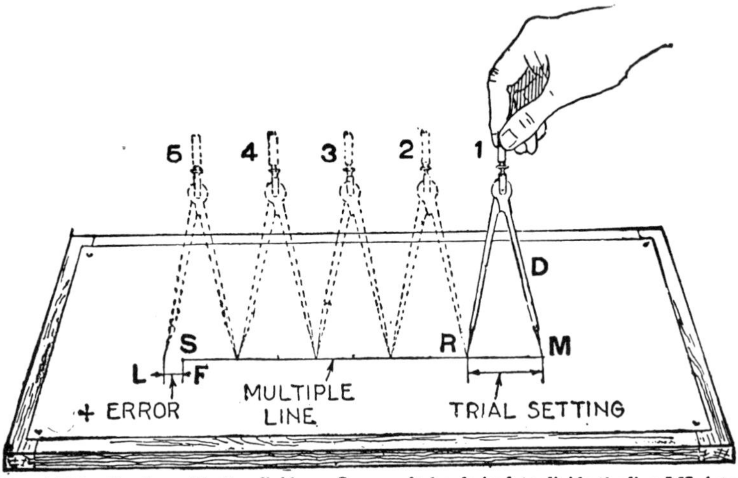

FIG. 1,573.—Spacing with the dividers. Suppose it be desired to divide the line MS, into 5 equal parts. Set dividers "by eye" to one-fifth the distance as MR. Space along the line by moving the dividers clockwise and counterclockwise to position 2, 3, 4, 5. As seen, MR, was taken too large. Adjust setting and respace until the correct setting has been obtained. Before setting the dividers "by eye," unloosen adjustment nut D, two or three turns, otherwise if the error LF, were negative (—), the hair spring could not be adjusted in the positive (+) direction.

in calculations. This can be avoided by simply dividing the rule itself by 2, from the beginning. Such a rule or *scale* as it is generally called, will be divided in ½ inches, each half inch representing one full inch divided into ½, ¼, ⅛, 1/16, each of these representing the same proportions of the actual sizes of the object to be drawn. From this contracted scale the dimensions and measurements are laid off on the drawing.

A quarter size scale is made by taking three inches to represent one foot. Each of the three inches will be divided into 12 parts representing inches, each one of these again will be divided in ½, ⅛, 1/16, etc.; each one of these representing to a quarter size scale the actual sizes of ½, ¼, ⅛, 1/16 of an inch.

It must be mentioned that in several instances, in this work, distances in one figure are said to be equal to corresponding distances in the same object in another view, while by actual measurement they are somewhat different; this is owing to the use of different scales—each scale separate should be marked on the drawing.

It must be understood that the scale on a drawing is not given

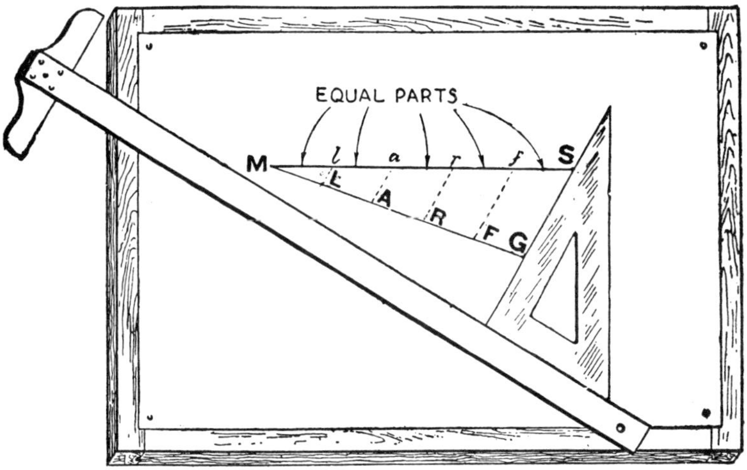

FIG. 1,574.—To divide a given line into any number of equal parts without use of dividers. Let MS, be the given line. From M, draw a diagonal line MG, of length in inches equal to the number of parts (say five) into which MS, is divided. Mark the inch divisions (L, A, R, F,) on MG, and join GS. With aid of T square and triangle, draw lines through L, A, R, F, parallel to GS, giving the points *larf* on MS. These points divide MS, into the desired number of equal parts. Note new use of T square by inclining it to any desired degree, the head not being in contact with the edge of the board.

for a shopman to take his dimensions from; such dimensions must all be taken from the dimension figures; the scale is given for the chief draughtsman's use, or whoever may check the drawing, and also for the use of other draughtsmen who may

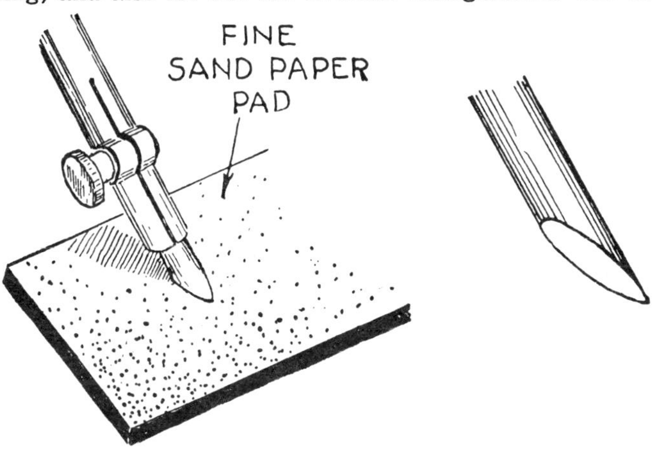

FIGS. 1,575 and 1,576.—Method of sharpening compass pencil by rubbing it over a pad of very fine sand paper (fig. 1,575), and appearance of pencil point after sharpening (greatly enlarged) fig. 1,576.

make at some future time alterations or additions to the drawing.

Dimensioning Drawings.—Every dimension necessary for the execution of the work indicated by the drawing should be clearly stated by figures on the drawing, *so that no measurements need to be taken in the shop by scale.* All measurements should be given with reference to the base or starting point from

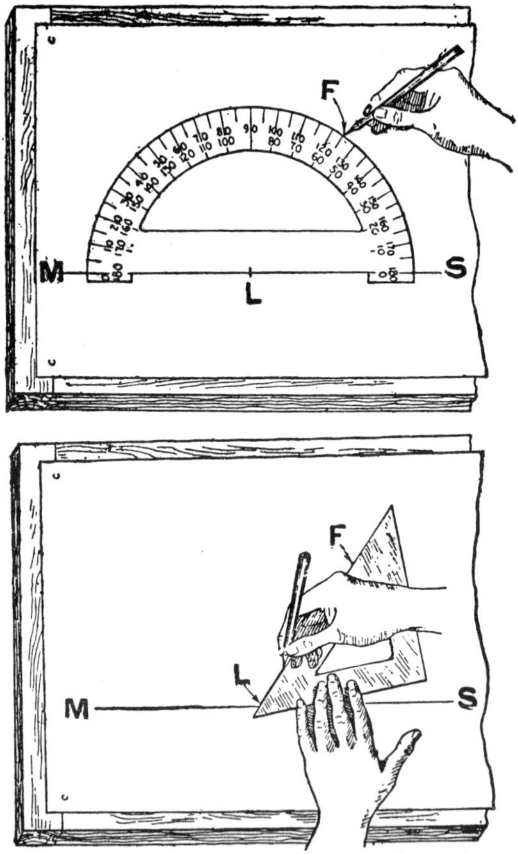

FIGS. 1,577 and 1,578.—Method of using the protractor to draw a line through a given point on a line and making a given angle with the line. Let MS and L, be respectively the given line and point on the line. Place protractor so that its center registers with point L, and its straight side with the line MS. Find angle on circular scale and mark same as at F. Remove protractor and join LF, by aid of triangle as in fig. 1,578, obtaining the required angle FLS.

which the work is laid out, and also *with reference to center lines.*

All figured dimensions on drawings must be in plain, round vertical figures, not less than one-eighth inch high, and formed by a line of uniform width and sufficiently heavy to insure printing well, omitting all thin, sloping or doubtful figures. All figured dimensions below two feet are best expressed in inches.

It is not necessary to put down a multiplicity of inch marks (″)—these can well be left off, using the foot and inch marks only when the dimension is expressed in feet and inches. This will save time and improve the appearance of the drawing.

It may be put down as a rule that the draughtsman must anticipate the measurements which will be looked for by the workman in doing the work, and these dimensions only must be put on the drawing.

The author objects to the usual style of dotted dimension and center lines and weak arrow heads, and prefers to make these lines and arrow heads solid, the lines being drawn very fine to distinguish them from the object as shown in fig. 1,579.

The dimensions written on the drawing should always give the actual finished sizes of the object, no matter to what scale the object may be drawn.

All dimensions which a shopman may require should be put on a drawing, so that no calculation be required on his part.

For instance, it is not enough to give the lengths of the different parts of the object, but the length over all, which is the sum of all these lengths, placed outside and the figures also be put outside, in which case an arrow should be put in to indicate the proper position of the figures.

The figure should be placed in the middle of the dimension line at right angles to that line, and so as to read either from the bottom, or from the right hand side of the drawing. The arrow heads should be put inside of the lines, from which the distance, as given in the dimension, is reckoned.

The dimension lines should also be put in the drawing, very near to the spaces or lines, to which they refer.

When "the view" is complicated, dimension lines drawn within it, might tend to make it still more obscure and difficult to understand; in such a case the dimension lines should be carried outside of the view and extension lines drawn from the arrow heads to the points, between which the dimension is given.

When the dimension includes a fraction, the numerator should be separated from the denominator preferably by a horizontal line instead of by an inclined line; care should also be taken to write the figures in a very clear and legible manner and crowding should be avoided.

Tracings.—Whenever it is desired to have more than one copy of a drawing, a "tracing" is made of it and from this as many blue prints can be obtained as are required.

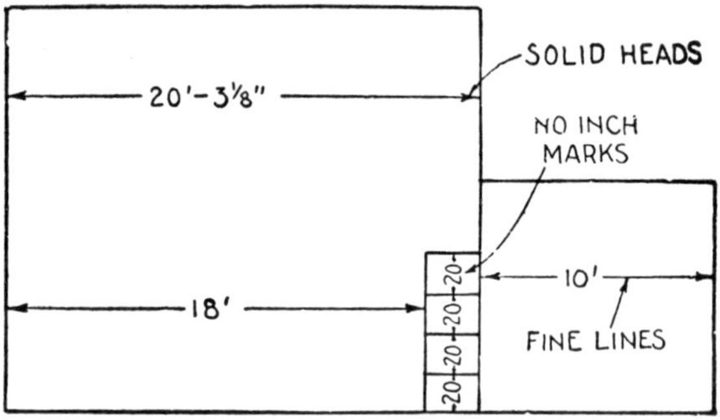

FIG. 1,579.—Method of dimensioning drawing as preferred by the author. Note, solid arrow heads that can be seen; fine dimension lines which by contrast are not confused with the lines of the drawing; no inch marks where dimension is in inches only.

When a tracing is needed for making blue prints, a piece of tracing paper or tracing cloth of the same size as the drawing is placed over the original drawing and fastened to the board. This tracing paper or cloth is almost transparent; the tracing is a mechanical copy of a drawing made by reproducing its lines as seen through a transparent medium such as has been described and the lines of the drawing can be seen through it.

The surfaces of the tracing cloth are called the "glazed side" and the "dull side," or "front" and "back"; the glazed side has a smooth polished surface and the dull side is like a piece of ordinary linen cloth.

Drawing on tracing paper or cloth is effected by pencil and drawing pen as in ordinary work.

The tracing cloth must be fastened to the board, over the drawing by pins or thumb tacks; moisture or dampness should be carefully avoided and the drawing done, preferably, on the smooth side of the cloth.

When tracing cloth will not take ink readily a small quantity of pounce may be applied to the surface of the cloth and distributed evenly with a piece of cotton waste, chamois, or similar material, but the pounce should be thoroughly removed before applying the ink.

In making tracings the order to be followed is as follows: 1, ink in the small circles and curves; 2, ink in the larger circles and curves; 3, then all the horizontal lines, beginning at the top of the drawing and working downward; 4, next ink in all the vertical lines, commencing at the left and moving back to the right; 5, draw in the oblique lines; 6, in finishing the figuring and lettering should be done with India ink, thoroughly black.

"Erasing," in case of mistakes or errors, should be done with an ink eraser or a sharp, round erasing knife; the surface of the tracing cloth must be made smooth in those places where lines have been erased; this is accomplished by rubbing the cloth with soapstone or powdered pumice stone, applied with a soft cloth or with the finger. When a mistake made is so serious that it cannot be corrected by erasing, a piece of the tracing cloth may be cut out and a new one inserted in its place.

A finished tracing should be provided with the title of the drawing, the date, scale and the initials of the draughtsman.

Lettering.—When the information necessary to the reading of a drawing cannot be expressed by lines and scale dimensions, it must be indicated in the form of printed explanations, remarks, etc.

To do good lettering is not an easy task, and unless the student is already experienced, he should devote considerable time to practicing the art. He should work slowly and bear in mind that much time is required to make well finished letters.

The character and size of the letters on all working drawings should be in harmony with the drawing on which they appear. It is desirable to have all lettering on a drawing made in the same style, only differing in size or finish of details.

Capital letters should always be sketched in pencil, especially by the beginner, and inked in afterwards: the lettering used on mechanical drawis usually of the simplest character, the letters being composed of heavy and light strokes only; for headings, titles of large drawings, where comparatively large lettering is required, it will be most appropriate to use large letters.

The title should be conspicuous, but not too much so; sub-titles should be made smaller than the main-title.

The "scale" and general remarks placed in the margin of the drawing or near the title should come next in size. All explanations and remarks on the views should not be larger than one-eighth inch.

After deciding on the size of the letters, lightly draw two parallel lines at a distance apart equal to the height of the letters. Good lettering requires

NOTE.—Many concerns have rules of their own, directing their draughtsmen to use either the smooth or the rough side for all purposes; if there be no such rules, it is left to the judgment of the draughtsman. While it is immaterial which side of the cloth is used in tracing, however, if any mistakes be made and have to be corrected this can be done easier on the glazed side; on the contrary, if any *additions* must be made to the tracing, which have to be drawn in pencil first, the dull side will be found most convenient, as the pencil marks show plainer on the dull side.

FIG. 1,580.—Winsor & Newton liquid Indian ink.

considerable practice. The beginner should first pencil in the letters before inking, but this is not necessary for the expert.

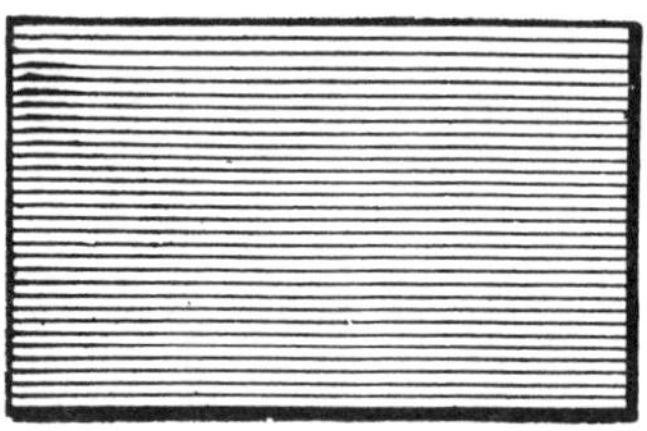

FIGS. 1,581 and 1,582.—Parallel line shading of flat and cylindrical surfaces.

Free hand lettering should only be taken up after the student is proficient in mechanical lettering; pencil guide lines for letters and words should be drawn; larger letters may first be penciled in very lightly, and an ordinary writing pen may be used for inking them in.

Letters should be so placed as not to interfere with the lines of the drawing and should clearly point out the part intended to be described. When single letters are used, they should be inked in before the shade or section lines are drawn; it is a good plan to start with the middle letter of the inscription and work in both directions.

Parallel Line Shading.—Plane surfaces are shaded by a number of parallel lines running parallel to the length of the plane which is to be shaded. If the plane is to be represented very light, it may be left blank or covered with very fine parallel lines, as shown in fig. 1,581.

A cylinder is shaded by a number of parallel lines, which are heaviest near to the side of the cylinder which does not receive

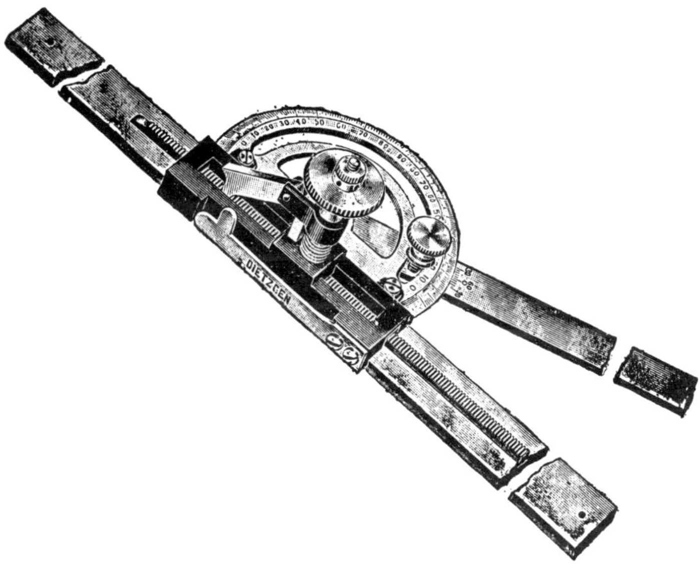

FIG. 1,583.—Dietzgen section liner and scale divider for drawing equally spaced parallel lines. *It consists of* a flat rack bar or base which carries a rack and carriage which slides on the rack bar. A semi-circular protractor with a ruler arm is attached to the carriage and by means of a simple mechanism from 4 to 200 parallel lines to the inch can be drawn. When the instrument is properly set, inch scales from ⅛ to 3 inches to the foot, decimal scales up to 1,000 to the foot can be produced.

any direct light, as in fig. 1,582. The heavy lines become lighter gradually and are drawn very fine near the middle of the cylinder; after this the lines are again drawn slightly heavier up to the side of the cylinder, which is nearest to the source of the light. The shading lines near the lighter side of the cylinder should never be as heavy as the heaviest lines on the dark side of the cylinder.

Section Lining is sometimes necessary to make use of a section, in order that certain details, which would otherwise be hidden, may be shown in a plain, clear manner. Such sections are usually indicated by drawing parallel oblique lines within the section, usually inclined 45°.

By changing the direction of these lines a clear distinction may be made between different pieces in the same view, which may be in contact.

Placing the lines too near together makes the work of sectioning much harder; the lines should not be drawn first in pencil, but only in ink, as the neat appearance of the drawing depends largely upon the uniformity of the lines in the section and these lines are to be spaced by the eye only. The process consists simply in ruling one line after another, sliding the triangle along the edge of the T square for an equal distance after drawing each section line.

Wooden beams are sectioned by a series of rings and radiating lines in imitation of the natural appearance of a cross section of the wood.

The side of a beam or board is represented by lines (drawn in free hand) running similarly to the grain of the wood. Sections of thin strips of metal as beam hanger straps, etc., are usually represented by filling in the whole sectional area solid black. In this case a white line must be left between adjoining sections.

CHAPTER 9

How to Read Plans

There are various ways of representing objects in drawings, and a knowledge of these different methods is essential in order to intelligently read a drawing or blue print.

The various methods of illustrating an object by a drawing are:

1. Perspective
2. Cabinet projection
3. Isometric projection
4. Orthographic projection
5. Development of surfaces

Of these methods, the first three may be classed as "pictorial" in that they show the entire visible portion of the object in one view, whereas the fourth requires several views to fully present the object and may be called "descriptive."

It is this latter method that is most generally used and which requires a little study to comprehend it.

A perspective drawing shows an object as it really appears to the eye, but presents so many difficulties of construction that the projection methods have been devised to overcome them. These projection methods will accordingly be considered first before taking up perspective.

Cabinet Projection.—In this system of drawing, the lines of an object are drawn parallel to three axes, one of which is horizontal, a second vertical, and the third, inclined 45° to the horizontal, as in fig. 1,584. The horizontal axis lie in the plane of the paper, and the vertical and inclined axes lie in a plane intended to appear to the eye as being at right angles to the plane of the paper.

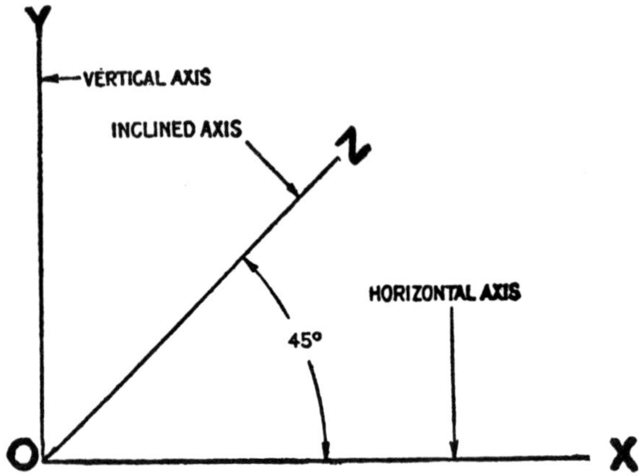

Fig. 1,584.—Cabinet projection axes.

These axes lie in planes at right angles to each other and known as the horizontal, vertical and profile planes.

In cabinet projection it is to be remembered that

1. All horizontal measurements, parallel to the length of the object must be laid off parallel to the horizontal axis, in their actual sizes.

2. All vertical measurements, parallel to the height of the object, must be drawn parallel to the vertical axis in their actual sizes.

3. All measurements parallel to the thickness of the object must be laid off on lines parallel to the 45° axis, in sizes of only one-half of the actual corresponding measurements.

It is not essential which side of the object should be considered its length and which side its thickness.

Problem 1.—To draw a cube in cabinet projection.

First draw the three axes, OX, OY, OZ, as in fig. 1,585. Lay off OA and OC, on OX and OY, equal to side of the given cube, and complete the side by drawing CB and AB. On OZ, lay off OG, equal to ½ OA. Through C, draw a line parallel to OZ, and through G, a dotted line parallel to OY, giving the lines CF and GF. Similarly through points G,F,A and B, draw parallels to the axes, thus completing the cube.

In the drawing the face ABCO, is regarded as lying in the plane of the paper, the face DEFG, as parallel and the other faces ABED and OCFG, as perpendicular to the plane of the paper. The edges which would be invisible if the cube were made of opaque material such as wood, are represented by dotted lines.

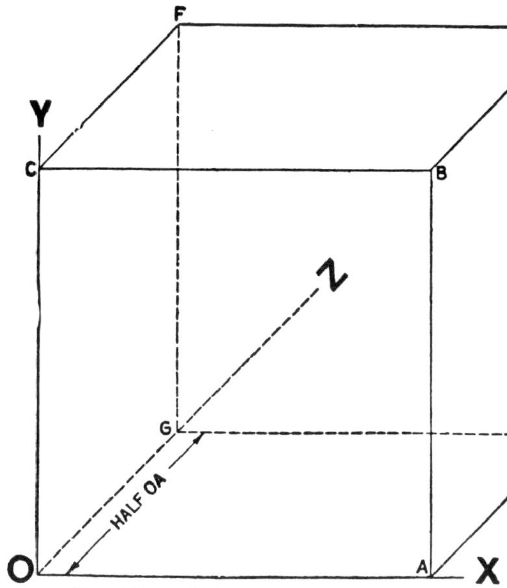

FIG. 1,585.—Cabinet projection of a cube.

Problem 2.—To draw a right cylinder with its bases in the XOY plane; length of cylinder 3 times the diameter.

This is the best position to draw a cylinder because the bases will be circles and the difficulty of describing ellipses avoided.

Draw the axes as usual. With O, as center and OA, equal to radius of cylinder, describe a circle. On OZ, lay off OM = 3 times OA (since length of cylinder = 3 times the diameter.)

With same radius describe a circle through M, and draw tangents BC, and DE, thus completing the outline of the cylinder.

The portion of the circle about M, between C and E, is shown by dotted lines because it would be invisible if the cylinder were made of opaque material.

Problem 3.—To draw a prism enclosing a cylinder with its bases parallel to the YOZ plane; length of cylinder 3 times the diameter.

Draw in the cube as directed in problem 1, making AD = ⅙ OA, as in fig. 1,587. Now, show half of the base ABED, in the plane of the paper as in fig. 1,588. Here, draw diagonals OB and OA, and describe the half circle tangent to the sides. Through the intersection of the circle with diagonals draw line 12.

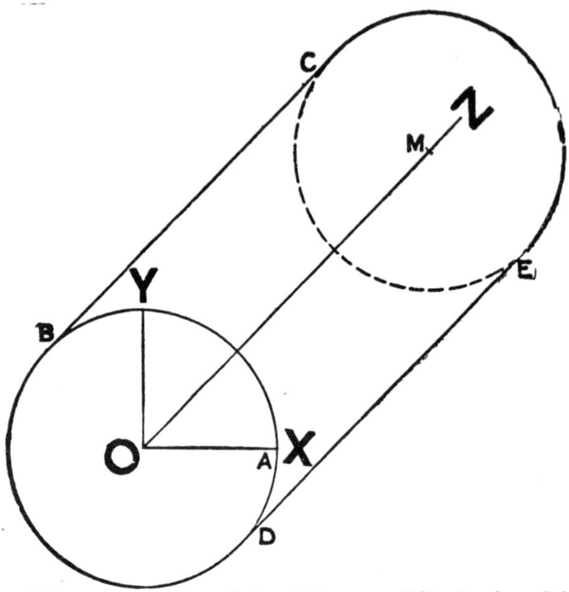

Fig. 1,586.—Cabinet projection of a cylinder with bases parallel to the plane of the paper.

In fig. 1,587 make B1 = ½ of B1 in fig. 1,588 and draw line 12, in fig. 1,587, and by similar construction line 34. Next draw diagonals AE and BD. The intersection of lines 12 and 34 with these diagonals will give four points together with points MLSF, through which to construct an ellipse representing the base of the cylinder as seen in profile constructing a similar ellipse at the other end and drawing the two tangents to the ellipses completes the outline of the cylinder.

Problem 4.—To draw a hexagonal prism inscribed in a right cylinder whose base equals twice diameter.

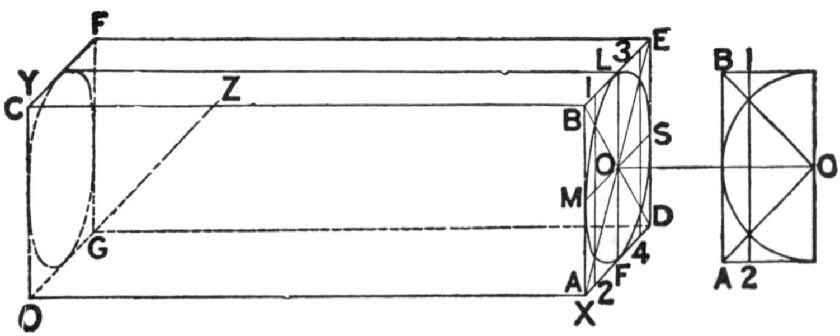

Figs. 1,587 and 1,588.—Cabinet projection of a prism enclosing a right cylinder with its bases parallel to the YOZ plane.

In construction fig. 1,589, describe a circle of diameter equal to diameter of the cylinder. Inscribe a hexagon. In fig. 1,590 lay off OF and OC, equal to *of* and *oc*.

Transfer points *m*, *s*, obtaining M, S, and through M and S, draw lines parallel to OZ, and on these lines lay off ME = ½ *me;* MA = ½ *ma*, etc. Through the points thus obtained draw in the ellipse ABCDEF.

Similarly construct upper ellipse A′B′C′D′E′F′ at elevation OO′ = ¾ FC, and draw tangents thus completing the cylinder. Join AB, A′B′, BC, B′C′, etc., and AA′, BB′, etc., thus completing outline of inscribed hexagonal prism.

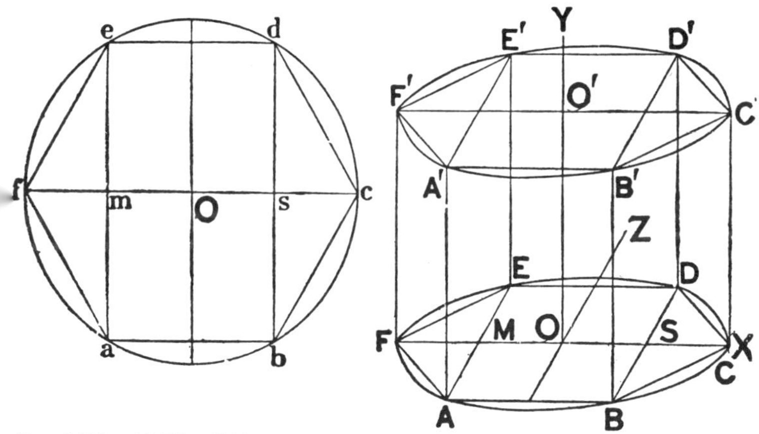

FIGS. 1,589 and 1,590.—Cabinet projection of a hexagonal prism inscribed in a right cylinder.

Modified Cabinet Projection.—There are various ways in which the 45° one-half foreshortened profile cabinet projection just described can be modified for convenience to suit special conditions. For instance, instead of inclining the OZ, axis to the right, as in fig. 1,591, it may be pointed to the left as in fig. 1,592.

Instead of foreshortening the profile dimension one half, as in fig. 1,593, in some cases, where dimensions are to be taken from the drawing, the profile may be made full dimension as in fig. 1,594.

Sometimes, because of the limited space available, it is desirable to take the OZ, axis at some angle other than 45°. In such cases it is usually taken at 30° or 60°, because these angles are obtained directly with the T square

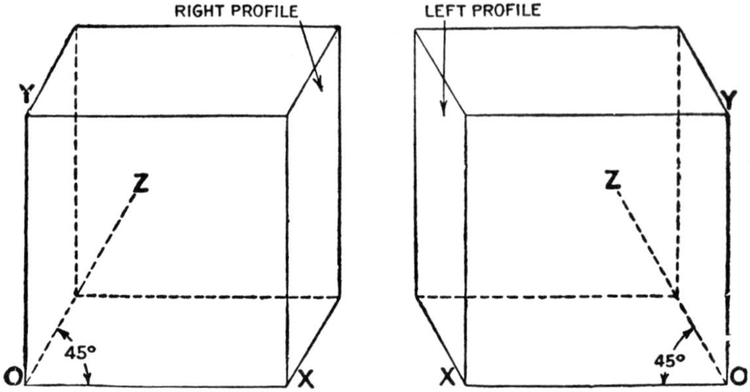

FIGS. 1,591 and 1,592.—45° cabinet projection with right and left profile.

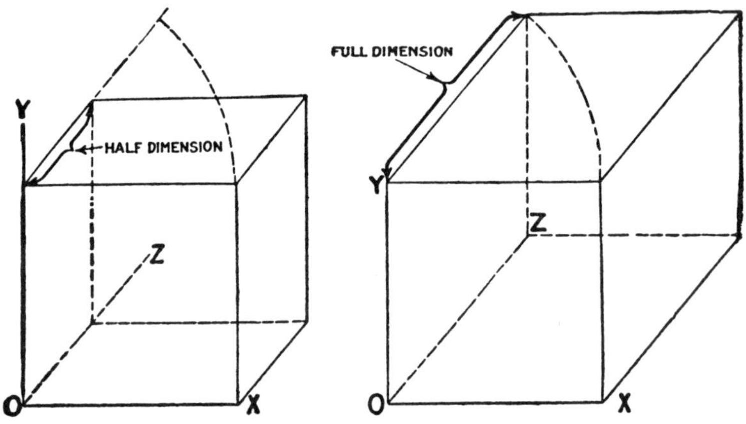

FIGS. 1,593 and 1,594.—Half and full profile dimension 45° cabinet projection. Evidently where all lines are drawn full dimension as in fig. 1,594, the drawing is made without calculating the profile dimensions, and especially in case of a complicated object time is saved.

and triangles—the object of the projection system being to use these instruments to conveniently and quickly execute drawings. In special cases, obviously any angle may be taken to suit the conditions.

These modifications of cabinet projection are shown in figs. 1,595 to 1,600. The first three figures show the object in full profile dimension for comparison. However, to save space and approach the natural appearance of the object, the following are the approved proportions for the profile.

OZ axis 45°, profile half dimension

OZ axis 30°, profile two-thirds dimension

OZ axis 60°, profile one-third dimension

The appearance of an object drawn to these proportions is shown in figs. 1,598 to 1,600.

Isometric Projection.—By definition the word *isometric* means *equal distances*, and as here applied, isometric projection is a system of drawing *with measurements on an equal scale in every one of three sets of lines* 120° *apart and representing the three planes of dimension.*

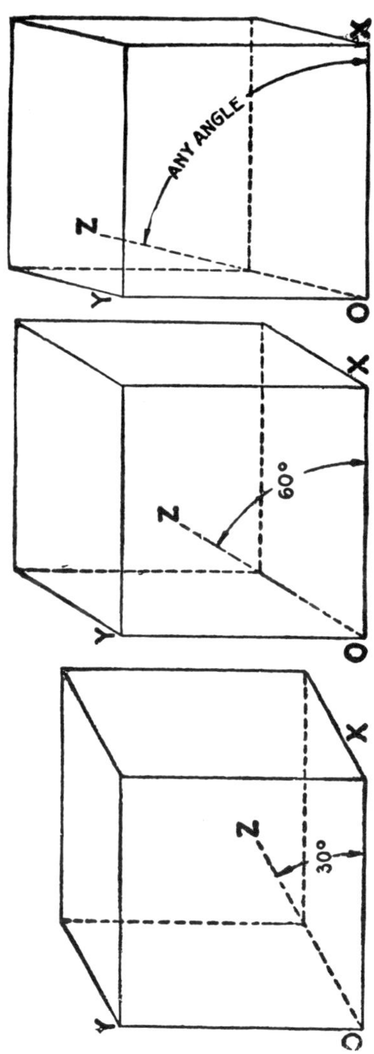

FIGS. 1,595 to 1,597.—Modified full profile dimension cabinet projection, with OZ axis at 30° (fig. 1,595); 60° (fig. 1,596), and at any angle (fig. 1,597). Evidently these modifications render the system flexible with respect to space and clear representation of any special part of an object.

In other words, the axes are taken 120° apart and there is no profile foreshortening as in cabinet projection, all lines being drawn full length.

Isometric projection further differs from cabinet projection in that none of the three planes lie in the plane of the paper.

Fig. 1,601 shows method of laying out the isometric axes using T square and 30° triangle. Figs. 1,602 and 1,603 show comparison of axes of the cabinet and isometric systems.

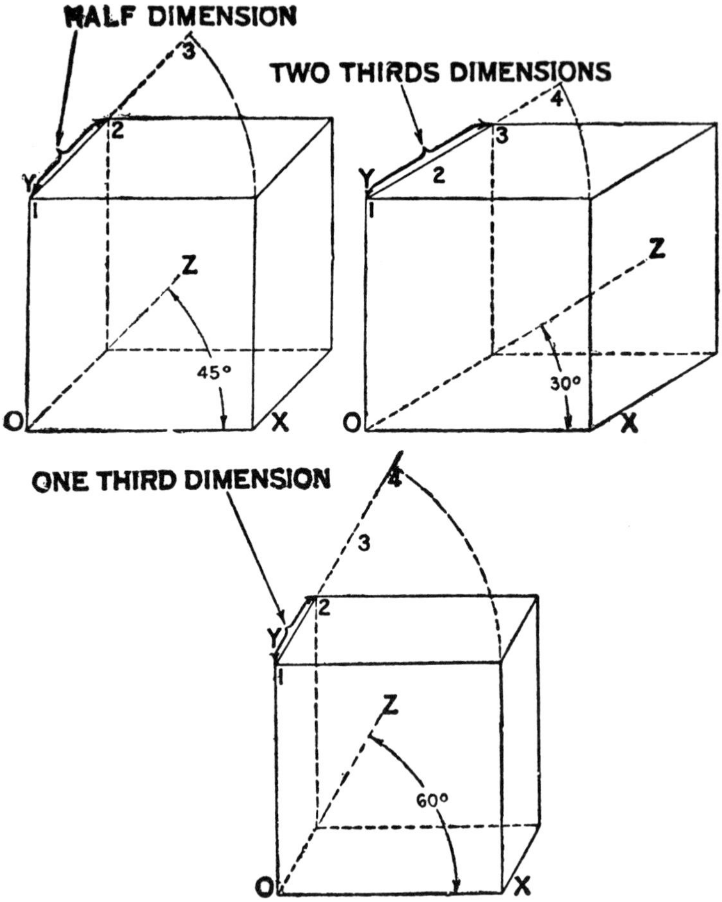

FIGS. 1,598 to 1,600.—Approved proportions for profile dimensions of 45°, 30° and 60° cabinet projection.

Problem 5.—To draw a prism in isometric projection.

First draw the axes OX, OY and OZ, at 120° as explained in fig. 1,601.

From O, fig. 1,604, lay off on the axes just drawn OA = OB = OC = length of side of the cube. Through points A,B,C, thus obtained, draw lines parallel to the axes, giving points D,E,F, thus completing visible outline of the cube.

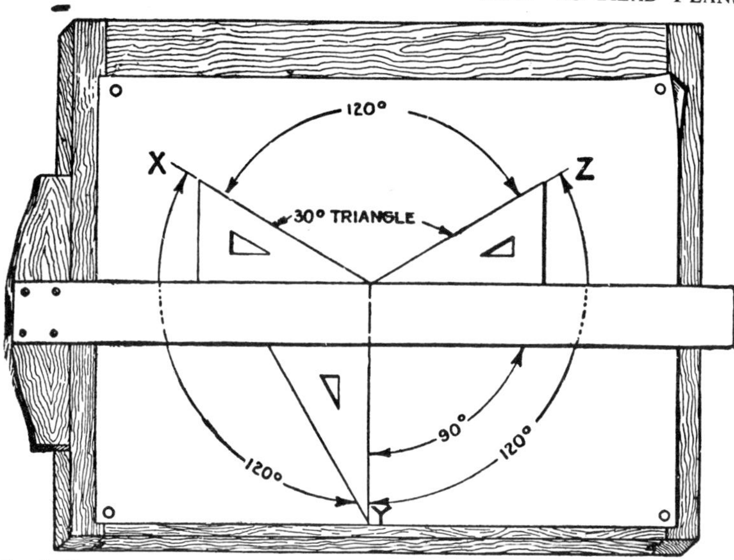

FIG. 1,601.—**Isometric axes** laid out at 120° to each other with 30° triangle and T square.

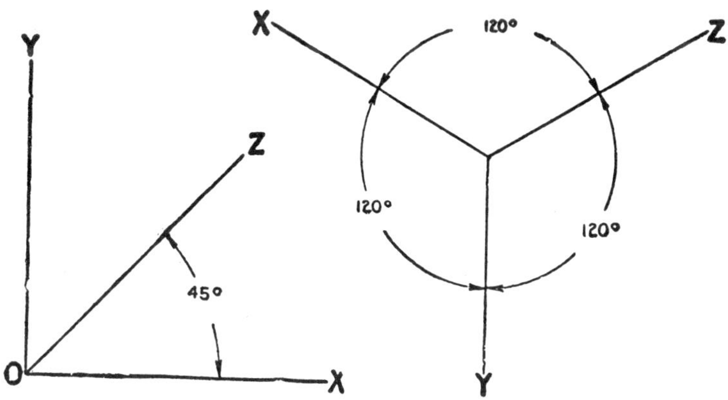

FIGS. 1,602 and 1,603.—Comparison of cabinet and isometric axes.

Through D, E and F, draw dotted lines intersecting at G, which gives the invisible outlines of the cube, assuming it to be opaque. An objection to this view is that the point G, falls behind the line OB, thus the outline of the invisible portion does not appear so well defined as it would in the case of a parallelopipedon as in the little fig. 1,605 at the right.

An objection to isometric projection is that, since no projection plane lies

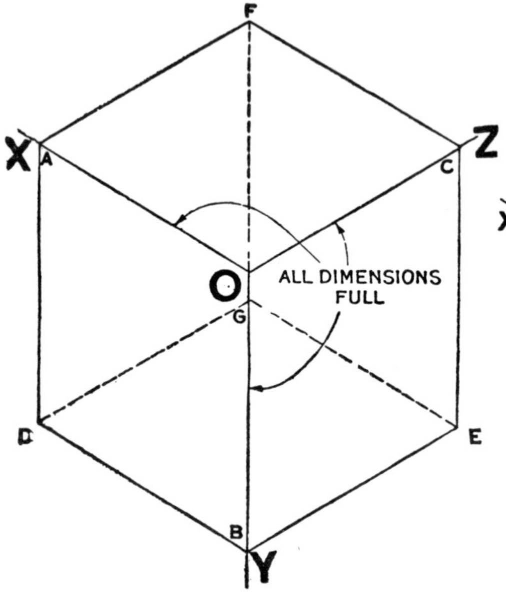

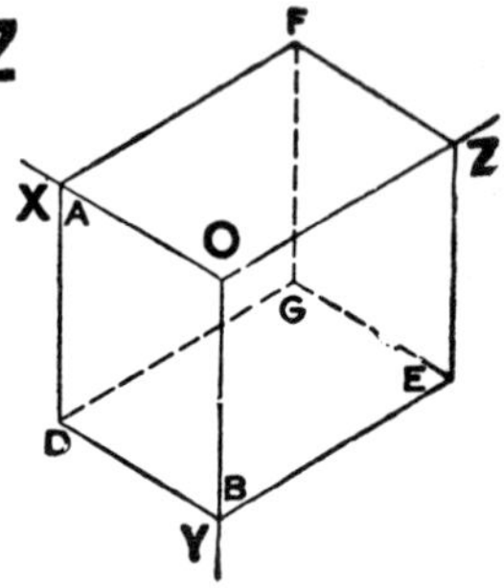

FIG. 1,604.—Isometric projection of a cube.

FIG. 1,605.—Isometric projection of a parallelopipedon showing completed dotted outline of invisible portion as compared with that of the cube fig. 1,604 in which part of the dotted line FG, falls behind OB. The reference letters are the same in both figures.

in the plane of the paper, it is necessary to construct ellipses to represent circular portions of an object and this requires time and skill.

Problem 6.—Draw a horizontal prism with inscribed cylinder; length of cylinder two times the diameter.

Draw the prism as explained in fig. 1,604, and drawn in fig. 1,606, making its length twice its side. Now construct the half end view in plane of paper (fig. 1,607); describe circle, diagonals and intersecting line 12.

Transfer from fig. 1,607 line 12 to fig. 1,606 and draw symmetrical line 34 and diagonals. The intersections, together with points MS, and LF, of axial lines through the center, give eight points through which construct ellipse.

Construct also a similar ellipse at other end of prism and join two ellipses with tangents, thus completing outline of inscribed cylinder.

An objection to isometric projection is the greater amount of space required as compared with cabinet projection; this point is shown in figs. 1,608 and 1,609.

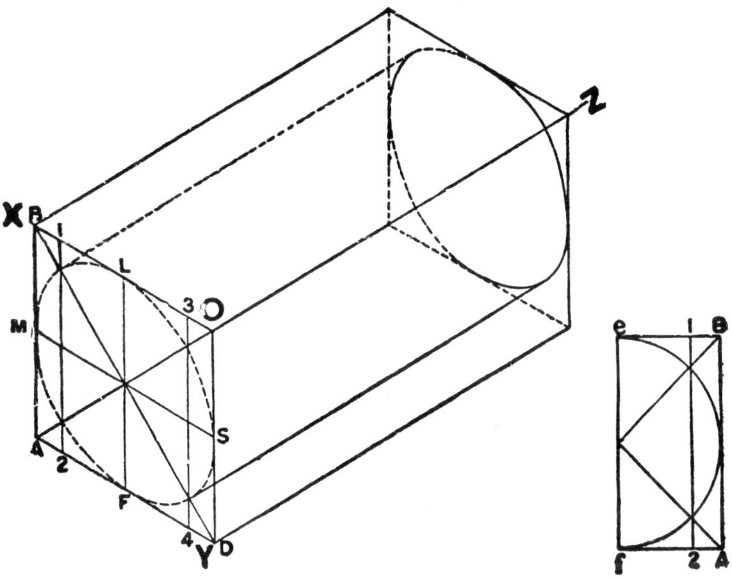

Figs. 1,606 and 1,607.—Isometric projection of a horizontal prism with inscribed cylinder.

Orthographic Projection.—Isometric drawing and cabinet projection, while showing the object as it really appears to the eye of the observer, are neither of them very convenient methods to employ where it is necessary to measure every part of the drawing for the purpose of reproducing it.

Drawings suitable for this purpose, generally known as *working drawings*, are made by the method known as *orthographic projection*.

In cabinet or isometric projections, three sides of the object are shown in one view, while in a drawing made in orthographic projection, but one side of the object is shown in a single view.

To illustrate this, a clear pane of glass may be placed in front of the object intended to be represented.

In fig. 1,610 a cube is shown on a table; in front of it, parallel to one face (the front face) of the cube, the pane of glass is placed.

Now, when the observer looks directly at the front of an object from a considerable distance, he will see only one side, in this case only the front side of the cube.

The rays of light falling upon the cube are reflected into the eyes of the observer, and in this manner he sees the cube. The pane of glass, evidently,

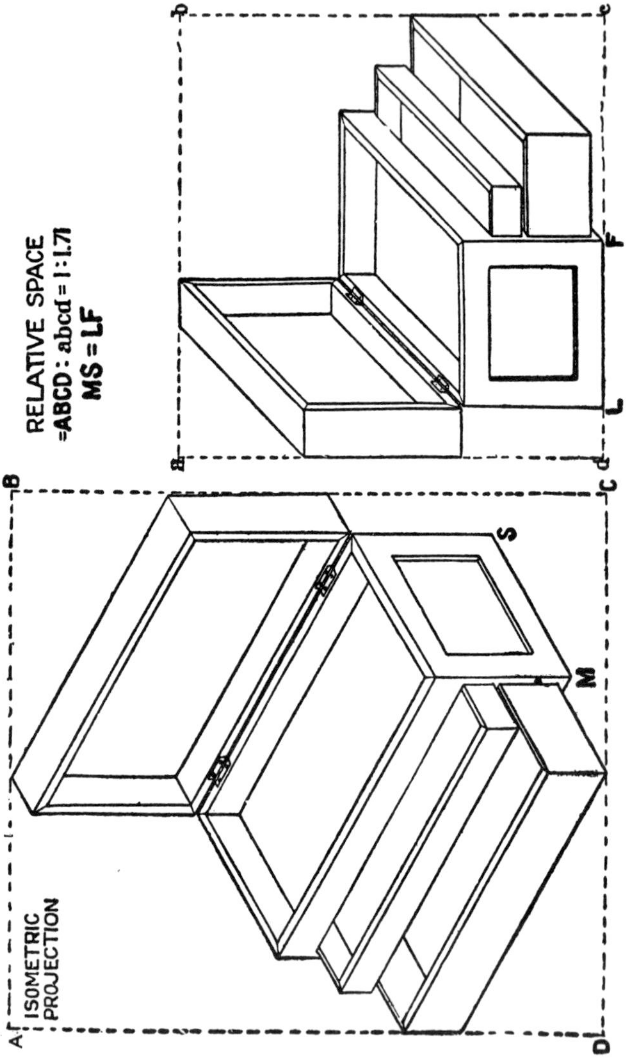

Figs. 1,608 and 1,609.—Comparison of isometric and cabinet projection showing relative space required to represent the same object drawn to same scale. Note that dimension MS = LF. The saving in space by cabinet projection is due to the position of the axes and the fore shortening of the profile dimensions.

is placed so that the rays of light from the object will pass through the glass in straight lines, to the eye of the observer. The front side of the object, by its outline, may be traced upon the glass, and in this manner a figure drawn on it (in this case a square) which is the view of the object as seen from the front which in this case is called the *front elevation*.

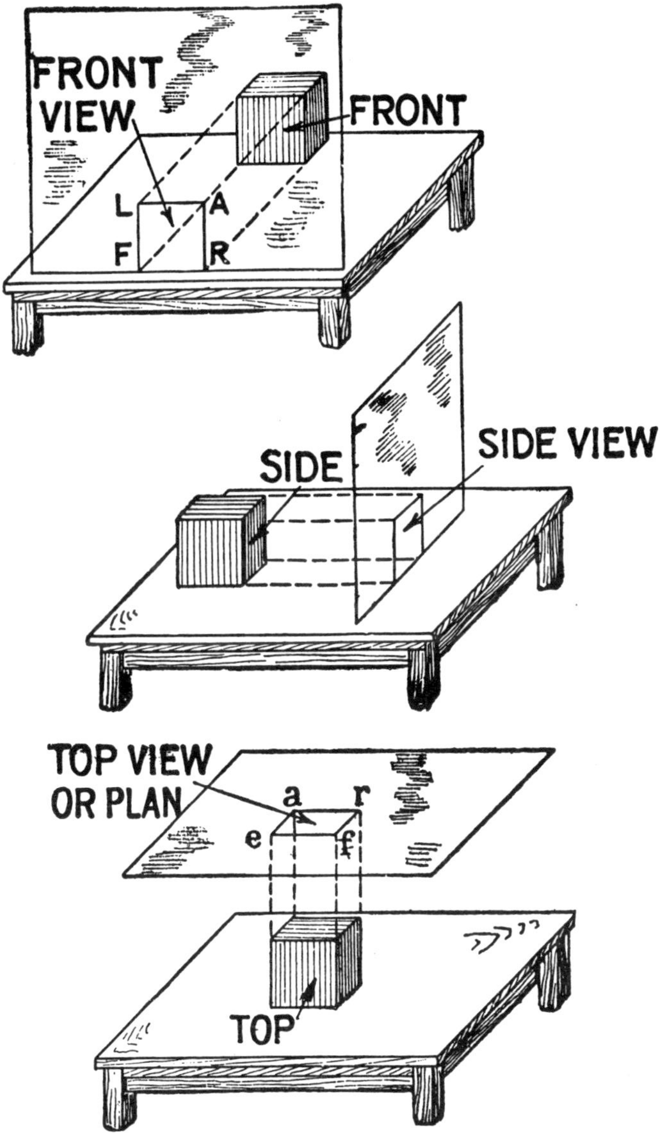

FIGS. 1,610 to 1,612.—Illustration of the various "views" of an object in orthographic projection by means of pane of glass placed in the different planes of projection.

One view, however, is not sufficient to show the real form of a solid figure. In a single view two dimensions only can be shown, length and height; hence the thickness of an object will have to be shown by still another view of it, as the top view or *plan*.

Now, place the pane in a horizontal position above the cube which is resting on the table, as in fig. 1,611, and looking at it from above, directly over the top face of the cube, trace its outline upon the pane; as a result, a square figure is drawn upon the glass, which corresponds to the appearance of the cube, as seen from above. This square on the glass is the top view of the cube, or its *plan*.

Fig. 1,612 shows the manner in which a side view of the cube may be traced; the glass is placed on the side of the cube, which rests on the table

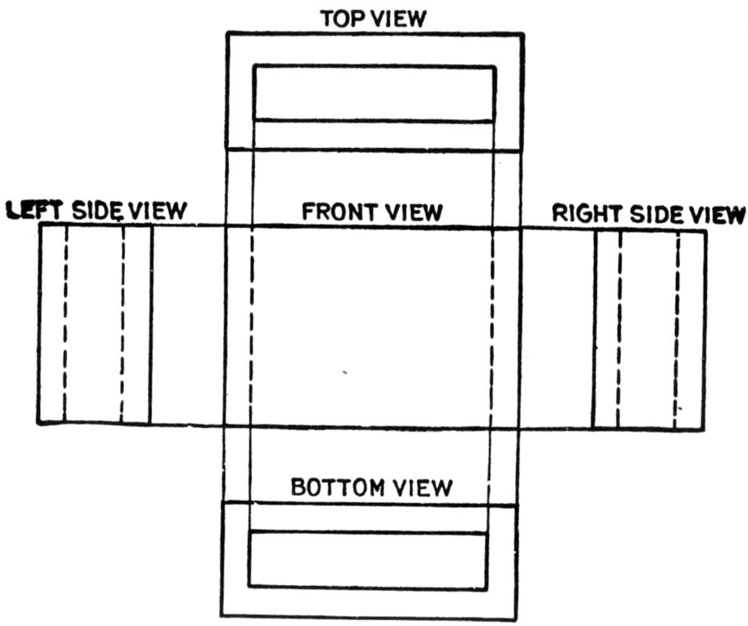

Figs. 1,613.—Five views of an object as drawn in orthographic projection.

as before, and the outline of the cube on the glass in this position, is called its *side elevation*.

Usually either two of the above mentioned views will suffice to show all dimensions and forms of the object, but to completely represent complicated objects, three or four views may be required.

In complicated pieces of machinery, however, more views, three and even more may be required to adequately represent the proportions and form of the different parts.

A drawing which represents the object as seen by an observer looking at it from the right side is called *the right side elevation* and a drawing showing the object as it appears to the observer looking at it from the left side is called *the left side elevation.*

In the case of a long object, a view at the end is called an *end view.*

A view of the object as seen from the rear is called the rear view or rear elevation, and a view from the bottom, the bottom view.

The different views of an object are always arranged on the drawing in a certain fixed and generally adopted manner, thus—

The front view is placed in the center; the right side view is placed to the right of the front view, and the left side view to the left; the top view is placed above the front view and the bottom view below it. The different views are placed directly opposite each other and are joined by dotted lines called *projection lines.*

By the aid of projection lines, leading from one view to the other, as in fig. 1,613, measurements of one kind may be transmitted from one view to the other; for instance, the height of different parts of an object may be transmitted from the front view to either one of the side views; in like manner the length of different parts of the object may be transmitted by the aid of projection lines, to the bottom view and top view.

It is often desirable to show lines belonging to an object, although they may not be directly visible. In fig. 1,613 the top view and the bottom view show plainly that the object is hollow; looking at the object from the front or from the sides, however, the observer could not see the inside edges of the object, except it were made of some transparent material.

In projection drawing it is assumed for convenience that all objects are made of such material, transparent enough to show all hidden lines, no matter from which side the object is observed; these hidden lines are represented in the drawing by dotted lines.

Problem 7.—Draw a plan, elevation and end view of the prism shown in fig. 1,614.

First draw the top view or plan as in fig. 1,615, by drawing the rectangle ABCD, to scale making AB, = 1 in. and AD, 4 ins. For the elevation, project the point AB, down to the parallel line obtaining line AB.

Lay off AE = 2 ins. and complete rectangle giving ABFE, or elevation. The end view BCGF, is drawn in a similar manner, side BF, being obtained by projection and BC, by measurement.

Problem 8.—Draw plan, end and front views of the barn shown in fig. 1,616.

The plan will consist simply of a rectangle CDFE (fig. 1,617), the length of whose sides being obtained from the dimensions in the orthographic projections. The end view is projected down from points C,A,E, of the plan, being identical with the end in fig. 1,616, because it is here drawn in the

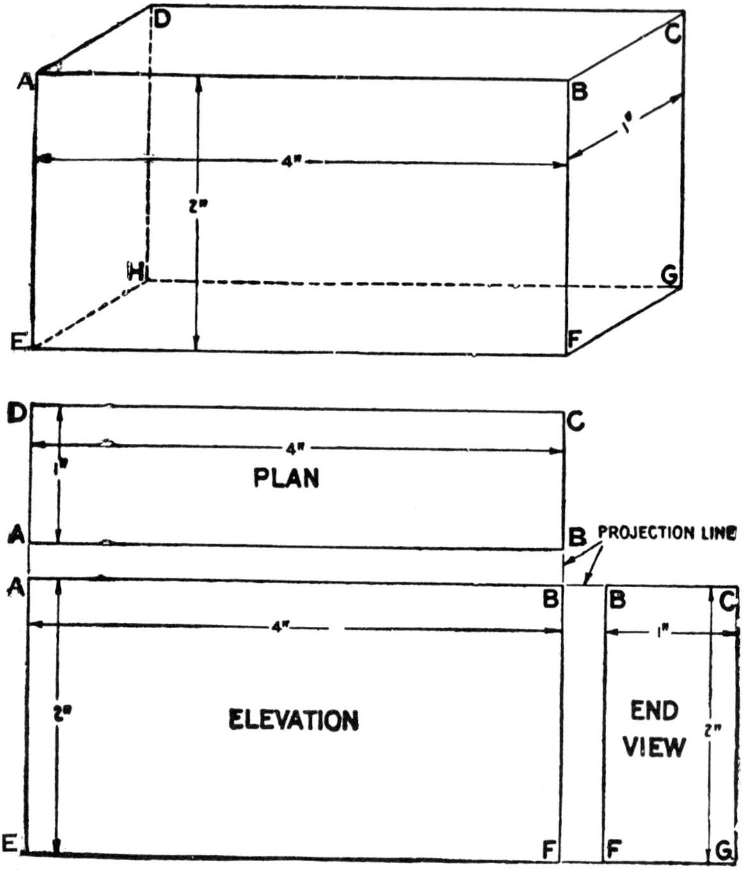

Figs. 1,614 and 1,615.—Cabinet projection of a prism, and orthographic views of same being shown in plan, elevation and end view

"OX, plane" which is the plane of the paper and accordingly is seen in true size.

Similarly for the front view project over the points A,B,I, of the end view and lay off AB, EF, and IJ, equal to 40 ft., the elevation of these lines being obtained from the given dimensions. The door is laid out in a similar manner.

Center Lines.—Objects which are symmetrical with respect to some axis drawn through the center are most easily drawn by first drawing such axis or *center line* and then drawing the object so that its center coincides with the center line. It is usual to

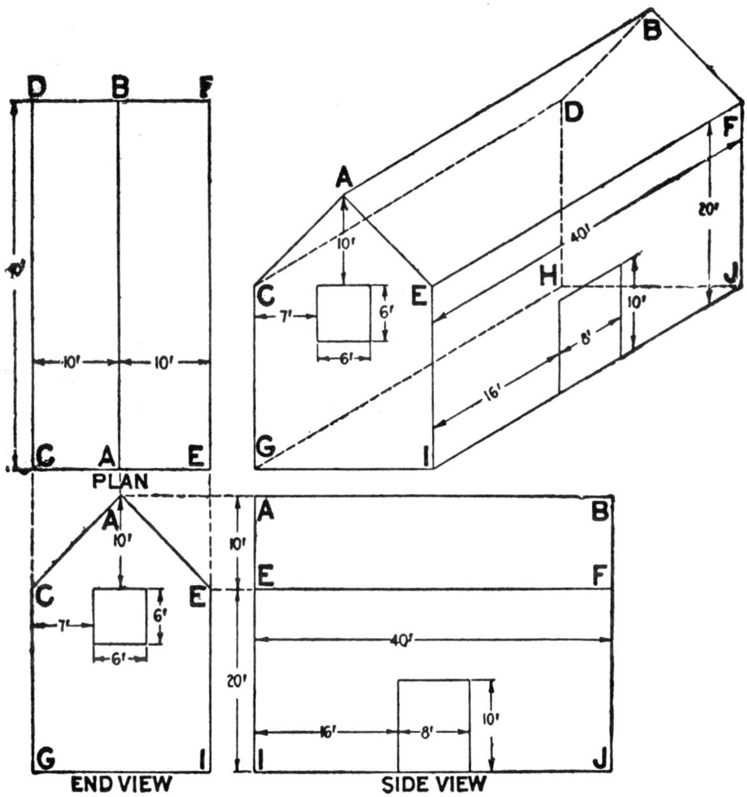

FIGS. 1,616 and 1,617.—Cabinet projection outline drawing of a barn and same drawn in orthographic projection.

make such lines broken by dot and dash, to distinguish them from the lines of the object. The author, however, prefers to draw solid center lines, obtaining the proper contrast to avoid confusion by drawing them much finer than those of the object, the same method being used also for dimension lines.

Figs. 1,618 and 1,619 show a rectangular built-up post drawn with center lines. Evidently since the figure is symmetrical with respect to these lines, equidistances are conveniently spaced off each side by aid of dividers and the drawing made quickly and with precision.

Since to show the entire length of the post would require considerable space, it is usual to "break it off" as at MS, by two ragged free hand lines, which indicate that part of its length is not shown. The construction of

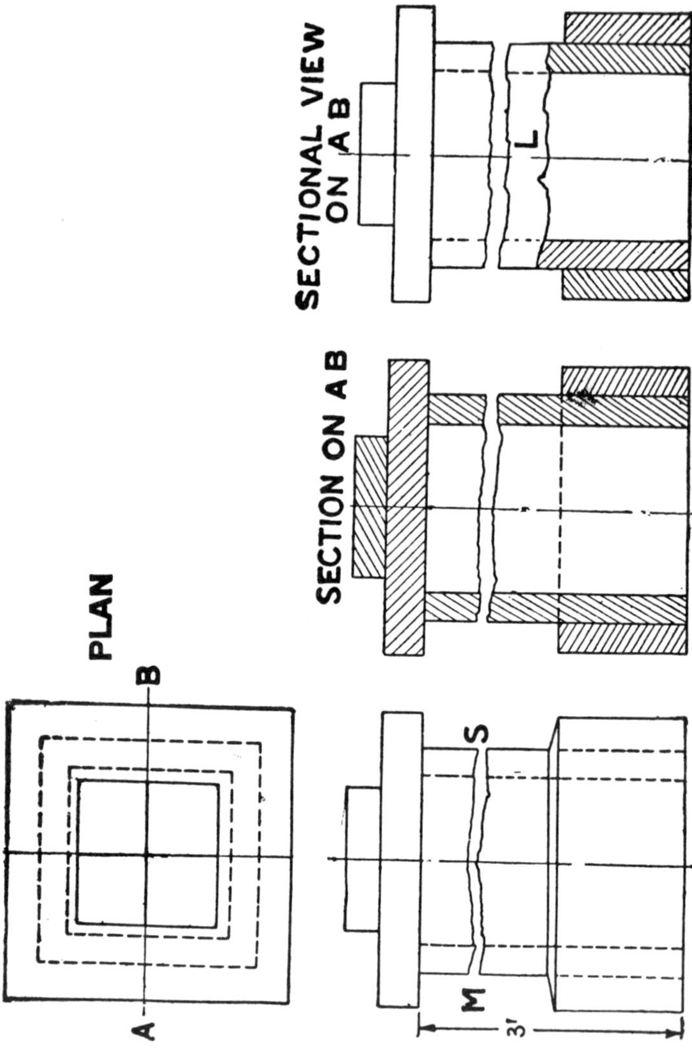

FIGS. 1,618 to 1,621.—Orthographic projection drawings of a built up post illustrating center lines, section, and sectional view; also the method of reducing space required for drawing of a long object by "breaking it off" as at MS.

the post may be more clearly shown by drawing it as though it were sawed through from end to end along the line AB; thus showing only the half back of AB; it would then have the appearance as shown in fig. 1,620, called a *section* on AB. The surface assumed to be sawed is "section lined." That is, covered with a series of parallel lines usually inclined 45°, 30° or 60°. It will be noted that the section lines in one plank run in a different direction from those in an adjacent plank so as to distinguish the separate parts.

To draw section lines consumes time, hence, time may be saved by showing in section not more than is sufficient for clearness, the rest of the drawing being seen "in full" as in fig. 1,621; this is called a sectional view. Here the post is shown in full down to the ragged line L, below this line the post is considered as being cut away to the axis AB.

Development of Surfaces.—The principles of projection already explained may be readily applied to the important problem of development of surfaces.

Whenever it is necessary to make an object of some thin material like sheet metal, as in the case of elbows or tees for leader or stove pipe, the surface of the desired object is laid out on sheet metal, in one or in several pieces; these are called the

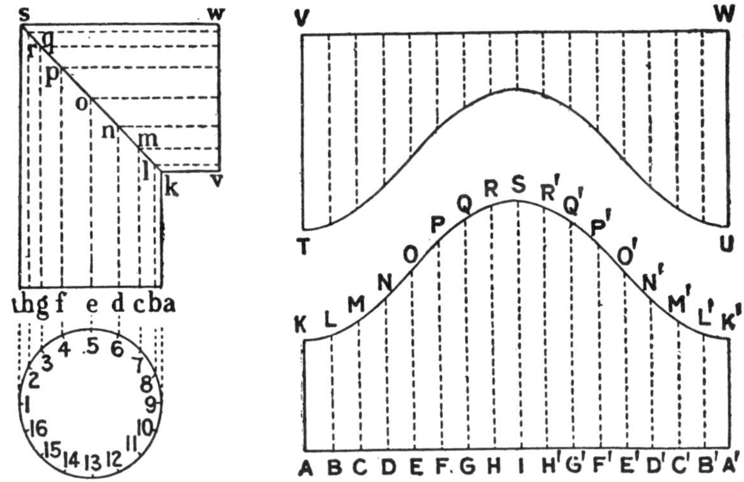

FIGS. 1,622 to 1,624.—Elevation and plan of 90° stove pipe elbow, and development of surfaces

Elbow Patterns.—In all elbow work the difficulty lies in obtaining the correct rise of the mitre line. By the use of a protractor this is overcome and thus the necessity of drawing a complete quadrant is avoided. Following the rule given in the illustration the rise can be easily found, when the throat and diameter of the pipe is known. The following table gives the rise of mitre line for elbows of various degrees and of various number of pieces.

patterns of the object; the pattern being first laid out on the sheet metal and then cut out; when this is done the separate pieces are ready to be fitted together to form the required object.

The method by which the surface of an object is laid out on a plane is called ***the development** of the object*. A few exercises

ELBOW TABLE

No. of pieces	Divide by	Degree of elbow	Rise of mitre line
2	2	105	52½°
2	2	90	45°
2	2	70	35°
3	4	90	22½°
4	6	90	15°
6	10	90	9°

will sufficiently acquaint the student with the methods used in problems of this character.

Problem 9.—To draw the development of a right, or 90° stove pipe elbow.

A right elbow is made by joining two pieces of pipe for the purpose of forming a right angle. It is really *an intersection of two cylinders of equal diameters;* the center lines of the two cylinders meeting at one point, and as the joint is to be a right elbow, the center lines must be perpendicular to each other.

To develop the surfaces, divide the circumference of the cylinder into any number of equal parts, and through the points of division draw lines parallel to the center line of the cylinder.

On these parallel lines, mark the points which belong to the curve of intersection with another cylinder, or any other figure as happens to be the case, and then roll out the surface of the cylinder into a flat plate.

The rolled out surface will be equal in length to the circumference of the cylinder, and it will contain all parallel lines, which were drawn upon the cylinder, with spaces between them just equal to the actual space between the parallel lines which were drawn upon the surface of the cylinder.

By marking the points of intersection on the parallel lines in the rolled surface, the development of the cylinder or its part is obtained.

In fig. 1,622, the circle showing the circumference of the pipe is divided into any number of equal parts by the divisions 1, 2, 3, etc. Lines are drawn through these divisions parallel to the center line of the vertical portion of the joint. These lines are *ak, bl, cm, dn,* etc.

The points *k, l, m, n, o* are the points on the parallel lines designating the curve of intersection.

The development of the two branches of the right elbow are shown in figs. 1,623 and 1,624; the length of the development, VW (or AA′) is equal to the circumference of the figure shown in fig. 1,622. To obtain this length all spaces, 1, 2, 3, 4, etc., laid out upon the circle in fig. 1,622 are

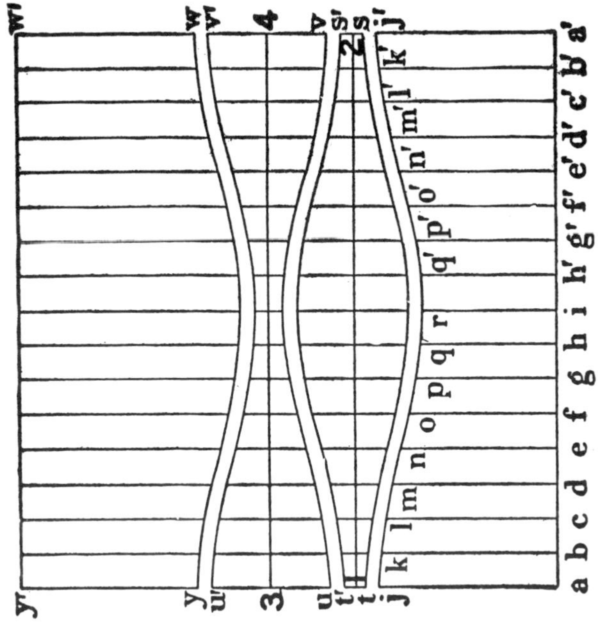

FIG. 1,626.—Development of surface of the four-part elbow, shown in elevation and plan in fig. 1,625.

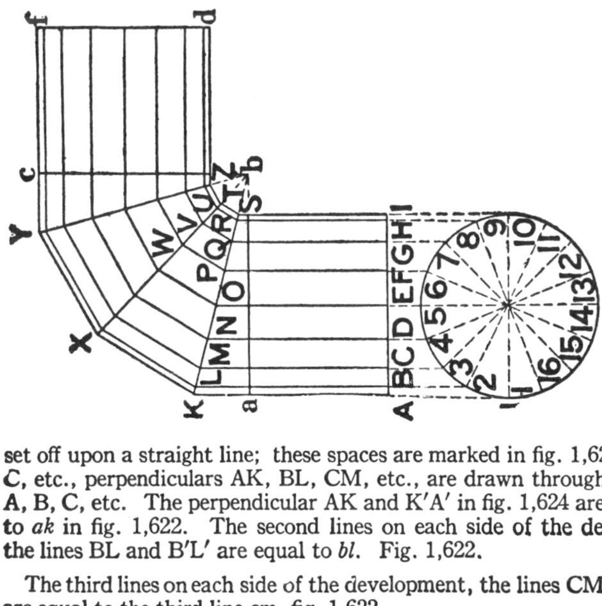

FIG. 1,625.—Elevation and plan of a four-part elbow, whose surfaces are developed as shown in fig. 1,626.

set off upon a straight line; these spaces are marked in fig. 1,624 by A, B, C, etc., perpendiculars AK, BL, CM, etc., are drawn through the points A, B, C, etc. The perpendicular AK and K′A′ in fig. 1,624 are each equal to *ak* in fig. 1,622. The second lines on each side of the development, the lines BL and B′L′ are equal to *bl*. Fig. 1,622.

The third lines on each side of the development, the lines CM and C′M′, are equal to the third line cm, fig. 1,622.

The fourth lines in the development are made equal to the fourth parallel in the elevation, fig. 1,622, and in the same manner all other lines in the development are made equal to the corresponding parallels in the elevation of the pipe in fig. 1,622.

The middle line, SI in the development is made equal to the line *si,* in the elevation; the points KLMSM′L′K′, etc.; thus found, define the position of the curve of intersection in the development of the cylinder.

The required curve is traced through these points; the development AA′K′K′ is the pattern for the part *aksi* of the right elbow shown in fig. 1,622.

The other part of the elbow is developed in fig. 1,623. It will be readily seen that the figure TVWU is laid out in the manner in which the first development was obtained; in this figure the shortest parallels are laid off above the longest parallels in the first development. This arrangement gives the advantage of cutting out both branches of the right elbow from one square piece of sheet metal without any waste of material.

It will be noticed that the patterns shown in figs. 1,623 and 1,624 do not provide for the lap by which the two branches are held together. A lap of any desired width may be added to the pattern, after it is constructed by drawing an additional curve, parallel to the curve of the above pattern, the distance between the two curves being equal to the width of the desired lap.

Problem 10.—To draw the development of a 90° four part elbow.

A four part elbow is a pipe joint made up of four sections such as is used for stove pipes where it is desired to obtain an easier bend than with the abrupt turn in fig. 1,622.

Fig. 1,625 shows an elevation and plan of the four part elbow. Here the four sections forming the elbow are AKSI, KXTS, XYZT, and Yf*d*Z. Of these four parts, the two larger parts, AKSI and Yf*d*Z are equal. The same is true of the two remaining smaller parts, KXTS and XYZT.

To lay out these parts in the elevation a right angle *abc* is drawn, the sides of which intersect at right angles, the two largest branches of the joint. It is evident that the point *b*, must be equidistant from both pipes.

The right angle *abc*, is divided first into three equal parts and then each one of these parts is divided in turn into two equal parts; the right angle is thus divided into six equal parts, of which Kb*a*, is one part, K*b*X, equals two parts, X*b*Y equals two parts and Y*bc* one part. It will be noticed that this construction does not depend on the diameter of the pipe.

The problem of developing the four part elbow resolves itself into developing two only of its parts, one large branch and one smaller part of the elbow, the remaining parts being correspondingly equal to these.

The circumference of the pipe, fig. 1,625, is divided into sixteen equal parts by the points 1,2,3,4,5, etc.

Through these points are drawn lines parallel to the center line of the pipe which is to be developed.

In fig. 1,626, the vertical branch of the elbow, AKSI (of fig. 1,625), will be taken up for the purpose. The parallels upon the surface of this branch are AK, BL, CM, DN, EO, FP, GQ, HR, and IS. Through the points K,L,M,N,O,P,Q,R, and S, draw parallels for the part KXTS, which will be next developed; some of these parallels are ST, RU, QV, PW.

To develop the vertical branch of the four-part elbow set off, upon a straight line *aa'*, fig. 1,626, sixteen equal parts, which altogether are equal to the circumference of the cylinder, which is to be developed.

Let the division points, *a,b,c,d,e,f,* etc., correspond to the division points, 1,2,3,4, etc., upon the circle, fig. 1,625. Through the points, *a,b, c,d,e,* etc., draw vertical lines equal to the parallel lines drawn upon the surface of the vertical branch of the joint; thus *aj*, is made equal to AK (fig. 1,626), *bk*, equal to BL; *cl*, equal to CM and so on until *ri* is made equal to SI (fig. 1,626).

The part laid out so far is *ajklmnopgri*. This is one-half of the development; the other half, *i'r'j'a'* being exactly the same as the first one, may be laid out in the same way.

The part *tt'ss'* is the development of the small part of the elbow. It is evident that its length, *ts*, must be equal to the circumference of the pipe in the elbow. The lines in the pattern, *tt'ss'* drawn at right angles to the center line of it, and bisected by it, are made equal to the parallel lines, ST, RU, QV, PW, etc., drawn upon the surface of the part, KXTS, fig. 1,626.

It is plain that the part, *uu'vv'* is equal to the part *tt'ss'*. with the difference that the small parallels in it are laid out above the large parallels in the other part; in the same manner, the part *yy'ww'* is equal to the part *aja'j'*.

Laying out the pattern in this manner makes it possible to cut out the complete elbow from one square piece of metal, *ay'w'a'*. The spaces between the patterns are left for laps, which are necessary for joining all parts.

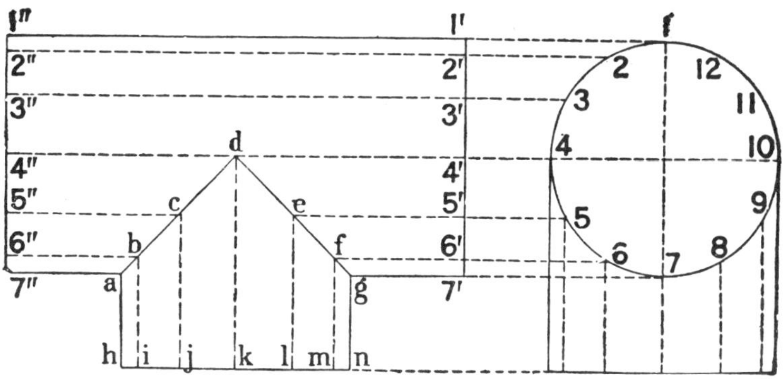

FIG. 1,627.—Side and end elevations of a tee pipe, where surfaces are developed as shown in fig. 1,628 to 1629; case, where diameter of all three outlets are the same.

Problem 11.—To draw the development of a tee pipe in which all branches are of equal diameter.

Front and side elevations of the tee are shown in fig. 1,627. As seen, it is made by the intersection of two cylinders of equal diameter. The section of the cylinders is represented in the front view by two 45-degree lines *ad* and *dg*.

To develop the pipes divide the circle in the end view of fig. 1,627, into any number of equal parts, in this case let it be twelve parts.

The greater the number of these divisions the more accurate will be the resultant pattern.

Through the divisions 1,2,3, etc., draw horizontal lines cutting the horizontal cylinder in the side view in the points 1″1′, 2″2′, 3″3′, 4″4′, 5″5′, 6″6′, 7″7′; the line 4″4′ just meets the lines of the section in the point *d*.

The line 5″5′ cuts the lines of the section in the points *e* and *c*, the line 6″6′ cuts the section lines in the points *f* and *b*, and the line 7″7′ cuts the lines of the section in the points *g* and *a*.

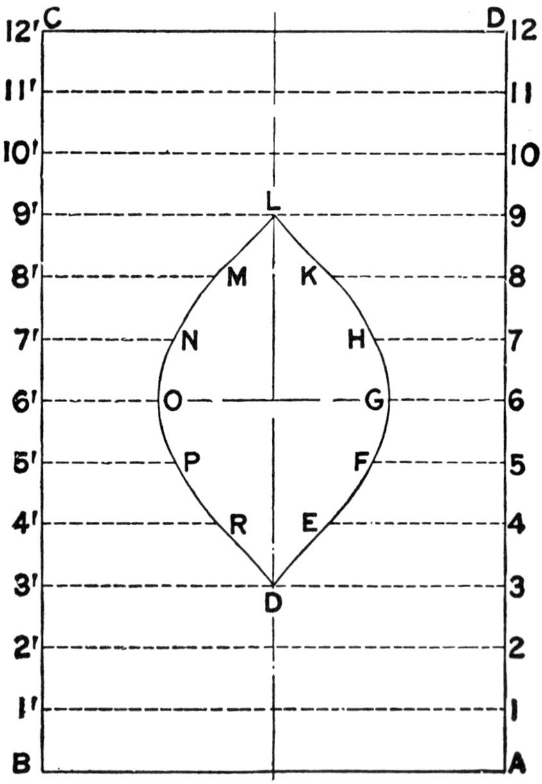

FIG. 1,628.—Development of run of tee pipe.

Draw vertical lines through the points *a,b,c,d,e,f,* and *g*. After all these lines are drawn we have all that is necessary to complete the development of the cylindrical surfaces.

Fig. 1,628 shows the development of the horizontal cylinder; the rectangle ABCD, is equal to the cylinder surface. The curve ODGL, is cut out within the rectangle for the joint which is the outline of the opening, into which the vertical cylinder will fit.

The rectangle ABCD, has one side AB, equal to the length of the horizontal cylinder, fig. 1,627; the other side AD, is equal to the circumference of the circle, show in the end view of the horizontal pipe, fig. 1,627 The twelve divisions marked on the circle are set off on the straight line AD. fig. 1,628, so that together they are equal to the circumference of the circle.

The outline of the opening for the intersection of the horizontal pipe with the vertical branch is laid out in the middle of the rectangle ABCD, in the following manner: On the middle line 6′6 are set off the distances 6′O and 6G each equal to *g*7′ (or *a* 7″) in fig. 1,627, on the lines 5′5 and 7′7 are set off the distances 5′P, 5F, 7′N and 7H each equal to the distance 6′*f*, fig. 1,627 (or *b*6″). The distances 4′R, 4E, 8K and 8′M are set off on the lines

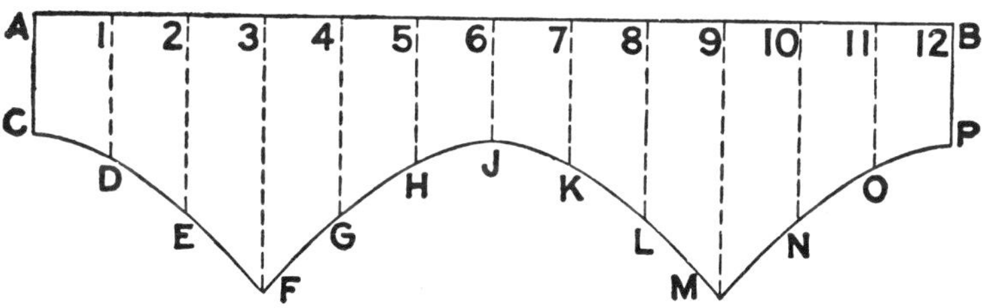

FIG. 1,629.—Development of outlet or branch of tee pipe.

8′8 and 4′4 to equal the distance *e*5′ (or *c*5″) of fig. 1,627. The lines 3′3 and 9′9 are touched by the curve of intersection in their center at points D and L.

There still remains to be drawn the development of the vertical branch of the tee-pipe; this is found in the same manner as the horizontal part, *i. e.*, by laying out the surface of the vertical cylinder; that is, by making it equal in length to the circumference of the circle showing the end view of the cylinder. The development is shown in fig. 1,629.

On the line AB, are set off the twelve parts of the circumference and in each one of these divisions is erected a perpendicular to the line AB; on these perpendiculars are laid off successively the length of the vertical lines drawn on the surface of the vertical branch; the lines AC, 1D, 2E, 3F, G4, 5H and 6J in fig. 1,629, are equal correspondingly to the lines *ah, bi, cj, dk, el, fm* and *gn*, in fig. 1,627.

Thus one-half of the development ACJ6 is constructed; the other 6JPB is exactly equal to the first part.

The method employed in these cases may be applied to nearly all developments of cylindrical surfaces; it consists in drawing on the surface of the

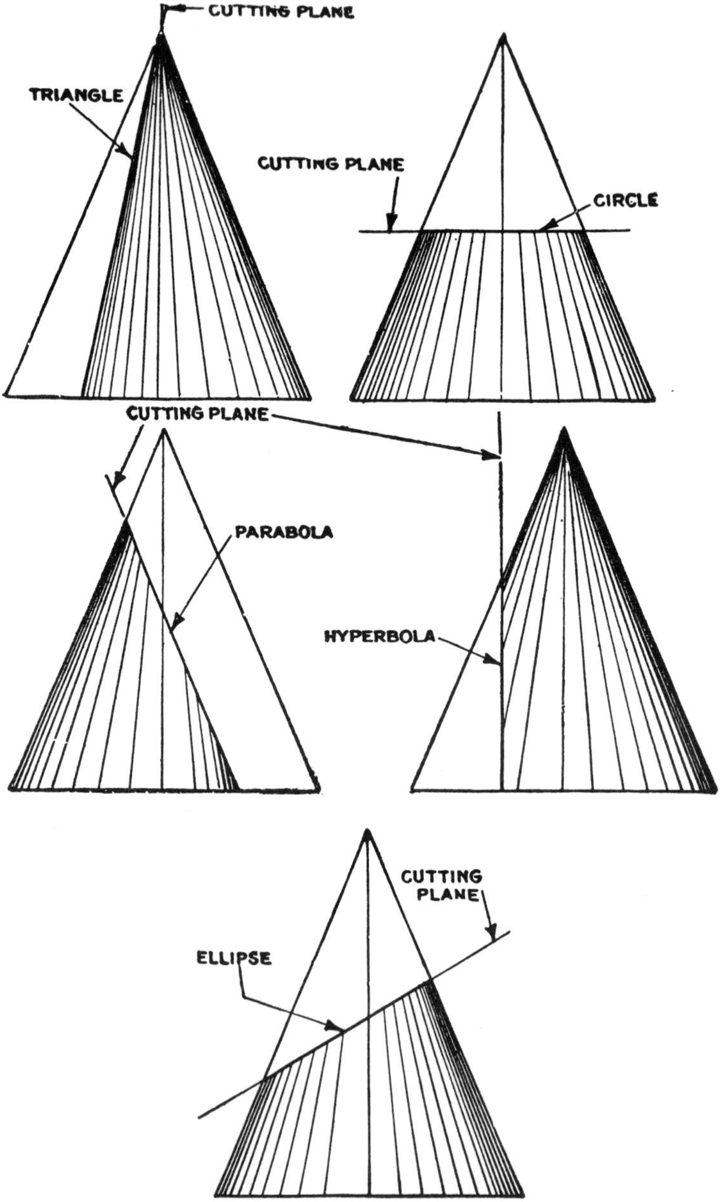

FIGS. 1,630 to 1,634.—The conic sections obtained by cutting a cone with a plane; showing position of plane for obtaining the various sections.

cylinder, which is to be developed, any number of equidistant parallel lines. The cylindrical surface is then developed and all parallel lines drawn in it. *By setting off the exact lengths of the parallel lines a number of points are obtained, through which may be traced the outline of the desired development.*

It has been noted in fig. 1,627, that the intersection of two cylinders of equal diameters—their arcs intersecting each other—will always appear in the side view as straight lines at right angles to each other. If one cylinder be of a smaller diameter than the other then the intersection will be a curve.

Conic Sections.—By definition a conic section is *a section cut by a plane passing through a cone.*

These sections are bounded by well known curves, and the latter may be any of the following depending upon the inclination or position of the plane with the axis of the cone, as shown in figs. 1,630 to 1,634.

1. Triangle

   Plane passes through apex of cone

2. Circle

   Plane parallel to base of cone

3. Ellipse

   Plane inclined to axis of cone

4. Parabola

   Plain parallel to one element of cone

5. Hyperbola

   Plane parallel to axis of cone

These sections appear as straight lines in elevation, while in plan they appear (with exception of the triangle) as curves.

Problem 12.—Find curve cut by a plane passing through apex of cone as in fig. 1,635.

Let ABC, be elevation of cone and MS. cutting plane passing through apex. Project point D, down to plan parallel to axis cutting base of cone at D′ and E′, obtaining line D′ E′, base of developed surface.

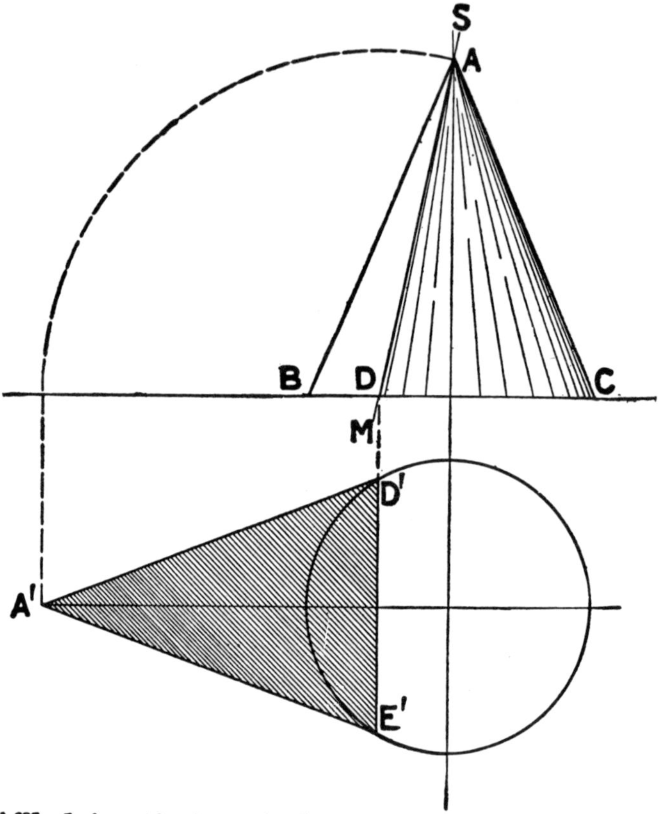

FIG. 1,635.—Surface cut by plane passing through apex of a cone—*triangle.*

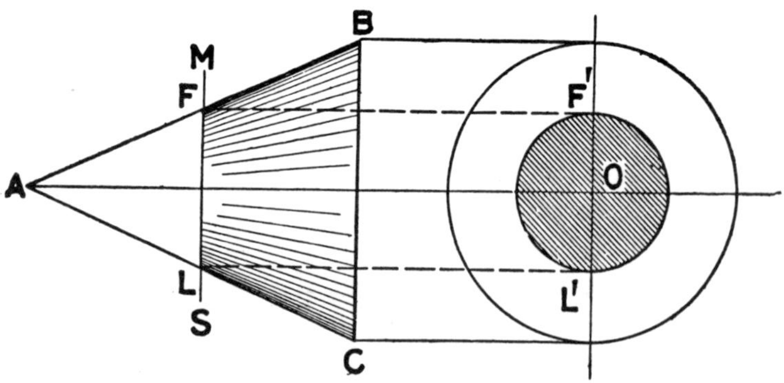

FIG. 1,636.—Surface cut by plane passing through a cone parallel to its base—*circle.*

With D, as center and radius DA, equal to element of cone swing A, around to base line and project down to A′. Join A′ with D′ and E′. Then, A′D′E′ is the developed surface or triangle cut by plane MS, with cone.

Problem 13.—Find surface cut by a plane passing through a cone parallel to its base as in fig. 1,636.

This may be found by simply projecting over to the plan. Where MS, cuts the element AB, as at F, project over to the axis of the plane and obtain point F′.

Similarly point L′ may be found. These points are equidistant from the center O, hence with radius = OF′ = OL′, describe a circle which is the curve cut by plane MS, when parallel to base of cone.

Problem 14.—Find the curve cut by a plane passing through a cone inclined to its axis as in fig. 1,637.

In fig. 1,637 let the plane cut the elements **OA** and **OB**, of the curve at M and S, respectively. Project S, down to *s′*, in plan giving one point on the curve. With S, as center swing M, around and project down to *m′*, in plan giving a second point on the curve, *m′s′*, being the major axis of the curve.

To find the minor axis of curve, bisect MS, at R, and swing R, around to horizontal with S, as center and project down to plan. Through R, draw radius 3, and describe arc with radius 3, about O′ as center. Where this cuts projection of R at *r′*, project over to plan, intersecting the vertical plan projection of R at *r′*. *O″ r′*, is half the minor axis.

To find the projection of any other point as L, or F, proceed in similar manner as indicated, obtaining *l′*, or *f′*. The curve joining these points and symmetrical points below the major axis is an ellipse.

Problem 15.—Find the curve cut by a plane passing through a cone parallel to an element of the cone as in fig. 1,638.

Let the plane MS, cut element AB, at L, and base at F. Project F, down to plan cutting base at F′ and F″, which are two points in the curve.

With F, as center and radius LF, swing point L, around and project down to axis of plan, obtaining point L′ in the curve.

Now any other point as R, may be obtained as follows: swing R around with F as center and project down to plan with line HG.

Describe a circle in plan with a radius (=radius *rr* of cone at elevation of point R), and where such cuts the projection of R at R′; project R′ over to line HG, and obtain point R″, which is a point in the curve.

Other points may be obtained in a similar manner. The curve is traced through points F′,R″,L′, and similar points on the other side of the axis, ending at F″. Such curve is called a *parabola*.

Problem 16.—Find the curve cut by a plane passing through a cone parallel to the axis of the cone, as in fig. 1,639.

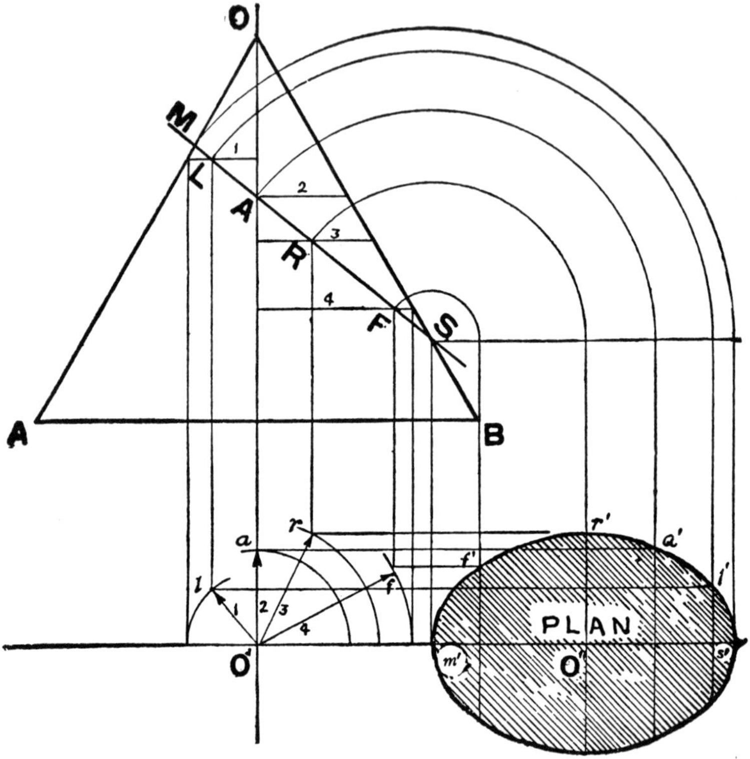

FIG. 1,637.—Surface cut by plane passing through a cone inclined to its axis—*Ellipse.*

Let plane MS, cut element AB, at L, and base at F. Project F, down to plan cutting base at F′ and F″, which are two points in the curve. With F, as center and radius FL, swing point L, around and project down to axis of plan obtaining point L′, in the curve.

Now any other point as R, may be obtained as follows: Swing R, around with F, as center and project down to plan with line HG. Describe a circle in plan with radius *rr′* (= radius of cone at elevation of point R) and where this circle cuts the projection of R at R′, project over to line HG, and obtain point R″, which is a point in the curve. Other points may be obtained in a similar manner.

The curve is traced through points F′,R″,L′, and similar points on the other side of the axis ending at F″. Such curve is called a *hyperbola*.

It will be noted that problems 1,638 and 1,639 are virtually worked out in the same way. In fact the text of one will apply

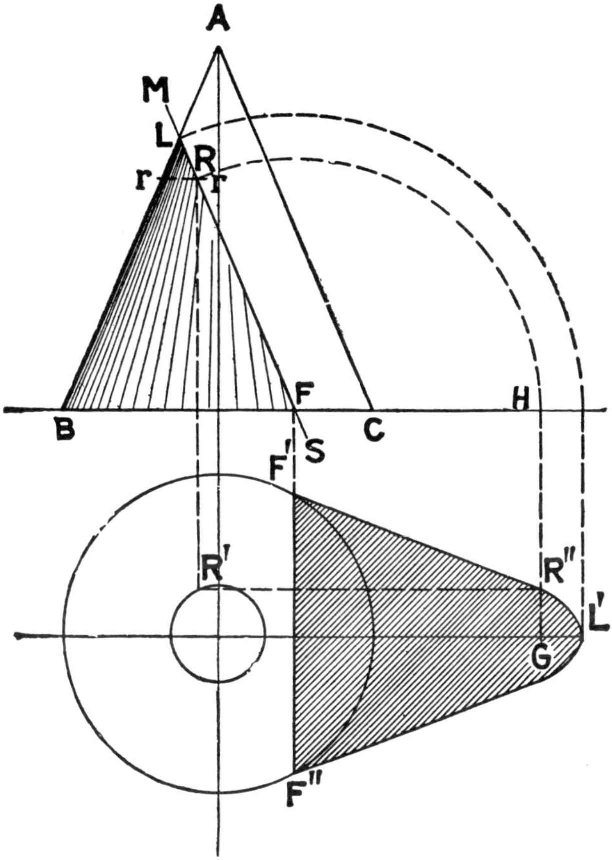

FIG. 1,638.—Surface cut by plane passing through a cone parallel to an element of the cone—***parabola.***

to the other, the two cuts being symmetrically lettered, the only difference being the shape of the curve.

Perspective Drawing.—This is the art of representing objects as they appear to the eye at a *definite* distance from the object. In orthographic (perpendicular) projection the views represent the object as seen when the eye is *infinitely* distant. By the perspective method then the lines drawn from points on the object to the eye converge and intersect at the point of sight.

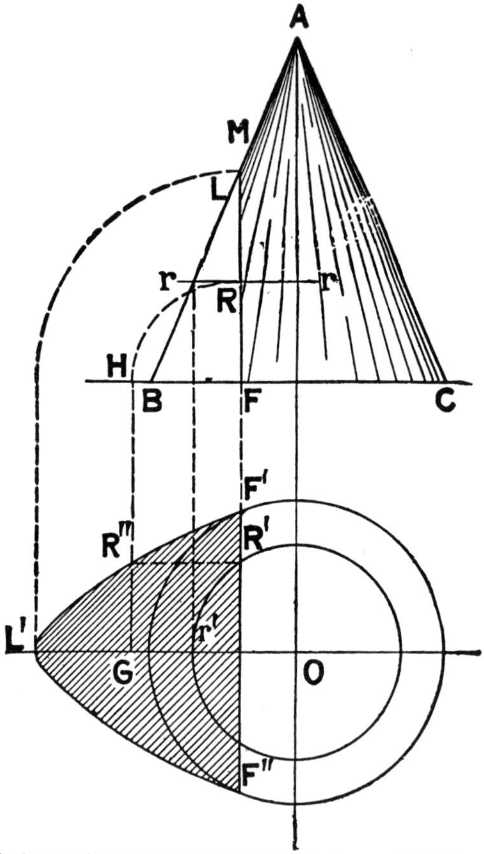

Fig. 1,639.—Surface cut by a plane passing through a cone parallel to the axes of the cone-*hyperbola.*

Before beginning the study of perspective projection it is well to consider some of nature's phenomena of perspective. These phenomena become more apparent when we attempt to sketch from nature. We notice that the size of an object diminishes as the distance between the object and the eye increases. If several objects of the same size be situated at different distances from the eye, the nearest one appears to be the largest and the others appear to be smaller as they are further and further away.

At last the distance between the lines becomes zero and the lines appear to meet in a single point. This point is called the

FIG. 1,640.—Perspective view with vanishing point at side of drawing.

FIG. 1,641.—Perspective view with vanishing point at center of drawing.

vanishing point of the lines, as shown in figs. 1,640 and 1,641.

By closer investigation of a perspective drawing it is found that

1. The limit of vision is a horizontal line called the ***horizon,*** situated at the height of the eye.

2. Objects of equal size appear smaller with increasing distance.

3. Parallel lines converge into one point, called ***vanishing point.*** For horizontal lines this point is situated at the height of the eye, that is, it lies in the horizon.

4. Vertical lines appear vertical.

5. The location of the observer's eye is called the ***point of sight*** and is located in the horizon.

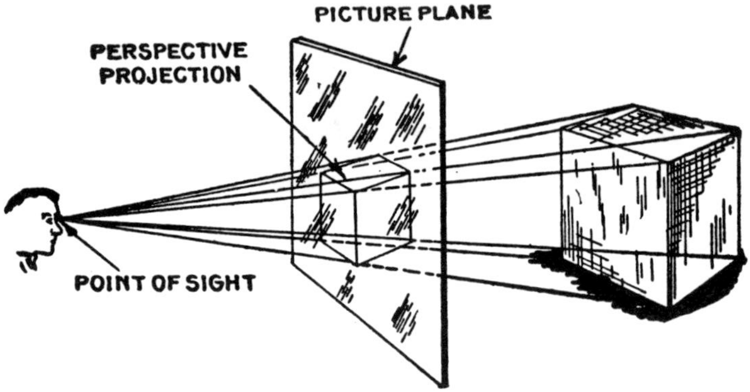

Fig. 1,642.—Picture plane illustrating principles of perspective.

When an object in space is being viewed, rays of light, called *visual rays*, are reflected from all points of its visible surface to the eye of the observers.

If a transparent plane, fig. 1,642, be placed between the object and the eye, the intersection of the visual rays will be a projection of the object upon the plane. Such projection is called the *perspective projection* of the object. The plane on which the projection is made is called the *picture plane*. The position of the observer's eye is the *point of sight*.

This principle is illustrated by models where red strings represent the rays, piercing a glass plate.

Perspective by Means of Plan and Elevation.—This method of perspective can be put to practical use if we obtain a perspective projection in plan and elevation and then proceed by orthographic projection to obtain the perspective.

Fig. 1,643 shows in plan and elevation, a prism, its front face making an angle with the picture plane. As a general rule, the object is placed behind the picture plane with one of its principal vertical lines lying in the picture plane. P, is the point of sight (the observer's eye). Its distance A, from the picture plane in plan depends a great deal on the size of the object and it is important that the best viewpoint is obtained.

If a house about 40′ high is to be sketched, the point of sight should be taken about 80′ from the picture plane. A good rule to follow is to make this distance about twice the greatest dimension. When large objects are to be represented the best results are obtained when the point is taken nearly in front of the object.

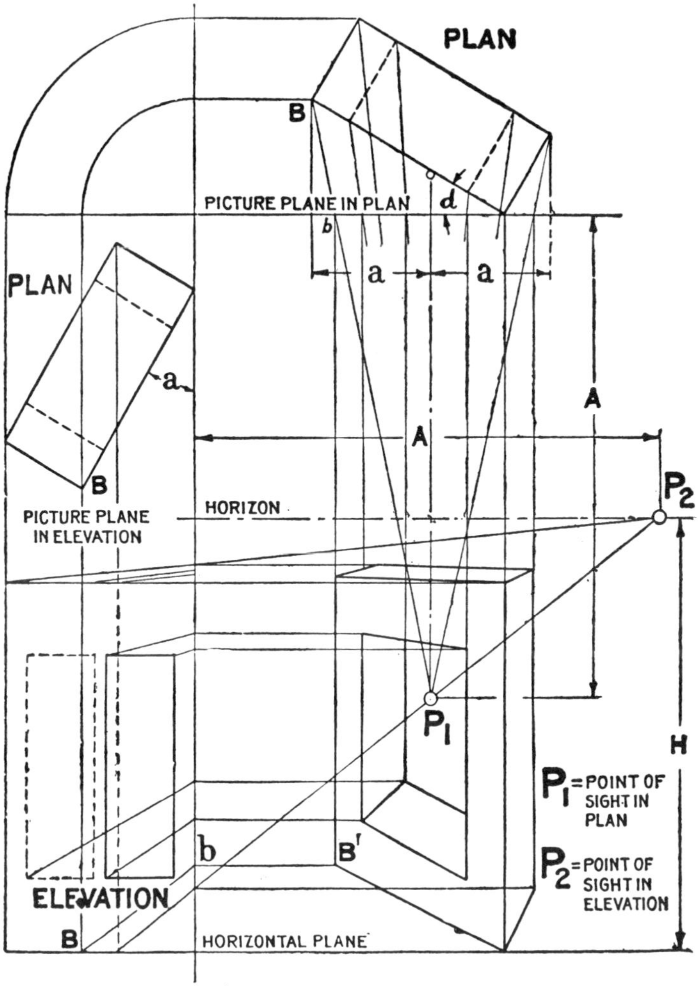

FIG. 1,643.—Perspective of a prism by means of plan and elevation. ***To obtain the perspective*** of any point of the object, for instance B, draw the visual ray in both plan and elevation to P_1 and P_2. respectively. From the point of intersection ***b***, in the picture plane (in plan and elevation) project perpendicularly and thus obtain the point B_1, as perspective picture of the point B of the object, In this manner all the other points of the perspective are obtained. This method of construction requires no further explanation and may be applied wherever plan and elevation is obtainable.

The distance of the horizon from the horizontal plane equals the height of the eye above ground and may be taken = 5′ — 3″. For high objects this distance may be increased and for low objects decreased. In our case it is shown slightly above the object. P_1 is assumed on a vertical line half way between two lines dropped from the extreme edges of the diagram. This is not necessary, but it usually insures a more pleasing perspective projection.

The method of obtaining the perspective is explained under the illustration.

Perspective by Means of Two Vanishing Points.—An example of this method of perspective is shown in fig. 1,644, which illustrates a rectangular prism in plan and elevation resting upon

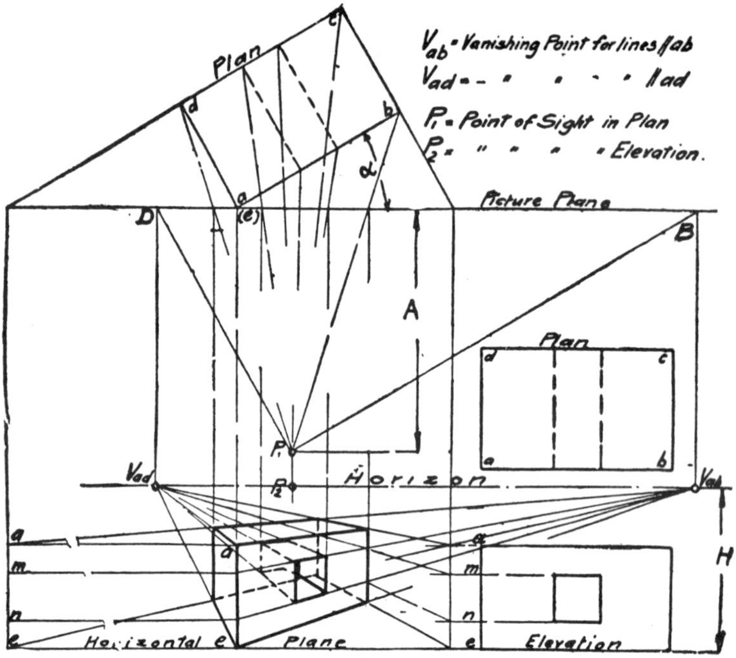

FIG. 1,644.—Perspective of rectangular prism by means of two vanishing points.

a horizontal plane.

The first step will be to redraw the plan, same as with the first method, behind the picture plane in plan, with the vertical line *ae*, lying in the picture plane and turned so that its long side makes an angle α (30°) with the picture plane. The point of sight P_2 is at a distance H, above the floor and is located at the same height as the horizon.

Next, find the vanishing points for the different systems of lines in the object. There are three systems of lines in the prism. V*ab* and V*ad*, are found by drawing lines P_1B and P_1D, through P_1 parallel to *ab* and *ad* of

the diagram and dropping vertical lines. from the intersection of these lines with the picture plane (B and D) to the horizon, giving the vanishing points V*ab* and V*ad*. The third system of lines embraces the vertical lines which are drawn actually vertical and not converging towards one another.

The edge *ae* of the diagram, being in the picture plane, is called the *line of measures*, as it appears in its true size in the perspective view, and from *a* and *e* in the perspective view the lines will vanish at V*ab* and V*ad*, respectively. establishing by intersection with the vertical edges all points desired.

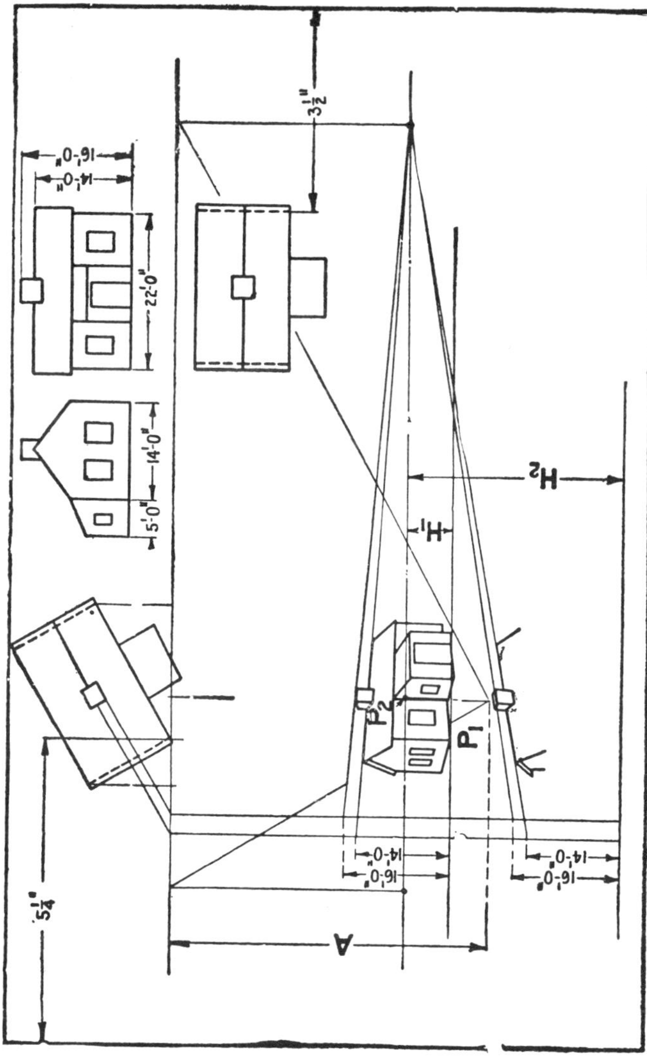

FIG. 1,645.—Perspective of a house. The projections are given, long side of house making an angle of 30° with the picture plane. Nearest vertical edge of house to lie in the picture plane. Two perspective views of the house shall be obtained, the house being viewed from two different points. Their common distance in plan A = 46'. The distances H_1 and H_2 of the point of sight above the horizontal shall be 6'-6" and 31'-6", respectively. The construction of both views is exactly the same. The fact the porch projects that in part in front of the picture plane makes no difference in the construction of the perspective projection.

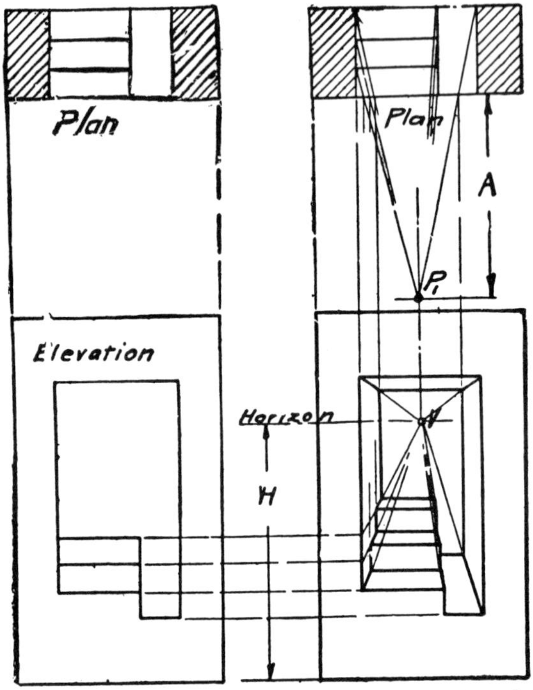

Fig. 1,646.—Perspective of a prism by means of one vanishing point.

Besides this principal line of measures other lines of measures may easily be established by extending any vertical plane in the object until it intersects the picture plane. This intersection, since it lies in the picture plane, will show in its true size and all points in it will show at their true height above the horizontal plane.

If no line in the object should lie in the picture plane there would not be any principal line of measures, and some vertical plane in the prism must be extended until it intersects the picture plane.

Instead of being some distance behind the picture plane, the prism might have been wholly or partly in front of the picture plane. In any case, find the intersection with the picture plane of some vertical face of the prism. This intersection will show the true vertical height of the prism.

Perspective by Means of One Vanishing Point.—In this method the plan is placed with one of its principal systems of horizontal lines parallel to the picture plane. This system therefore has no vanishing point, and as the vertical system has no vanishing point, only the third system of lines will have a vanishing point.

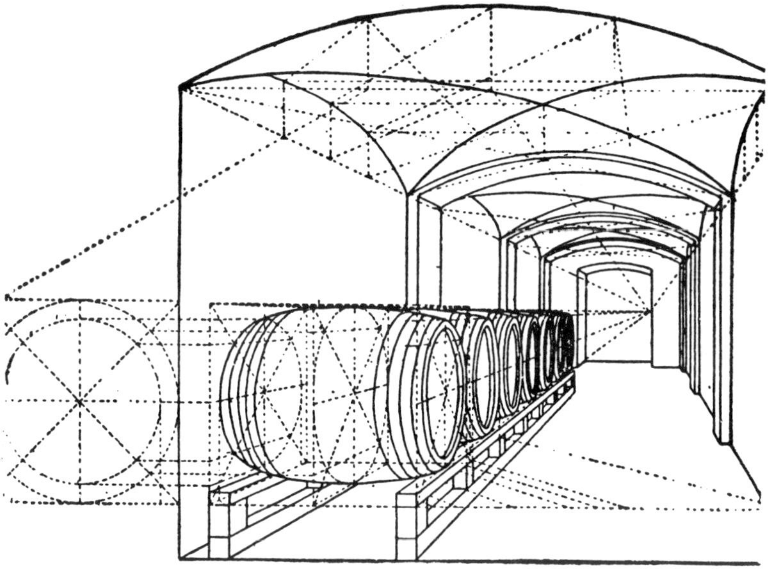

Fig. 1,647.—Perspective of a row of barrels by means of one vanishing point.

In fig. 1,646, the vertical face of the prism lies in the picture plane and shows in its true size. Its edges are lines of measures.

The construction of the perspective is easily apparent from the illustration.

Another example of perspective by means of one vanishing point is shown in fig. 1,647.

CHAPTER 10

Geometrical Problems

Geometrical Problems.—The following problems illustrating how various geometrical figures are constructed, are to be solved by the use of pencil, dividers, compasses, and scale.

Many of these problems are such as are encountered in sheet metal work in laying out patterns. Proficiency in the solution of these problems will be of value to draughtsmen.

Problem 1.*—To bisect or divide into two equal parts a straight line of arc of a circle.*

In fig. 132, from the ends AB, as centers, describe arcs cutting each other at C and D, and draw CD, which cuts the line at E, or the arc at F.

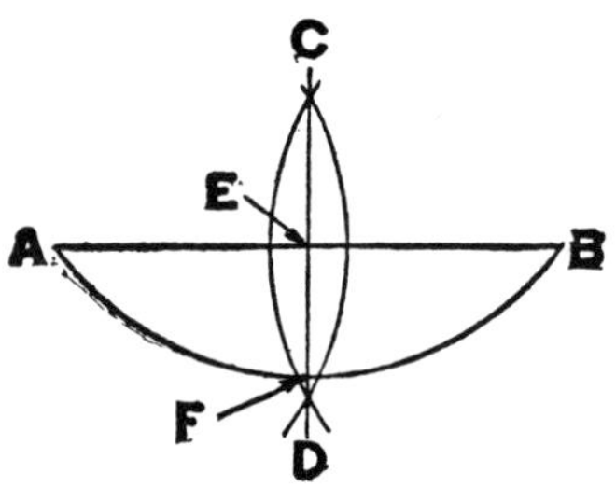

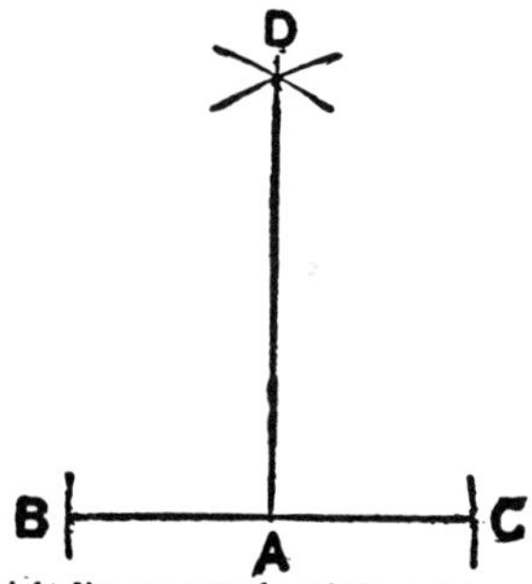

FIG. 132.—*Problems 1 and 2. To bisect a straight line or arc of a circle and to erect a perpendicular to the line which is a radial line to the arc.*

FIG. 133.—*Problem 3 To erect a perpendicular to a straight line, from a given point in that line.*

Problem 2.*—To draw a perpendicular to a straight line, or a radial line to an arc.*

In fig. 132 the line CD is perpendicular to AB, moreover, the line CD, is radial to the arc AFB.

Problem 3.*—To erect a perpendicular to a straight line, from a given point in that line.*

In fig. 133 with any radius from any given point A, in the line BC, describe arcs cutting the line at B and C. Next, with a longer radius describe arcs with B and C, as centers, intersecting at D, and draw the perpendicular DA.

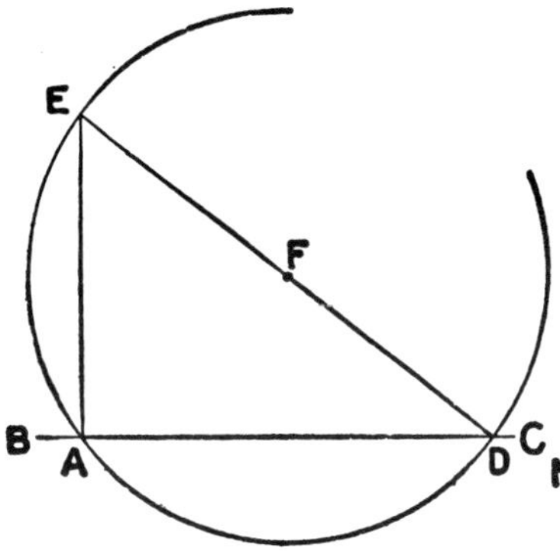

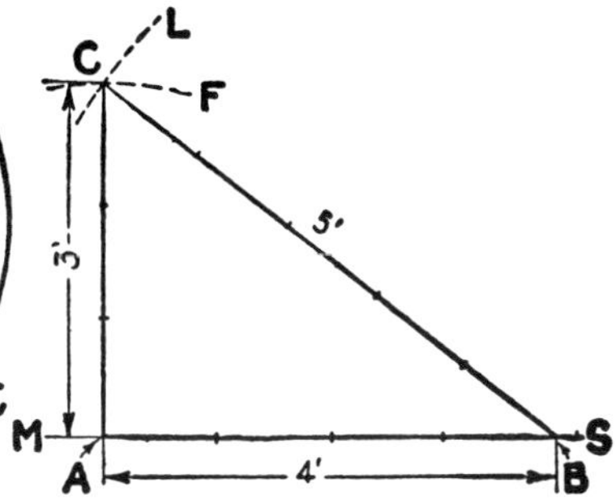

FIG. 134.—*Problem 3. Second method.*

FIG. 135.—*Problem 3. Third method (boat builder's laying down method).*

Second method. In fig. 134. from any center F, above BC describe a circle passing through the given point A, and cutting the given line at D; draw DF, and produce it to cut the circle at E; now draw the perpendicular AE.

Third method (boat builders' laying down method)—In fig. 135 let MS be the given line and A, the given point. From A, measure off a distance AB, say 4 ft. With centers A and B, and radii of 3 and 5 ft. respectively, describe arcs F and L, intersecting at C. Draw a line through A and C. which will be the perpendicular required.

Fourth method.— In fig. 136, from A, describe an arc EC, and from E, with the same radius, the arc AC. cutting the other at C; through C, draw a line ECD, and set off CD, equal to CE, and through D, draw the perpendicular AD.

***Problem 4.**—To erect a perpendicular to a straight line from any point without the line.*

In fig. 137, from the point A, with a sufficient radius, cut the given line at F and G; and from these points describe arcs cutting at E. Place triangle on points A and E, and from A, draw perpendicular to line GF

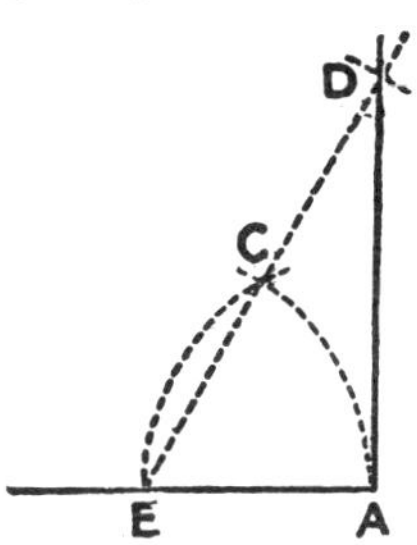

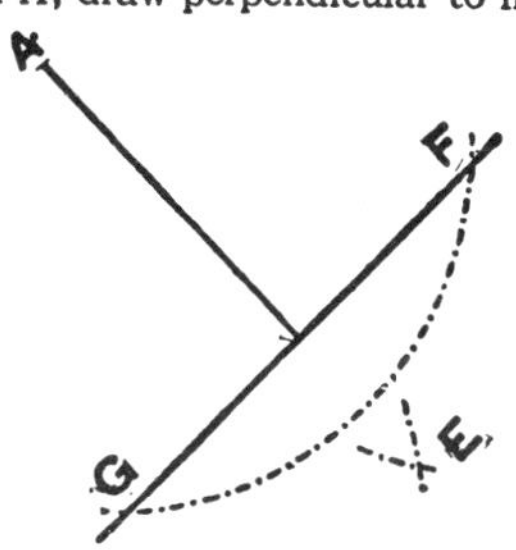

FIG. 136.—*Problem 3. Fourth method.*

Second method.— In fig. 138, from any two points, B,C, at some distance apart, on the given line, and with the radii BA, CA, respectively, describe arcs cutting at A and D. Place triangle on points A and D, and draw the perpendicular AD.

FIG. 137.—***Problem 4.*** *To erect a perpendicular to a straight line, from any point without the line.* If there be no room below the line, the intersection may be taken above the line, that is to say, between the line and the given point.

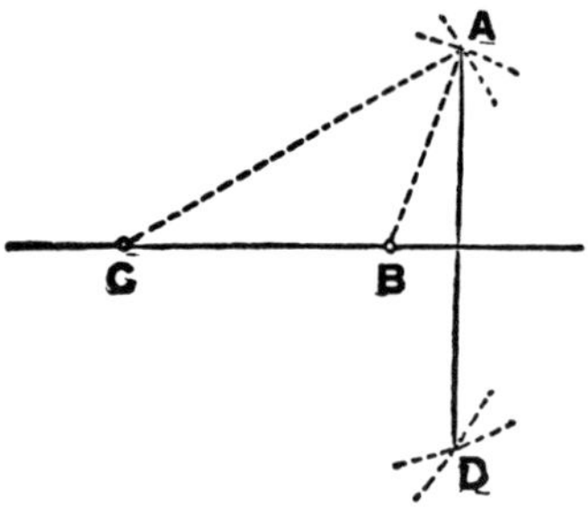

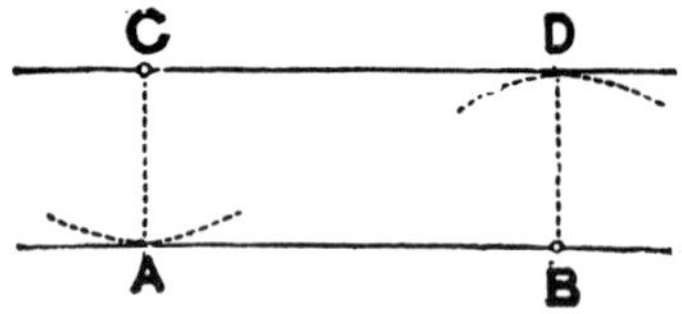

FIG. 138.—***Problem 4.*** *Second method.*

FIG. 139.—***Problem 5.*** *Through a given point to draw a line parallel with a given line.*

Problem 5.—*Through a given point to draw a line parallel with a given line.*

In fig. 139, with C, as center describe an arc tangent to the given line AB; the radius will then equal distance from given point to the given line. Take a point B, on line remote from C, and with same radius, describe an arc. Draw a line through C, tangent to this arc and it will be parallel to the given line AB.

Second method.— In fig. 140, from A, the given point, describe the arc FD, cutting the given line at F; from F, with the same radius, describe the arc EA, and set off FD, equal to EA. Draw the parallel through the points AD.

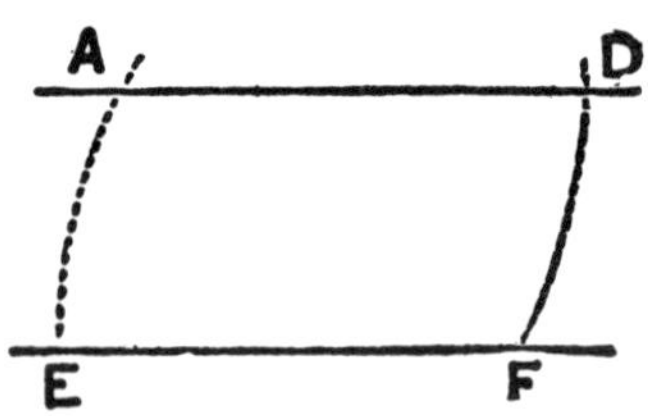

FIG. 140.—***Problem 5.*** *Second method.*

Problem 6.—*To divide a line into a number of equal parts.*

In fig. 141 assuming line AB, is to be divided into say 5 parts, draw a diagonal line AC, and space off 5 unit lengths. Join B5, and through the points 1,2,3,4, draw lines 1*l*, 2*a*, etc., parallel to B5, then will AB, be divided into five equal parts, A*l*, *la*, *ar*, *rf*, and *f*B.

Problem 7.—*Upon a straight line to draw an angle equal to a given angle.*

In figs. 143 and 144 let A, be the given angle and FG, the line

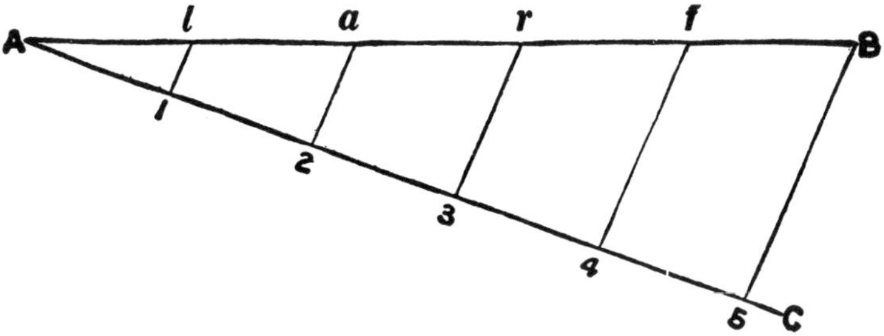

FIG. 141.—*Problem 6. To divide a line into a number of equal parts.*

With any radius from the points A and F, describe arcs DE, and IH, cutting the sides of the angle A, and the line FG.

Set off the arc IH, equal to DE, and draw FH. The angle F, is equal to angle A, as required.

Problem 8.*—To bisect an angle.*

In fig. 145, let ACB, be the angle; with center C, describe an arc

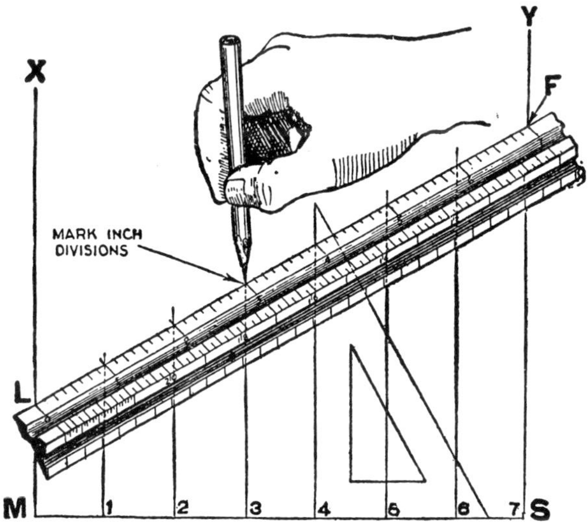

FIG. 142.—To divide a given line into any number of equal parts without dividers. Let M S, be the line and say it is to be divided into seven equal parts. Erect perpendiculars M X and S Y. Lay the O, mark of the scale on the line M X, and place scale at such angle that coincides with line S Y. Draw a light line L F, and mark the inch divisions as shown. With a triangle and T square draw lines from the points on L F, to M S, cutting M S, at 1, 2, 3, etc., which divide M S, into seven equal parts.

cutting the sides at A and B. On A and B, as centers describe arcs cutting at D. A line through C and D will divide the angle into two equal parts

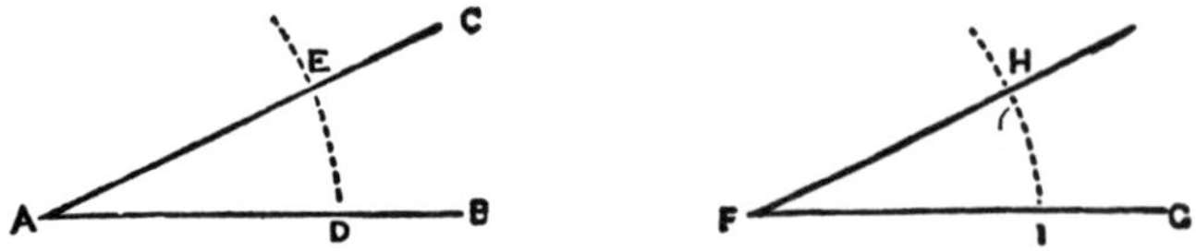

FIGS. 143 and 144.—*Problem 7. Upon a straight line to draw an angle equal to a given angle.*

Problem 9.—To find the center of a circle

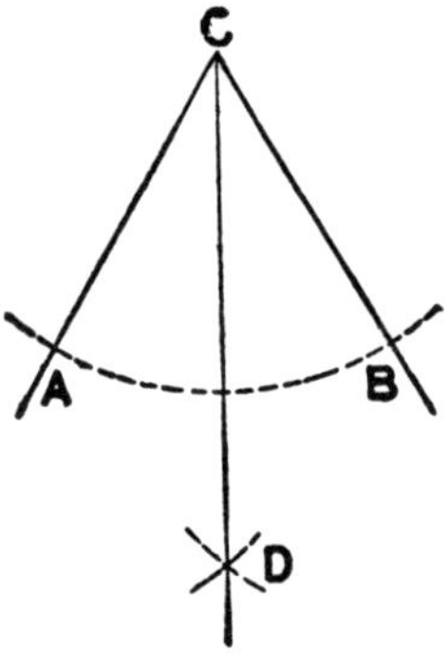

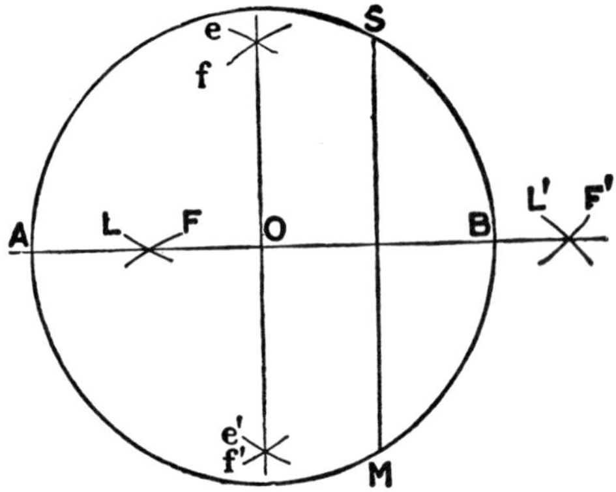

FIG. 145.—*Problem 8. To bisect an angle.*

FIG. 146.—*Problem 9. To find the center of a circle.*

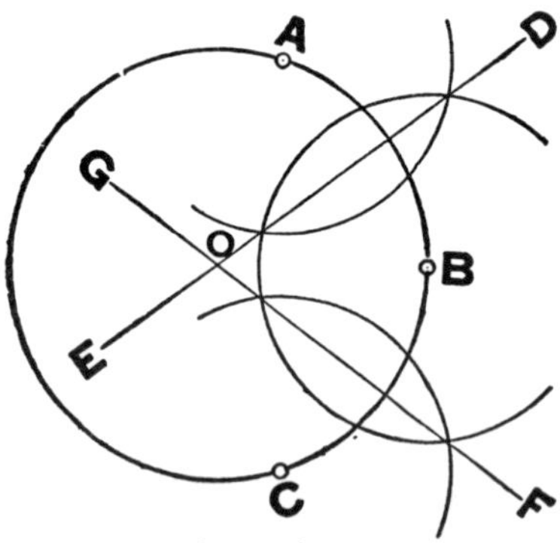

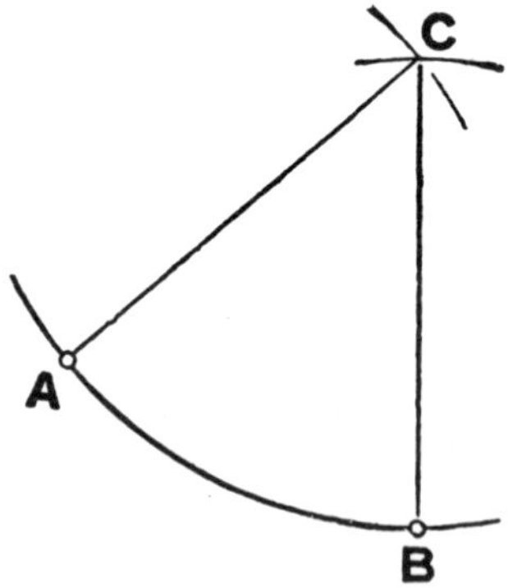

FIG. 147.—*Problem 9. Second method.*

FIG. 147.—*Problem 10. To describe a circle passing through three given points.*

FIG. 148.—*Problem 11. Through two given points to describe an arc of a circle with a given radius.*

In fig. 146, draw any chord as MS. With M and S, as centers and any radius, describe arcs L,F, and L′,F′, and a line through their intersection, giving a diameter AB. Applying same construction with centers A and B, describe arcs *ef*, and *e′f′*. A line drawn through the intersections of these arcs will cut AB, at O, the center of the circle.

Problem 9.—Second method. To find the center of a circle.

In fig. 147, select three points, A,B,C, in the circumference, well apart; with the same radius describe arcs from these three points cutting each other, and draw two lines DE, FG through their intersections. The point O, where they cut is the center of the circle or arc.

Problem 10.—To describe a circle passing through three given points.

In fig. 147, let A,B,C, be the given points and proceed as in last problem to find the center O, from which the circle may be described. This problem

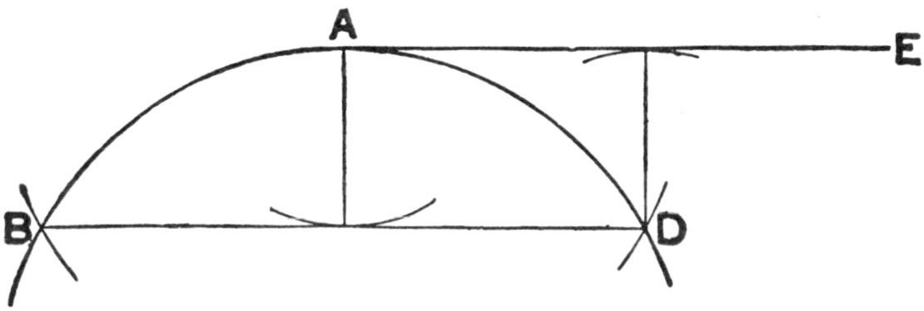

FIG. 149.—*Problem 12. To draw a tangent to a circle from a given point in the circumference.*

is useful in such work as laying out objects of large diameter as an arch, when the span and rise are given.

Problem 11.—Through two given points to describe an arc of a circle with a given radius.

In fig. 148, take the given points A and B, as centers, and, with the given radius, describe arcs cutting at C; from C, with the same radius, describe the required arc AB.

Problem 12.—To draw a tangent to a circle from a given point in the circumference.

In fig. 149 from point A set off equal segments AB, AD; join BD, and draw AE, parallel with it, for the tangent.

Problem 13.—On a given straight line A5, to construct any regular polygon say a pentagon.

In fig. 150 produce the given side A5, say to the left. With center A, and radius A5, describe a semi-circle. Divide the semi-circle into as many equal parts as the polygon is to have sides; in this case 5 equal parts, by trial with compasses. From A, draw A2, which gives another side of the polygon; and no matter how many sides the polygon is to have, always draw from A, to the second division on the semi-circle. Bisect the sides 2 A, A 5, by

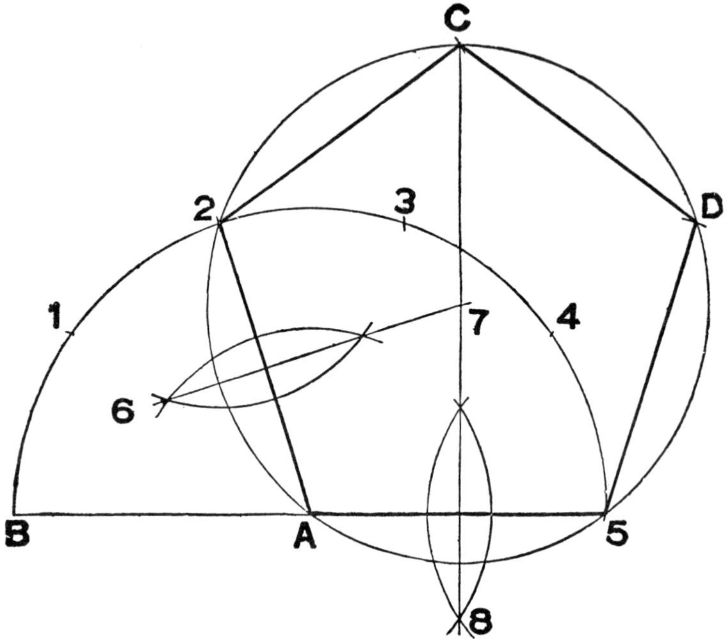

FIG. 150.—*Problem 13. On a given straight line, to construct a regular polygon.*

lines 6 7, and 8 7, intersecting at point 7, which is the center of the polygon. With center 7, and radius 7 A, describe the circle. Mark off, on the circumference, the divisions 2C, CD, equal to A5. Joint 2C, CD, D5. Then A2CD5, is the required regular polygon.

Problem 14.—To ascertain approximately, the length of the circumference of a given circle.

In fig. 151, draw a diameter AB. Find center C. Draw AD, perpendicular to AB, and 3 times the length of the radius. Draw BE, perpendicular to AB. With 30° triangle, draw angle BCH = 30°. Mark joint J, on BE. Join J D. Then line J D, is (approximately) equal to half the circumference; and twice J D = the whole circumference. This method is sufficiently accurate for all practical work, because the result is wrong only by about a $\frac{1}{100,000}$ part. This problem helps to ascertain approximately, the length of certain portions of the circumference. Thus $\frac{1}{3}$ of JD = $\frac{1}{6}$ of the circumference. Archimedes demonstrated that the diameter is to the circumference, within a minute fraction, as 7 is to 22, or 1 to $3^1/_7$. Thus, for all practical purposes, it may be assumed that if the diameter = 1 in., the circumference = $3^1/_7$ ins. To describe a circle having a circumference equal to the circumferences of any number of given equal or unequal circles: *Draw a line equal to the sum of the diameters of the given circles. This line is the diameter of the required circle.*

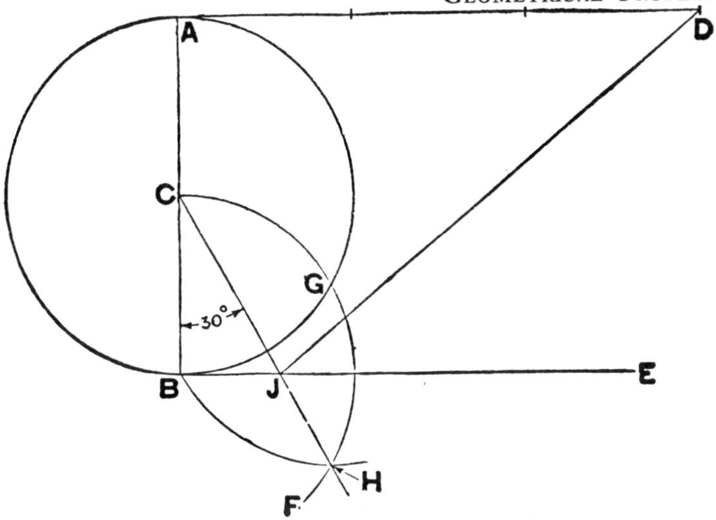

FIG. 151.—*Problem 14. To ascertain approximately, the length of the circumference of a given circle.*

Problem 15.—To find the center of a given circle, or arc of a circle.

In fig. 152, draw any two chords, 12 and 23. Bisect these chords by perpendiculars 45, and 67, intersecting at A. Point A, is the center of the circle or arc. The chords are not obliged to meet at 2. They may be drawn anywhere in the circle or arc, but it is better, when possible, to let them be at about right angles to each other. The chords may intersect. They should not be made too short.

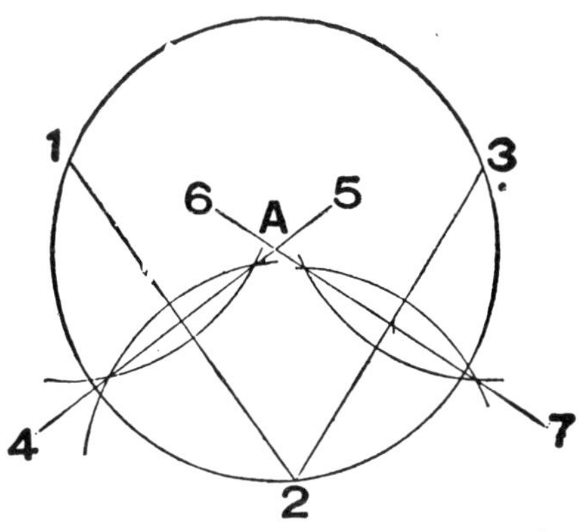

FIG. 152.—*Problem 15. To find the center of a given circle, or arc of a circle.*

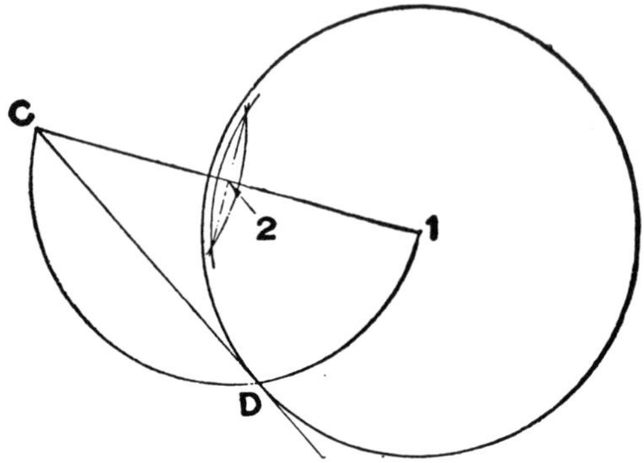

FIG. 153.—*Problem 16. To draw a tangent to a given circle from any given point.*

Problem 16.—To draw a tangent to a given circle from any given point C, outside the circle.

In fig. 153, 1 is the center of the given circle. Join points C1. Bisect C1, at 2; and with center 2, and radius 2C, or 21, describe a semi-circle, cutting the circle at D. Point D, is the point of contact. Through D, draw CD, which is the required tangent.

CD, is tangent because a line through the point of contact D, and center 1, of the circle makes a right angle with CD. Why?

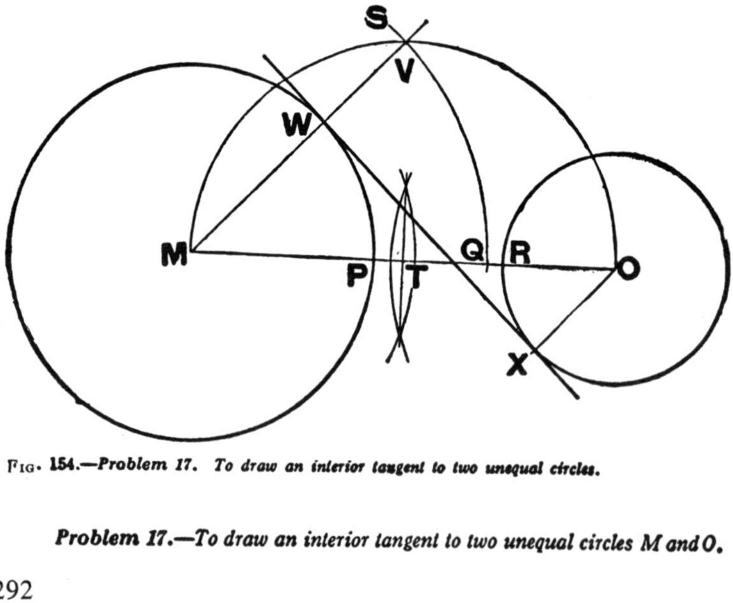

FIG. 154.—*Problem 17. To draw an interior tangent to two unequal circles.*

Problem 17.—To draw an interior tangent to two unequal circles M and O.

In fig. 154, join centers M and O. Bisect MO, at T, and describe a semi-circle on MO. From P, on the larger circle, mark off PQ = OR, the radius of the smaller circle. With center M, and radius MQ, describe arc QS, cutting the semi-circle at V. Join MV, and mark point W. Draw OX. parallel with MV. Through the points of contact W and X, draw the interior tangent WX.

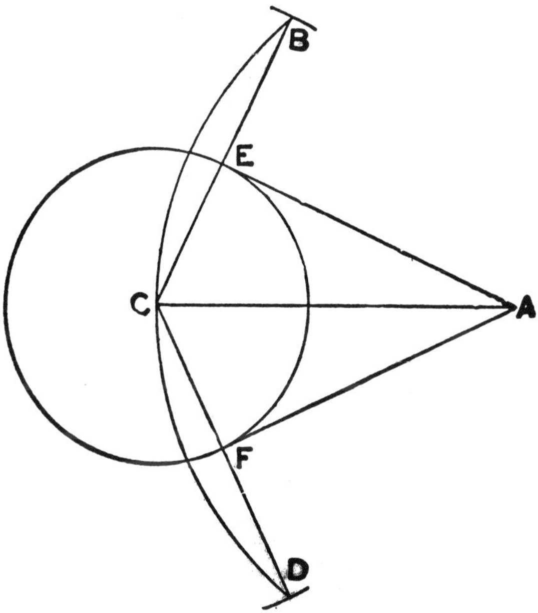

FIG. 155.—*Problem 18. To draw tangents to a circle from points without.*

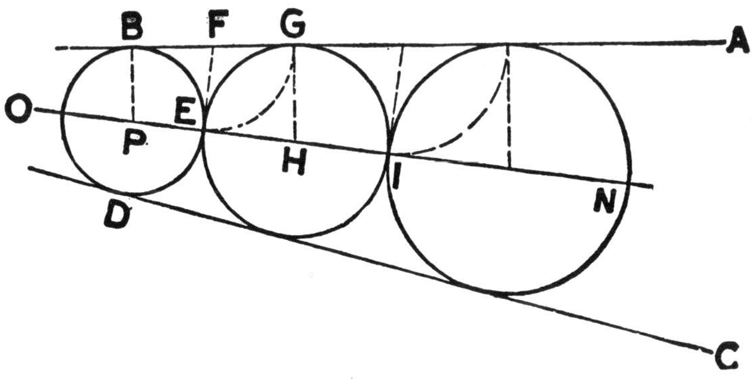

FIG. 156.—*Problem 19. To describe tangent circles tangent to two inclined lines.*

***Problem 18.**—To draw tangents to a circle from points without.*

In fig. 155, from A, and with the radius AC, describe an arc BCD, and from C, with a radius equal to the diameter of the circle, cut the arc at BD; join BC, CD, cutting the circle at EF, and draw the tangents, AF, AE.

***Problem 19.**—Between two inclined lines to describe a series of circles tangent to these lines and tangent to each other.*

In fig. 156, bisect the inclination of the given lines AB, CD, by the line NO. From a point P, in this line, draw the perpendicular PB, to the line AB, and about P, describe the circle BD, touching the lines and cutting the center line at E. From E, draw EF, perpendicular to the center line, cutting

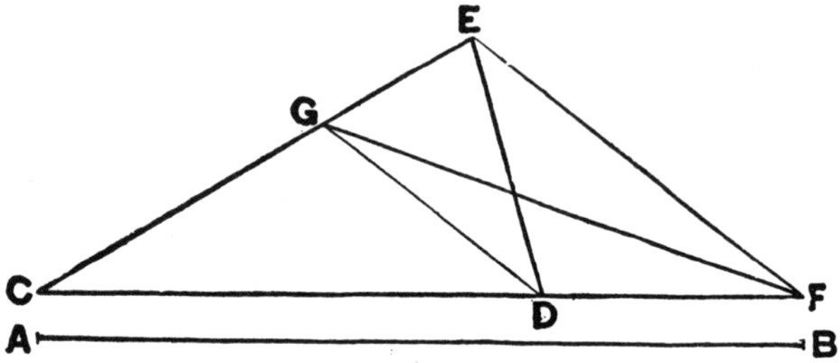

FIGS. 157 and 158.—*Problem 20. To construct a triangle having a given base and equivalent to any rectilineal figure*

AB, at F, and about F, describe an arc EG, cutting AB, at G. Draw GH parallel with BP, giving H, the center of the next circle, to be described with the radius HE, and so on for the next circle IN.

***Problem 20.**—To construct a triangle, having a given base A B, and equivalent to any rectilineal figure, say equal in area to the triangle CDE.*

In figs. 157 and 158, produce one side CD, to F, making CF, equal to the given base AB. Join FE. Draw DG, parallel to FE. Join FG. Then CFG, is the required triangle.

***Problem 21.**—To construct a rectangle, when each of the diagonals is equal to A B, and each of one pair of opposite sides is equal to C D.*

In figs. 159 and 160, bisect AB at 1, and with center 1, and radius 1A, describe a circle. With centers A and B, and radius CD, obtain points 2 and 3. Join A2, 2B, B3, 3A. Then A2B3 is the required rectangle. If the longer side A2 be given, instead of the shorter side, then describe arcs at 2 and 3, with the longer side as radius.

***Problem 22.**—To construct a square, whose diagonal is given*

In fig. 161 Bisect R S, by a perpendicular 2 3. Cut off 14, and 15, equal to 1R or 1S. Join R5, 5S, S4, 4R. Then R5S4 is the square required, having a given diagonal RS.

Problem 23.—To construct a square equal in area to any number of given squares.

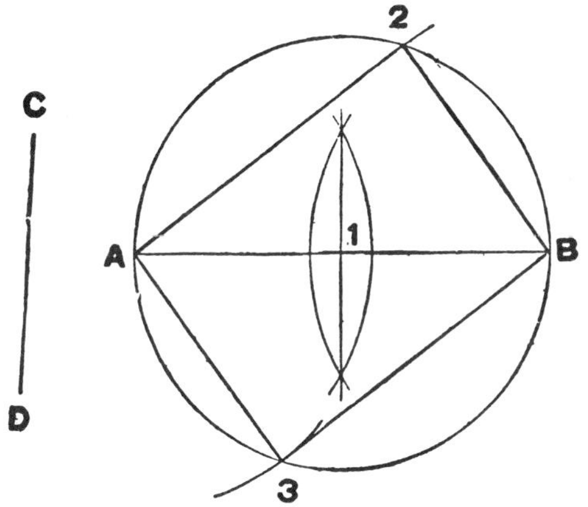

FIGS. 159 and 160.—*Problem 21. To construct a rectangle when each diagonal is equal to a given line and each of one pair of opposite sides is equal to another given line.*

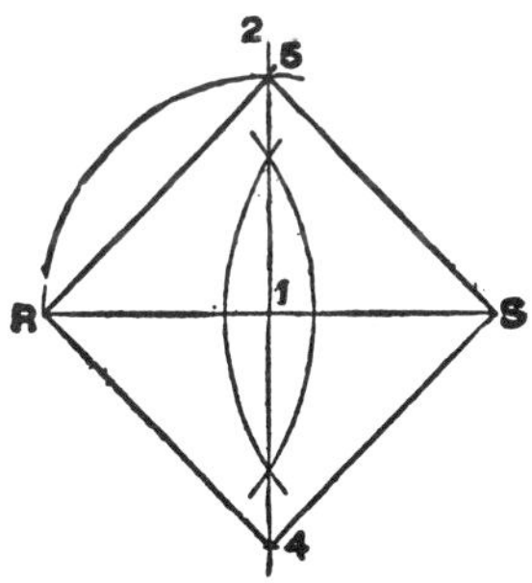

FIG. 161.—*Problem 22. To construct a square with given diagonal.*

In figs. 162 and 163:

Let A, B, C be the side of the three given squares. Make D G = A and D H = B. Join G H. Then the square upon G H equals the squares upon A and B. Make D K = G H and D L = C. Join K L. Then the square K L N M, equals the three squares upon A, B and C.

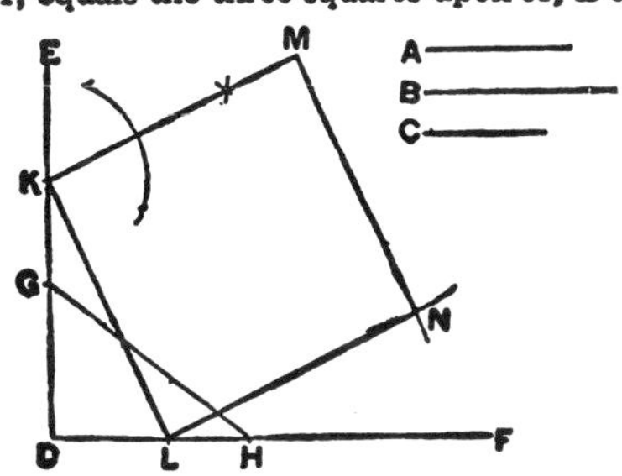

FIGS. 162 and 163.—*Problem 23. To construct a square equal in area to any number of given squares.*

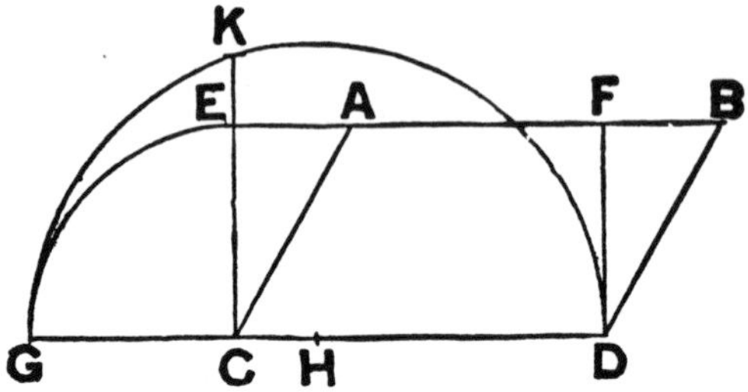

FIG. 164.—*Problem 24. To construct a square equal in area to any parallelogram.*

Problem 24.—To construct a square, equal in area to any parallelogram. Thus, construct a square equivalent to the rhomboid C D B A.

In fig. 164, make the rectangle C D F E, equal to C D B A, by producing E F, and erecting perpendiculars C E, D F. Produce D C. Make C G, = C E. Bisect G D, at H. With center H, and radius H G, describe a semicircle. Produce C E, to K. Then C K, is the mean proportional if G C, C D, and a square constructed with C K as a side is equal in area to the rhomboid C D B A.

Problem 25.—Describe a catenary having given the span and versed sine.

In fig. 165, divide half span, as AB, into any required number of equal parts, as 1, 2, 3 and let fall BC and AO, each equal to versed sine of curve;

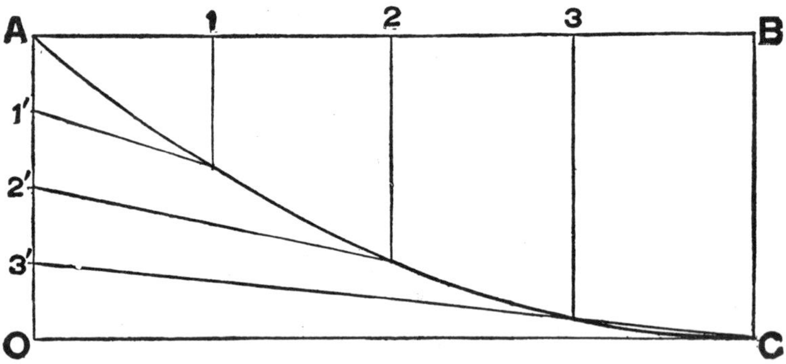

FIG. 165.—*Problem 25. To describe a catenary having given the span and versed sine.*

divide AO into like number of parts, 1′, 2′, 3′, as AB. Connect C′1, C2′ and C3′ and points of intersection of perpendiculars let fall from AB will give points through which curve is to be drawn.

The catenary is the curve assumed by a perfectly flexible cord when its ends are fastened at two points, the weight of a unit length being constant.

Problem 26.—To construct a circle equal in area to a given square.

In fig. 166, let LF be side of the given square. Through L, draw

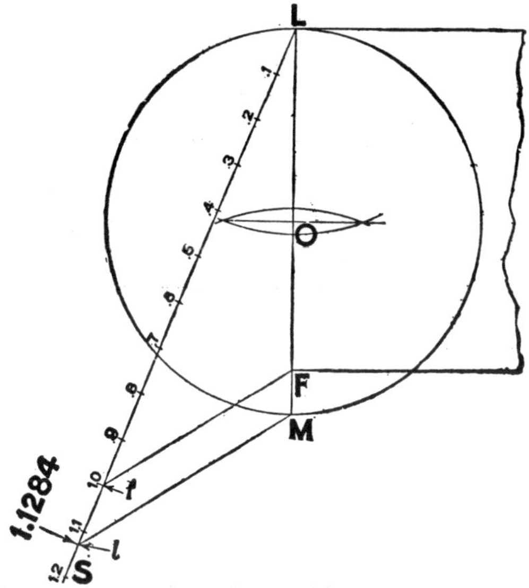

FIG. 166.—*Problem 26. To construct a circle equal in area to a given square.*

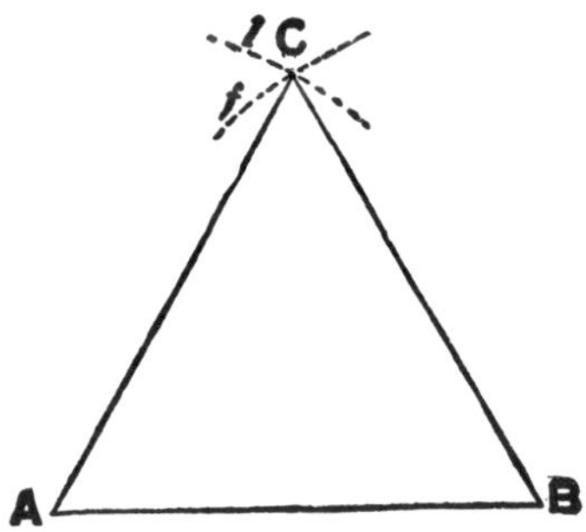

FIG. 167.—*Problem 27. To construct an equilateral triangle on a given base.*

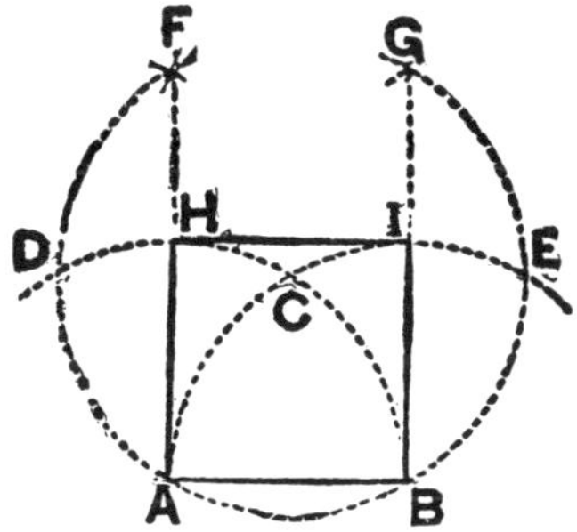

FIG. 168.—*Problems 28 and 29. To construct a square and a rectangle on a given base.*

proportional line L S, and with any convenient scale divide it into 12 equal parts. At the point *f*, or 10th division, draw line *f* F, and at a point *l*, between 11 and 12 draw *l* M parallel with *f* F. LM is the diameter of the required circle. Bisect this diameter at O and with radius O L describe the required circle.

***Problem 27.**—To construct an equilateral triangle on a given base.*

In fig 167, with A, and B, as centers and radius equal to AB, describe arcs *l* and *j*. At their intersection C, draw lines CA, and CB, sides of the required triangle.

***Problem 28.**—To construct a square on a given base.*

In fig. 168, with end points A and B, of base as centers and radius equal to AB, describe arcs cutting at C; with C as center, describe arcs

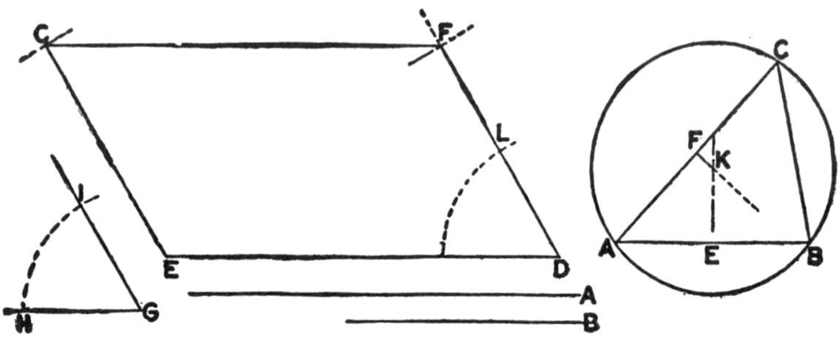

FIGS. 169 to 171.—*Problem 30. To construct a parallelogram having given the sides and an angle.*

FIG. 172.—*Problem 31. To describe a circle about a triangle.*

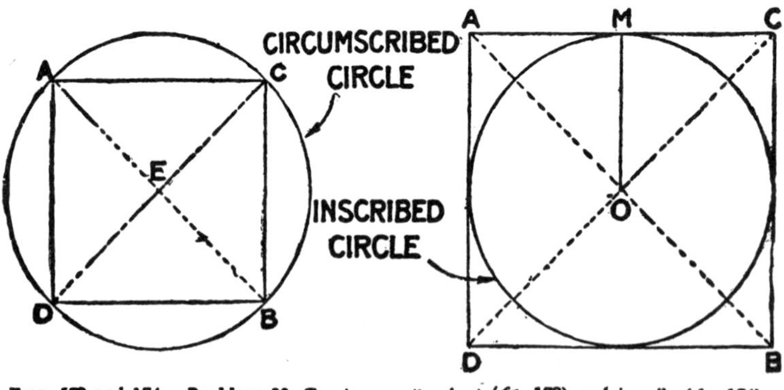

FIGS. 173 and 174.—*Problem 32. To circumscribe about (fig. 173) and inscribe (fig. 174) a circle in a square.*

cutting the others at DE; and with D and E, cut these at FG. Draw AF, and BG, and join the intersections HI, then ABIH is the required square.

***Problem 29.**—To construct a rectangle on a given base.*

In fig. 168, let AB, be given base. Erect perpendiculars at A and B, equal to altitude of the rectangle, and join their ends H and I, by line HI, ABIH, is the rectangle required.

***Problem 30.**—To construct a parallelogram having given the sides and an angle.*

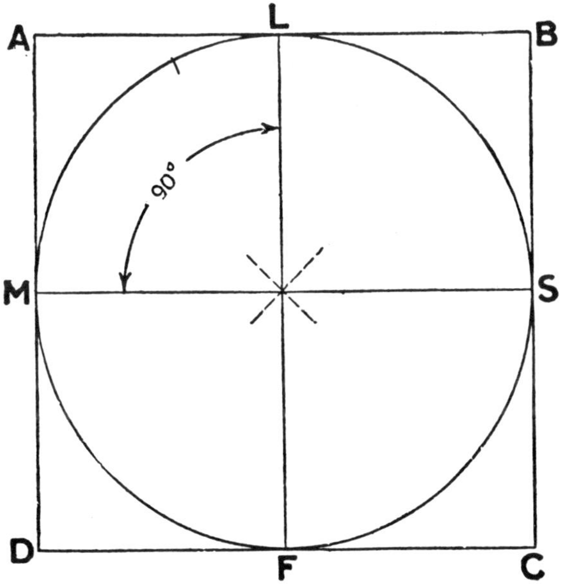

FIG. 175.—*Problem 33. To circumscribe a square about a circle. Second method*

In figs 169 to 171, draw side DE, equal to the given length A, and set off the other side DF, equal to the other length B, forming the given angle IGH. From E, with DF, as radius, describe an arc, and from F, with the radius DE, cut the arc at C. Draw FC, EC. Or, the remaining sides may be drawn as parallels to DE, DF

***Problem 31.**—To describe a circle about a triangle.*

In fig. 172, bisect two sides AB, AC, of the triangle at E and F, and from these points draw perpendiculars intersecting at K. From the center K, with the radius KA, describe the circle ABC.

***Problem 32.**—To circumscribe about and inscribe a circle in a square.*

In fig. 173, draw the diagonals AB and CD, intersecting at E. With radius EA, circumscribe the circle. To inscribe a circle let fall from the center (as just found) a perpendicular to one side of the square as OM, in fig. 174. With radius OM, inscribe the circle.

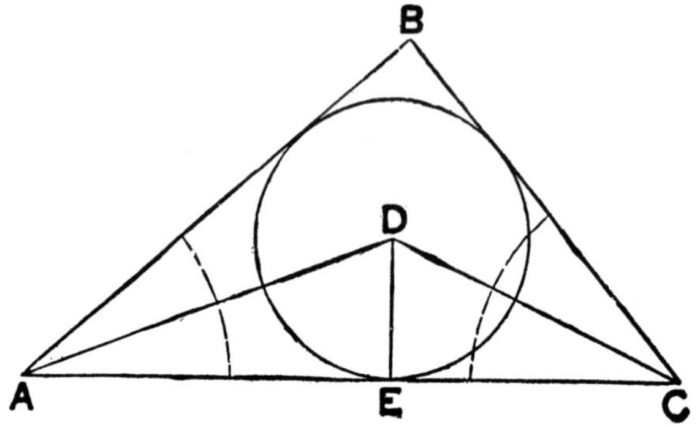

FIG. 176.—*Problem 34. To inscribe a circle in a triangle.*

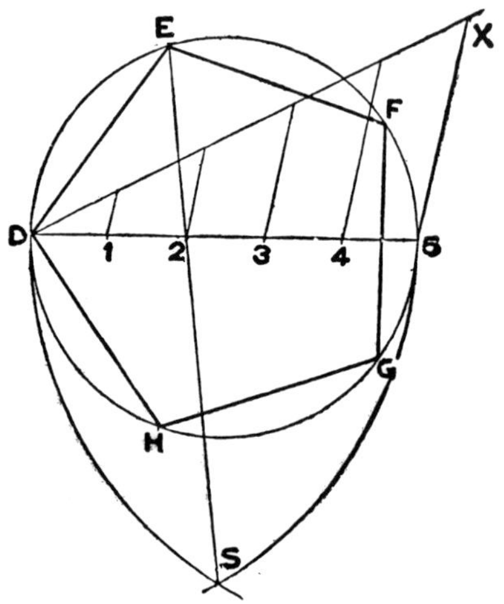

FIG. 177.—*Problem 35. To inscribe any regular polygon in a given circle.*

Problem 33.—To circumscribe a square about a circle.

In fig. 175, draw diameters MS and LF, at right angles to each other. At the points M, L, S, F, where these diameters cut the circle, draw tangents that is, lines perpendicular to the diameter, thus obtaining the sides of the circumscribed square ABCD.

Problem 34.—To inscribe a circle in a triangle.

In fig. 176, bisect two of the angles A and C, of the triangle by lines cutting at D; from D, draw a perpendicular DE, to any side, and with DE, as radius, describe a circle.

Problem 35.—To inscribe any regular polygon in a given circle.

In fig. 177, draw a diameter D 5. Divide D 5, into as many equal parts

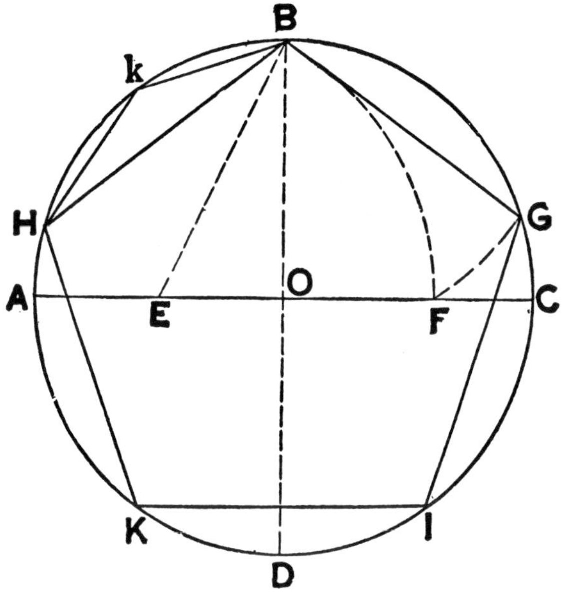

FIG. 178.—*Problem 36. To inscribe a pentagon in a circle.*

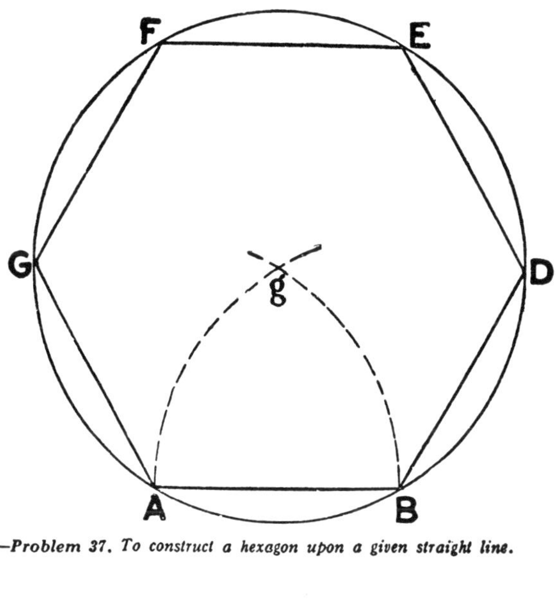

FIG. 179.—*Problem 37. To construct a hexagon upon a given straight line.*

as the polygon is to have sides, in this case, five equal parts. With points D and 5, as centers, and the diameter D 5, as radius, describe arcs intersecting at 6. From 6, draw a line through Point 2 to E. Join D E, which is one side of the required polygon. Make E F, F G, G H, each equal D E. Join E F, F G, G H, H D. Then D E F G H, is the required polygon.

This method is only approximately correct. It is however, sufficiently accurate for all practical work. On the same principle, an arc can (approximately) be divided into any number of equal parts, or a circle into equal sectors. By this method, a regular polygon having any number of sides can be inscribed, (approximately) within a given circle. If a nonagon is to be inscribed, divide the diameter into nine equal parts, and then proceed as above. To get the first side of the polygon, always draw a line from point 6; through the 2nd division on the diameter, no matter how many sides the polygon is to have. In a polygon, that has an even number of sides, a line drawn from one angle to the opposite angle (a diagonal) passes through the center. When there is an odd number of sides, a line from an angle through the center, bisects the opposite side. Note these facts as tests for accuracy in the work.

Problem 36.—To inscribe a pentagon in a circle.

In fig. 178, draw two diameters, AC, BD, at right angles intersecting at O; bisect AO, at E, and from E, with radius EB, cut AC, at F, and from B, with radius BF, cut the circumference at G,H, and with the same radius step round the circle to I and K; join the points so found to form the pentagon.

Problem 37.—To construct a hexagon upon a given straight line.

In fig. 179, from A and B, the ends of the given line describe arcs intersecting at *g*; from *g*, with the radius *g*A, describe a circle. With the same radius set off the arcs AG, GF and BD, DE. Join the points so found to form the hexagon.

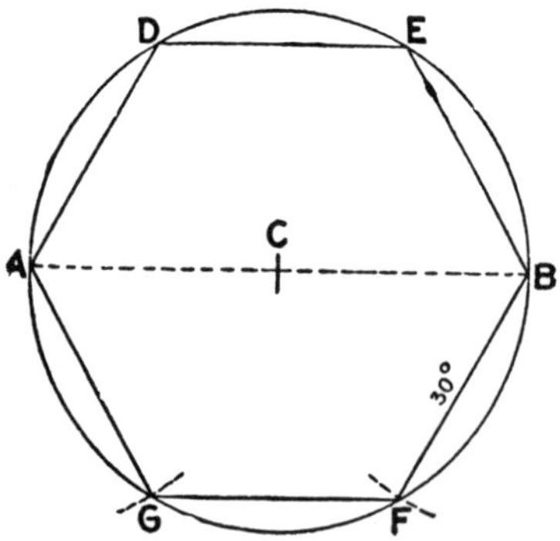

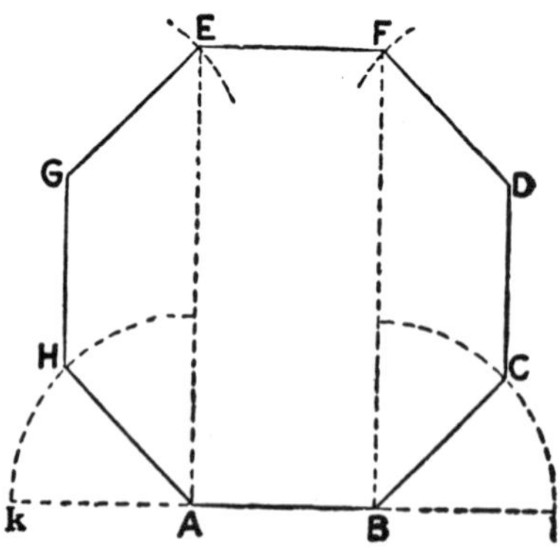

FIG. 180.—*Problem 38. To inscribe a hexagon in a circle.*

FIG. 181.—*Problem 39. To construct an octagon on a given straight line.*

Problem 38.—To inscribe a hexagon in a circle.

In fig. 180, draw a diameter ACB; from A and B, as centers with the radius of the circle AC, cut the circumference at D, E, F, G, and draw AD, DE, etc., to form the hexagon.

The points DE, etc., may be found by stepping the radius (with the dividers) six times round the circle.

Problem 39.—To construct an octagon on a given straight line.

In fig. 181, produce the given line AB, both ways, and draw perpendiculars AE, BF; bisect the external angles A and B, by the lines AH, BC, which make equal to AB. Draw CD and HG parallel with AE and equal to AB; from centers G, D, with the radius AB, cut the perpendiculars at EF, and draw EF, to complete the octagon.

Problem 40.—To inscribe an octagon in a square.

In fig. 182 draw the diagonals of the square intersecting at *e;* from the

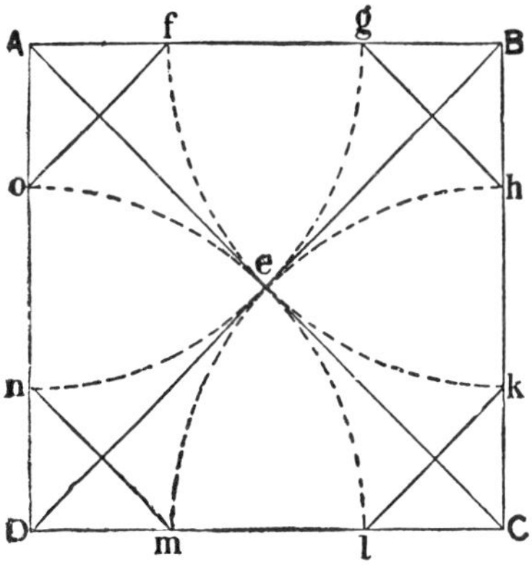

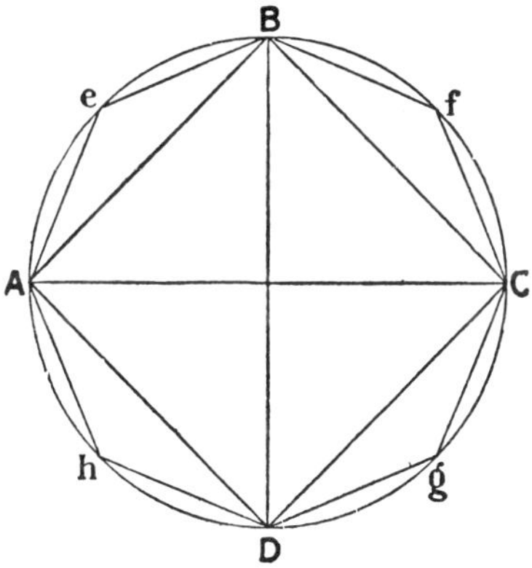

FIG. 182.—***Problem 40.*** *To inscribe an octagon in a square.*
FIG. 183.—***Problem 41.*** *To inscribe an octagon in a circle.*

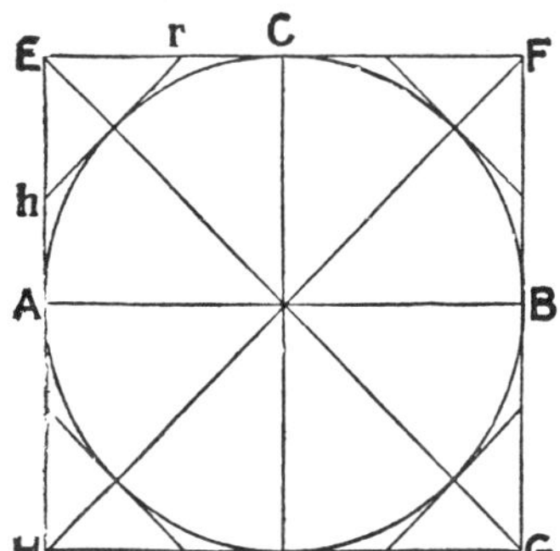

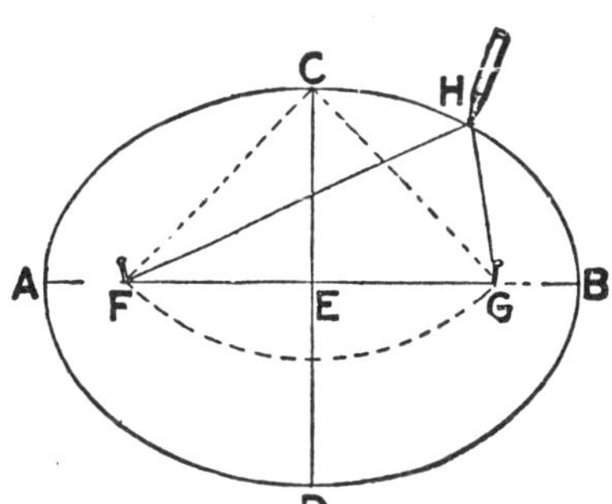

FIG. 184.—***Problem 42.*** *To circumscribe an octagon about a circle.*
FIG. 185.—***Problem 43.*** *To describe an ellipse when the two axes are given.*

corners A,B,C,D, with Ae, as radius, describe arcs cutting the sides at g, h, etc.; and join the points so found to complete the octagon.

Problem 41.—To inscribe an octagon in a circle

In fig. 183, draw two diameters AC, BD, at right angles; bisect the arcs AB, BC, etc., at e, f, etc., and join the points of division to form the octagon.

Problem 42.—To circumscribe an octagon about a circle.

In fig. 184, circumscribe a square EFGH, about the given circle Draw

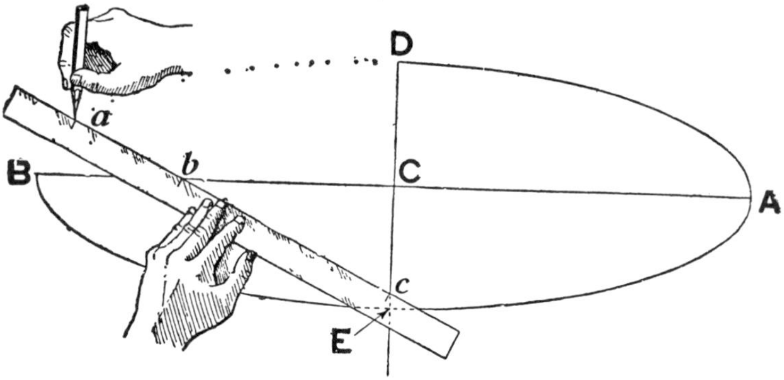

FIG. 186.—*Problem 43. Second method.*

diagonals HF and EG, and tangents h r, etc., through points where the diagonals cut the circle to form with the intercepts, the octagon.

Problem 43.—To describe an ellipse when the two axes are given

In fig. 185, draw the major and minor axes AB and CD, at right angles, intersecting at E. On the center C, with AE, as radius, cut the axis AB, at F and G, the *foci;* insert pins through the axis at F and G, and loop a thread or cord upon them equal in length to the axis AB, so that when stretched it reaches the extremity C, of the *minor axis*, as shown in dotted lines. Place a pencil inside the cord, as at H, and guiding the pencil in this way, keeping the cord equally in tension, carry the pencil round the pins FG, and so describe the ellipse.

Second Method.— In fig. 186 along the edge of a piece of paper, mark off a distance ac, equal to AC, half the major axis, and from the same point, a distance ab, equal to CD, half the minor axis. Place the slip so as to bring the point b, on the line AB, or major axis, and the point c, on the line DE, or minor axis. Set off the position of the point a. Shifting the slip, so that the point b, travels on the major axis, and the point c, on the minor axis, any number of points in the curve may be found, through which the curve may be traced.

Problem 44.—To construct a spiral or volute, by means of tangential arcs of circles.

Construct a square 1 2 3 4, and produce the sides, fig. 44. With center

2, and radius 2 1, describe arc 1 A; center 3, and radius 3 A, describe arc A B; center 4, and radius 4 B, describe arc B C; center 1, and radius 1 C, describe C D; center 2, and radius 2 D, describe D E. In the same way describe any number of arcs, E F, F G, G H. The curve obtained is a spiral or volute. 1 2 3 4, is the eye of the spiral. The eye can be formed by any regular or irregular rectilineal figure, not having a re-entrant angle. In every case, proceed as above.

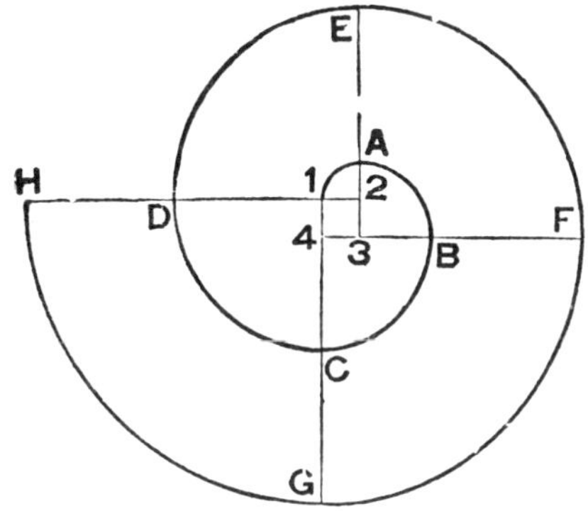

FIG. 187.—*Problem 44. To construct a spiral or volute, by means of tangential arcs of circles.*

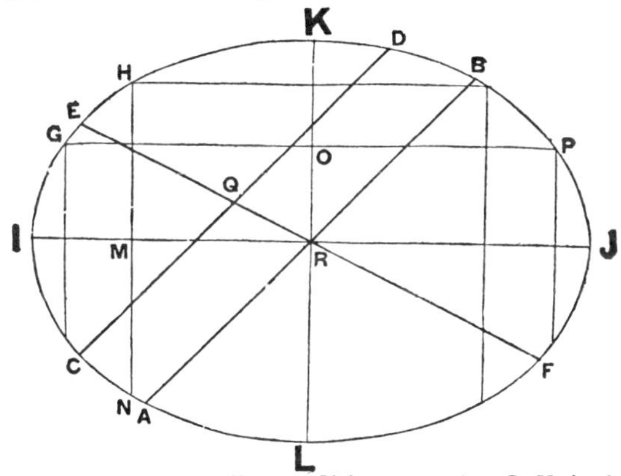

FIG. 188. ***The General notes about ellipses.*** If from any points G, H, in the curve of an ellipse, lines parallel to the major axis I J, be drawn, or to the minor axis K L, be drawn, and the distance M N, be made equal to M H, or O P, be made equal to O G, other points, N, P, in the elliptic curve are obtained. A line Q C, or Q D, drawn from any point Q, in a diameter E F, and parallel to a conjugate diameter A B, is called an ordinate. M H, O P, are also ordinates. The whole line C D, H N, or G P, is a double ordinate. Draw any cord C D, parallel to A B. Bisect A B, C D, at Q R. Then E F, drawn through Q R, is a *conjugate diameter* to A B. The minor axis is called "the conjugate axis," because of its relationship to the major axis. The major and minor axes are a *pair* of conjugate diameters.

Problem 45.*—To find the foci of an ellipse and then to draw the elliptic curve by means of intersecting arcs, the major axis P Q and minor axis T V being given.*

In fig. 189, with T, one end of the minor axis, as center and X Q, half the major axis as radius, describe arc Y, cutting the major axis at F′, F″. These points are the required foci. Between F′ and X, mark any number of points 1, 2, 3, 4. With centers F′, F″, and radius P1, describe arcs *a, a, a, a*. With the same centers and radius Q 1, cut arcs *a, a, a, a*, at *b, b, b, b*. With each focus as center and radius P2, describe arcs, *c, c, c, c*. With the

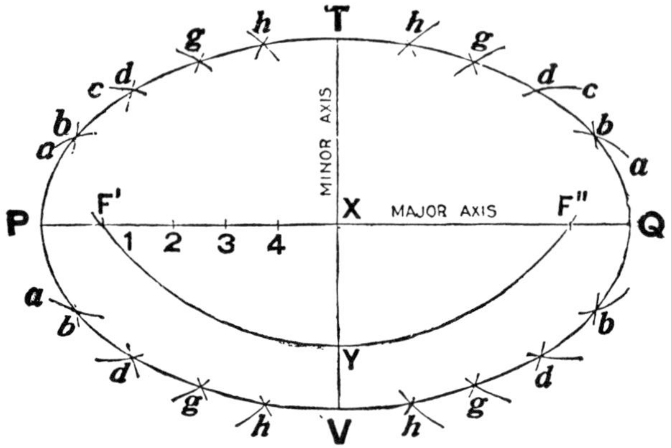

FIG. 189.—*Problem 45. To construct an ellipse having given minor and major axes.*
FIG. 189.—*Problem 46. The major axis and foci of an ellipse being given to find the minor axis.*

same centers and radius Q 2, cut these arcs at *d, d, d, d*. In the same way use points 3 and 4, to get *g, g, h, h*. Through points *b, d, g, h*, draw the curve of the ellipse. The points 1, 2, 3, 4 may be at any distance apart, but it is more convenient to let the divisions decrease in length toward F'. Do not make the arcs too long, as this causes confusion.

***Problem 46.**—The major axis and foci of an ellipse being given to find the minor axis.*

In fig. 189, bisect P Q, at X. With X P, or X Q, as radius and the foci as centers, strike arcs cutting at T and V. Join T V. Then T V, is the minor axis.

***Problem 47.**—Draw the curve of an ellipse by means of intersecting lines. The lengths of the major and minor axes AB, and CD, being given.*

In figs. 190 and 191 bisect CD at E. Bisect AB, at X, by a perpendicular. Make X G, X H, each equal to E C, or E D. With centers G, H, and radius X A, describe arcs at K, K,K,K. With centers A, B, and radius X G, cut these arcs at L, M, N, O. Join L M, M O, O N, N L. Divide A L, A N, B M, B O, A X, B X, each into the same number of equal parts, say four. Draw lines from G, to 1, 2, 3, on A L, B M. From H, through 1 (on A X), draw a line to meet 1 G, at *x*. Through 2, draw a line from H, to meet 2 G, at *x*. Through 3, draw a line to meet 3 G, at *x*. In the same way get points *x,x,x*, on the other side. Also get similar points for the lower half of the ellipse, as shown by dotted lines at R and S. Through *x,x,x*, R, S, draw the curve of the ellipse. The divisions on A L, A X, may be unequal, provided those on A X, be proportional to those on A L. A French curve may be used for drawing the elliptic curve, through the points *x,x,x*. By this method, an ellipse may be inscribed in any rectangle. By joining A G, G B, B H, H A, a rhombus is obtained. Therefore an ellipse can be circumscribed about a rhombus, or a rhombus can be inscribed in an ellipse.

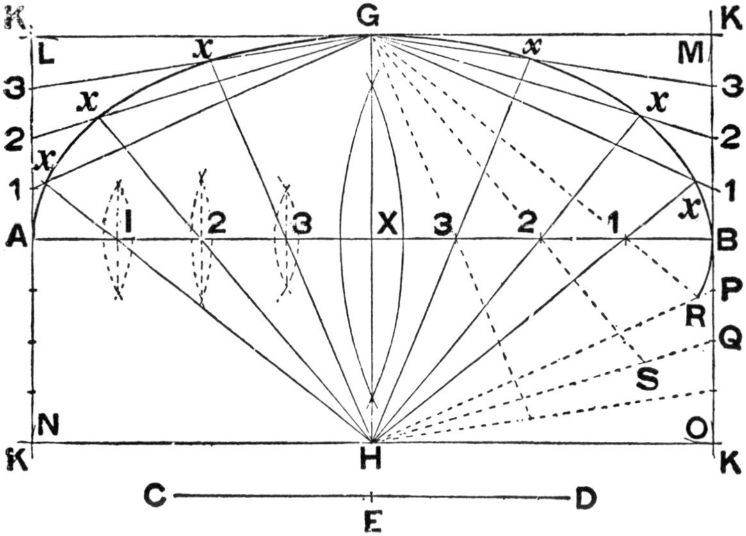

FIGS. 190 and 191.—*Problem 47. Draw the curve of an ellipse by means of intersecting lines.*

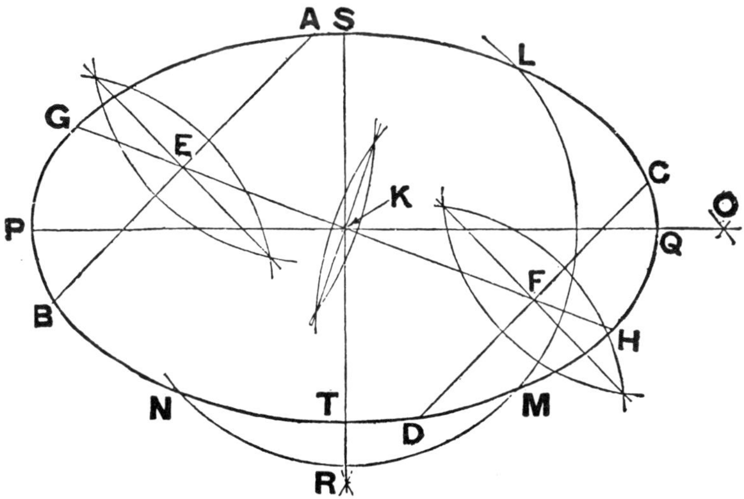

FIG. 192.—*Problem 48. The curve or portion of the curve of an ellipse being given, to find the center and the major and minor axes.*

Problem 48.*—The curve or portion of the curve of an ellipse being given, to find the center and the major and minor axes.*

In fig. 192, draw two parallel chords A B, C D. Bisect them at E and F, Through E, F, draw G H, which is a diameter. Bisect it at K, which is the center of the ellipse. With center K, and any convenient radius, describe the arc L M N. With centers L and M, and any radius, describe arcs cutting at O From O, through K, draw P Q, which is the major axis. With centers M and N; and any radius, describe arcs cutting at R. From R, through K, draw T S, which is the minor axis. Instead of describing arcs at O and R, L M, M N, may be joined and the axes through K, parallel with L M, M N, drawn. For convenience, the given ellipse may be drawn with a piece of thread as shown in fig 185. If a small portion of the curve be given, the chords A B, C D, must be drawn closer together, If only one end of G H, meet the curve, draw another pair of parallel chords, and get another diameter, then the intersection of the two diameters gives the center. The portion of the curve given should contain at least one end of each axis.

***Problem 49.**—At any point A, in the curve of an ellipse, to draw a normal; and through any point B in the curve to draw a tangent.*

In fig. 193, draw the ellipse with a piece of thread. From each focus, draw a line through A, to D, and C. Bisect angle D A C, by A E. The line A E, is the required normal (or perpendicular). From the foci draw lines to B. Produce one of the lines, say to G. Bisect the angle G B F″, by H K. Then the line H K is the required tangent. The normal may also be obtained by bisecting the angle F′ A F″. To draw a normal at either

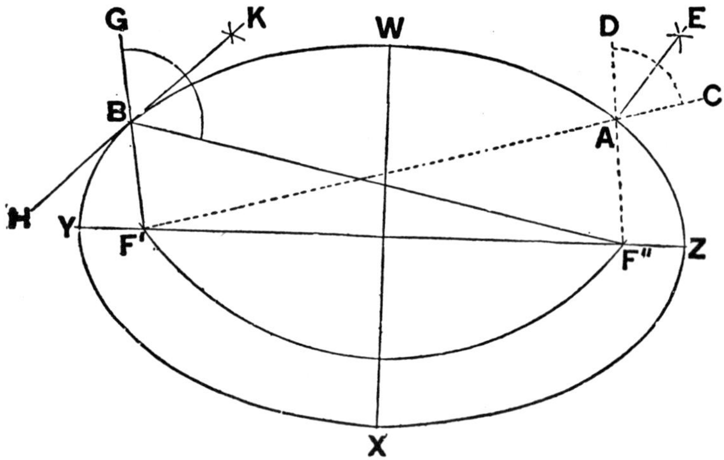

FIG. 193.—*Problem 49. At any point in the curve of an ellipse to draw a normal, and through any point in the curve to draw a tangent*

extremity of the major or minor axis, simply produce the axis. A tangent at either extremity of the major or minor axes must be drawn at right angles to the axis.

NOTE.—*To draw an ellipse* when the foci and one point in the curve are given. Draw a line of indefinite length through the foci. Draw a line from each focus to the given point. The sum of these two lines gives the length of the major axis. With half the major axis as radius, and the foci as centers, describe arcs intersecting at points, which give the ends of the minor axis. Obtain the curve of the ellipse.

Problem 50.—To get the curve of an ellipse approximately with arcs of circles, and by the use of a paper trammel.

In figs. 194 to 196, lines A B, C D, are major and minor axes. Drew E 1, E 2, at any angle. Make E G = X C, and E H = X A. Join G H. With center E, and radii E G, E H, strike the arcs G F, H J. Draw F L, J K, parallel with G H. Make D M, C N, equal to E K, and A O, B P, equal to E L. With centers N and M, and radius N C, describe arcs passing through C and D. With centers O, P, and radius O A, describe arcs at A and B. These four arcs give approximately parts of the ellipse. On one edge of a straight

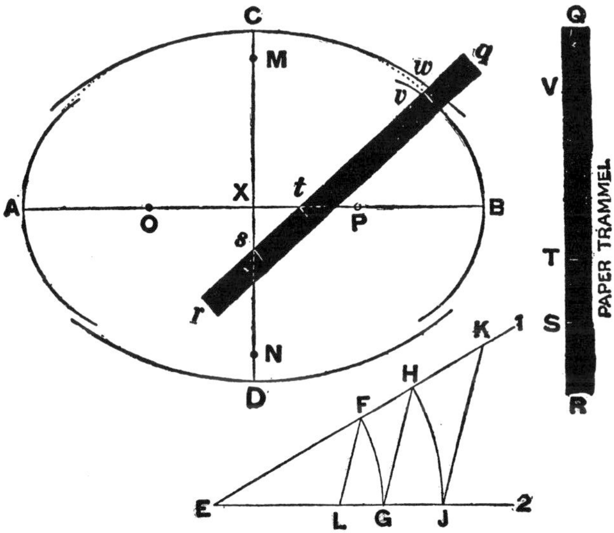

FIGS. 194 to 196.—***Problem 50.*** *To get the curve of an ellipse approximately with arcs of circles and by the use of a paper trammel.* ***This method applies only when the minor axis is more than about ⅗ of the major axis.*** *In making a narrow ellipse M and N will fall outside the ellipse.*

slip of paper Q R, set off V S, equal to A X, and V T, equal to C X. Then use Q R, as a trammel. Adjust the trammel Q R, in such a manner, that

NOTE.—***To describe an ellipse,*** having one diameter given, similar to any given ellipse. In two similar ellipses, any two conjugate diameters of one ellipse have the same proportion to each other as the corresponding conjugate diameters of the other ellipse have to each other. Therefore find a fourth proportional to the given diameter, and the two diameters of the given ellipse. This fourth proportional gives the length of the other diameter of the required ellipse. Place the two diameters bisecting each other, and at the required angle and describe the ellipse.

NOTE.—***To draw tangential lines to an ellipse from a given point outside the curve.*** Call given point 1, and place it in any position with regard to ellipse. With 1, as center and the distance to the nearer focus, as radius, describe about half of a circle cutting the ellipse in two places. With the further focus as center and the major axis as radius, cut the arc in points 2 and 3. From points 2 and 3, draw lines to the further focus. These lines cut the ellipse in two points. Call these points 4 and 5; they are the required points of contact. Draw two lines from the given point 1, through points 4 and 5; and these lines are the required tangents.

point *t*, rests somewhere on the major axis; and point *s*, on the minor axis. Wherever point *v*, comes, will be a point situated in the curve of the ellipse. Mark several points as at *w*, and through these points draw curves connecting the arcs. E L, is a third proportional less, and E K, is a third proportional greater, to the lines E G, E H. A French curve may be used to connect the arcs through the points at *w*. The entire curve can be drawn by means of points obtained with a trammel. When an ellipse has a short minor axis, the points M and N, fall outside the ellipse, on the minor axis produced. This method is exceedingly useful when representing circles in perspective, and also in mechanical drawing when describing ellipses.

TEST QUESTIONS

1. *Draw a perpendicular to a straight line.*
2. *Give the boat builder's laying down method.*
3. *Divide a line into a number of equal parts.*
4. *Bisect an angle.*
5. *Find the center of a given circle.*
6. *Describe a circle passing through three given points.*
7. *Find approximately the length of the circumference of a given circle.*
8. *From a given point draw a tangent to a given circle.*
9. *Construct a square having its diagonal given.*
10. *Construct a square equal in area to any number of given squares.*
11. *Construct a circle equal in area to a given square.*
12. *How is an equilateral triangle constructed on a given base?*
13. *Erect a rectangle on a given base.*
14. *Describe a circle about a triangle.*
15. *Inscribe a circle in a triangle.*
16. *Inscribe any regular polygon in a given circle.*
17. *Inscribe a hexagon in a circle.*

18. *Construct an octagon on a given straight line.*

19. *Describe an ellipse when the two axes are given.*

20. *What is the method of constructing a spiral or volute by means of tangential arcs of circles?*

21. *Give the method of describing an ellipse by means of intersecting arcs.*

22. *Describe an ellipse by the method of intersecting lines.*

23. *Draw a tangent to an ellipse at a given point in the curve of the ellipse.*

CHAPTER 11

Mensuration

Mensuration is *the process of measuring.*

It is that branch of mathematics that has to do with finding the length of lines, the area of surfaces, and the volume of solids. Accordingly the problems which follow will be divided into three groups, as:

1. Measurement of lines.
   a. One dimension, *length*
2. Measurement of surfaces (*areas*).
   a. Two dimensions, *length* and *breadth*
3. Measurement of solids (*volumes*).
   a. Three dimensions. *length, breadth,* and *thickness*

1. Measurement of Lines

(length)

Problem 1.—To find the length of any side of a right triangle, the other two sides being given.

Rule.—*Length of hypotenuse equals square root of the sum of the squares of the two legs; length of either leg equals square root of the difference of the square of the hypotenuse and the square of the other leg.*

Example.—The two legs of a right triangle measure 3 and 4 ft.; find length of hypotenuse. If the length of hypotenuse and one leg be 5 and 3 ft. respectively, what is the length of the other leg?

In fig 197

$$AB = \sqrt{3^2 + 4^2} = \sqrt{25} = 5$$

In fig. 198 $BC = \sqrt{5^2 - 3^2} = \sqrt{25 - 9} = \sqrt{16} = 4.$

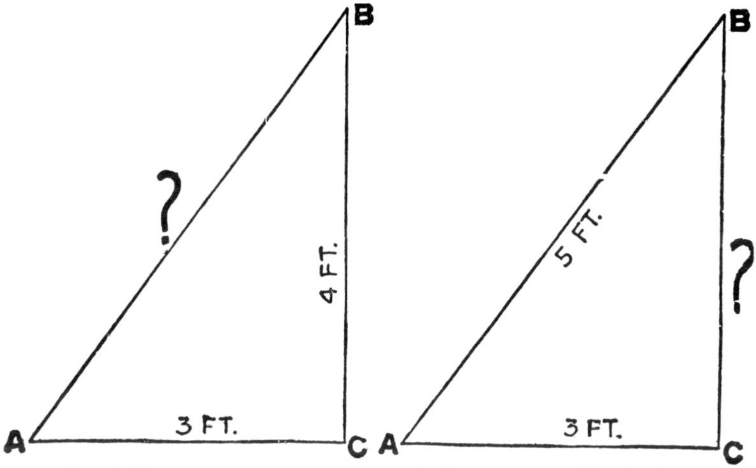

FIGS. 197 and 198:—*Problem 1. To find the length of any side of a right triangle.*

Problem 2.—To find length of circumference of a circle.

Rule.—*Multiply the diameter by* 3.1416.

Example.—What length of moulding strip is required for a circular window 5 ft. in diameter?

$$5 \times 3.1416 = 15.7 \text{ ft.}$$

As the mechanic does not ordinarily measure feet in tenths, the .7 should be reduced to inches; it corresponds to 8⅜ ins. from the table below. That is, the length of moulding is 15 ft. 8⅜ ins. (approx).

Problem 3.—To find the length of an arc of a circle.

Rule.—*As* 360° *is to the number of degrees of the arc so is the length of the circumference to the length of the arc.*

Example.—If the circumference of a circle be 6 feet, what is the length of 60° arc?

Let X = length of the arc, solving for X.

$$360 : 60 = 6 : X = \frac{60 \times 6}{360} = \frac{360}{360} = 1 \text{ ft.}$$

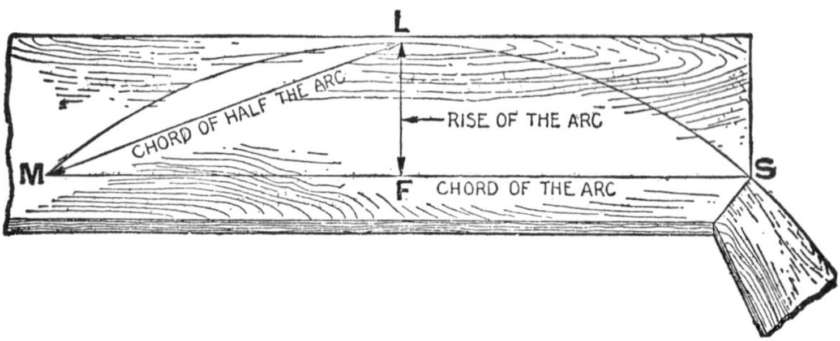

FIG. 199 —*Problem 4 To find width of board required for plate form of circular pattern.*

Decimals of a Foot and Inches

Inch	0"	1"	2"	3"	4"	5"	6"	7"	8"	9"	10"	11"
0	.0	.0833	.1677	.2500	.3333	4167	.5000	.5833	.6667	.7500	.8333	.9167
1-16	.0052	.0885	.1719	.2552	.3385	.4219	.5052	.5885	.6719	.7552	.8385	.9219
1-8	.0104	.0937	.1771	2604	.3437	.4271	.5104	.5937	.6771	.7604	.8437	.9271
3-16	.0156	.0990	.1823	.2656	.3490	.4323	.5156	.5990	.6823	.7656	.8490	9323
1-4	.0208	.1042	.1875	.2708	.3542	.4375	.5208	.6042	.6875	.7708	.8542	.9375
5-16	0260	1094	.1927	.2760	.3594	.4427	.5260	.6094	.6927	.7760	.8594	.9427
3-8	0312	.1146	.1979	2812	.3646	.4479	.5312	.6146	.6979	7812	.8646	.9479
7-16	.0365	.1198	.2031	.2865	.3698	.4531	.5365	.6198	.7031	.7865	.8698	9531
1-2	.0417	.1250	2083	.2917	.3750	.4583	.5417	.6250	.7083	.7917	.8750	.9583
9-16	.0469	.1302	.2135	.2969	.3802	.4635	.5469	.6302	.7135	.7969	8802	.9635
5-8	.0521	.1354	.2188	3021	.3854	.4688	.5521	.6354	.7188	.8021	.8854	9688
11-16	.0573	.1406	.2240	.3073	.3906	.4740	.5573	.6406	.7240	.8073	.8906	.9740
3-4	.0625	.1458	.2292	3125	.3958	.4792	.5625	.6458	.7292	8125.	.8958	.9792
13-16	.0677	.1510	.2344	.3177	4010	.4844	.5677	6510	.7344	.8177	.9010	.9844
7-8	.0729	.1562	.2396	.3229	4062	4896	5729	.6562	7396	.8229	.9062	.9896
15-16	.0781	.1615	.2448	.3281	4115	.4948	.5781	6615	7448	.8281	9115	.9948

***Problem* 4.**—To find the rise of an arc.

Rule 1.—*The rise of an arc is equal to the square of the chord of half the arc divided by the diameter.*

Rule 2.—*Length of chord subtending an angle at the center is equal to twice the radius times the sine of half the angle.*

Example.—A circular pattern 10 ft. in diam. has six plate forms. Find width of board required for these forms allowing 3 ins. margin for joints as in fig. 199.

Each plate will subtend an angle of 360 ÷ 6 = 60°
The "chord of half the arc" (mentioned in rule 1) will subtend 60 ÷ 2 = 30°.

Applying rule 2, "half the angle" = 30° ÷ 2 = 15°.

From table of "trigonometrical functions", sine of 15° = .259, which with radius of 5 ft., becomes

$$\sin 15° \text{ (on 10-ft. circle)} = 5 \times .259 = 1.295$$

Applying rule 2 length of chord MS. = $2 \times 1.295 = 2.59$

Applying rule 1 rise of arc MS, = $2.59^2 \div 10 = .671$ ft. or $8^1/_{16}$ ins. (approx.)

Add to this 3 ins. margin for joints and obtain

width of board $8^1/_{16} + 3 = 11^1/_{16}$ Use 12 in. board

2. Measurement of Surfaces

(areas)

***Problem* 5.—To find the area of a square.**

Rule.—***Multiply the base by the height.***

Example.—What is the area of a square whose side is 5 ft. as in fig. 200?

$$5 \times 5 = 25 \text{ sq. ft.}$$

***Problem* 6.**—To find the area of a rectangle.

Rule.—*Multiply the base by the height (i. e., width by length).*

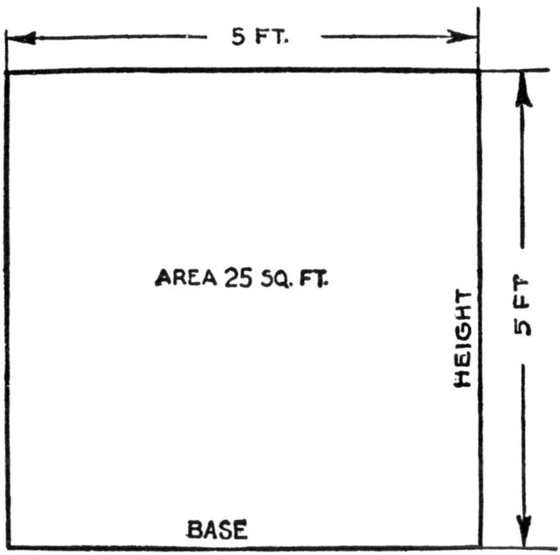

FIG. 200.—***Problem 5.*** *Area of square.*

Example.—What is the area of a rectangle 5 ft. wide and 12 ft. long, as in fig. 201?

$$5 \times 12 = 60 \text{ sq. ft.}$$

Problem 7.—To find the area of a parallelogram.

Rule.—*Multiply base by perpendicular height.*

Example.—What is the area of a parallelogram 2 ft. wide and 10 ft. long?

$$2 \times 10 = 20 \text{ sq. ft.}$$

Problem 8.—To find the area of a triangle.

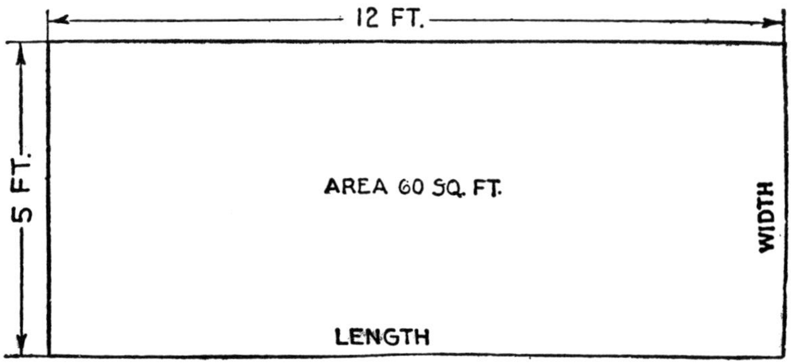

FIG. 201.—*Problem 6. Area of rectangle.*

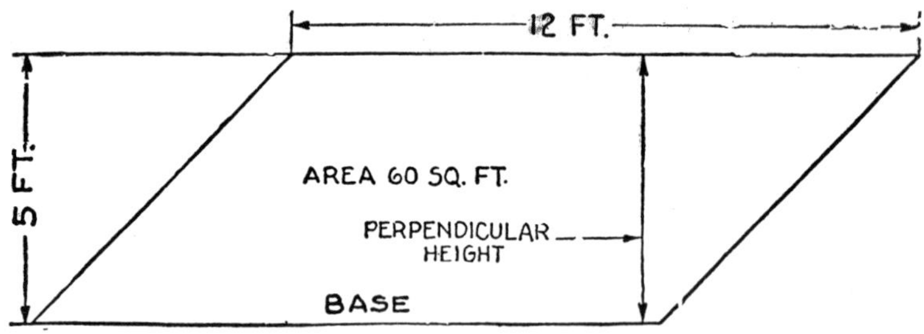

FIG. 202.—*Problem 7. Area of parallelogram.*

Rule.—*Multiply the base by half the altitude.*

Example.—How many sq. ft. of sheet tin are required to cover a church steeple having four triangular sides, measuring 12 ft. (base) × 30 ft. (altitude) as in fig. 203?

$$\tfrac{1}{2} \text{ of altitude} = 15 \text{ ft.}$$

$$\text{area of one side} = 12 \times 15 = 180 \text{ sq. ft.}$$

Total area (four sides) $4 \times 180 = 720$ sq. ft.

Problem 9.—To find the area of a trapezoid.

Rule.—*Multiply one-half the sum of the two parallel sides by the perpendicular distance between them.*

Example.—What is the area of the trapezoid shown in fig. 205?

Here LA and FR, are the parallel sides and MS, the perpendicular dis-

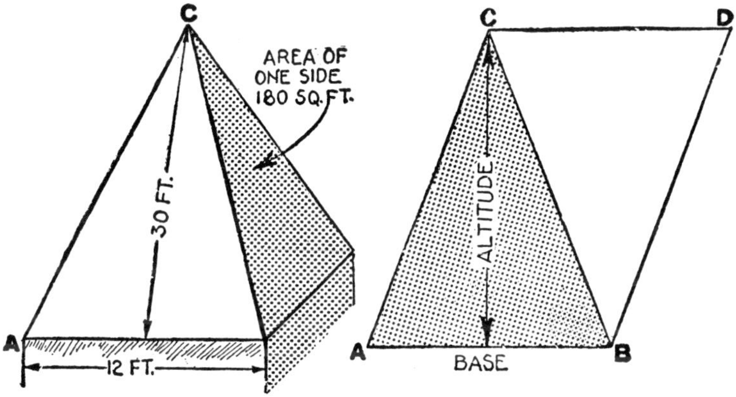

FIGS. 203 and 204.—***Problem 8.*** *Area of triangle.* An inspection of fig. 204 will show that area of triangle = base $\times$ ½ altitude because constructing a parallelogram ABCD, it is made up of two equal triangles and its area = base $\times$ altitude. Hence ½ altitude is taken in finding area of a triangle.

tance between them. Applying rule

$$\text{area} = \tfrac{1}{2}(LA + FR) \times MS$$
$$= \tfrac{1}{2}(8 + 12) \times 6 = 60 \text{ sq. ft.}$$

Problem 10.—To find the area of a trapezium.

Rule.—*Draw a diagonal, dividing figure into triangles; measure diagonal and altitudes and find area of the triangles.*

Example.—What is the area of the trapezium shown in fig. 206, for the dimensions given? Draw diagonal LR, and altitudes AM and FS.

$$\text{area triangle ALR} = 12 \times {}^{6}/_{2} = 36 \text{ sq. ft.}$$
$$\text{area triangle LRF} = 12 \times {}^{9}/_{2} = 54 \text{ sq. ft.}$$

area trapezium LARF = ...90 sq. ft.

Problem 11.—To find the area of any irregular polygon.

Rule.—*Draw diagonals dividing the figure into triangles and find the sum of the areas of these triangles.*

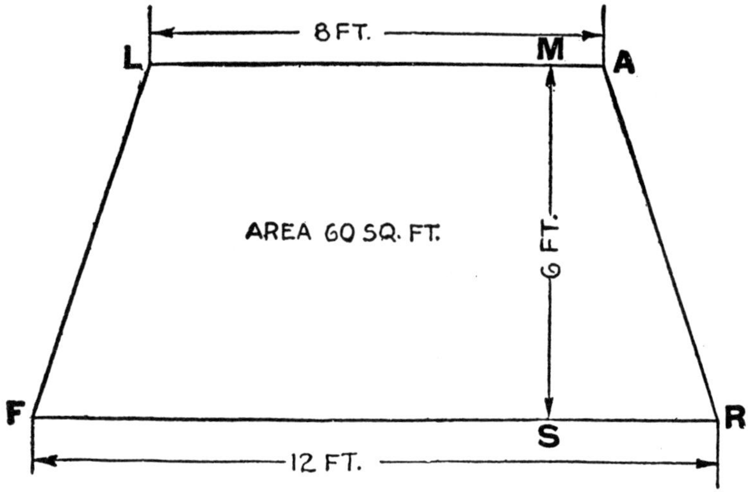

FIG. 205.—*Problem 9. Area of trapezoid.*

Problem 12.—To find the area of any regular polygon when length of side only is given.

Rule.—*Multiply the square of the sides by the figure for "area, when side = 1" in the table following:*

Number of sides	3	4	5	6	7	8	9	10	11	12
Area when side =1.........	.433	1.	1 721	2.598	3 634	4.828	6 181	7 694	9 366	11 196

Example.—What is the area of an octagon (8-sided polygon) whose sides measure 4 ft.

In the above table under 8, find 4.828. Multiply this by the square of one side.

$$4.828 \times 4^2 = 77.25 \text{ sq. ft.}$$

Problem 13.—To find the area of a circle.

Rule.—*Multiply square of diameter by .7854.*

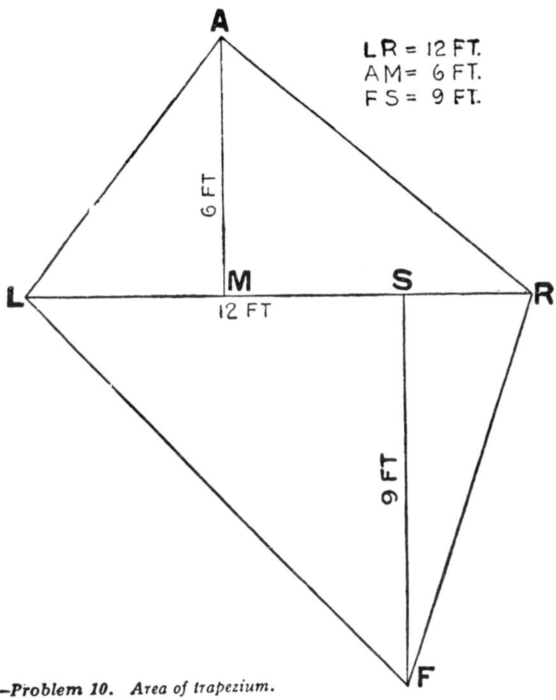

FIG. 206.—*Problem 10. Area of trapezium.*

Example.—What is the area of a circle 10 ft. in diameter?

$$10^2 \times .7854 = 78.54 \text{ sq. ft.}$$

Figs. 207 and 208 show why the decimal .7854 is used in finding the area of a circle.

Problem 14.—To find the area of a sector of a circle.

Rule.—*Multiply the arc of the sector by half the radius.*

Example.—How much tin is required to cover a 60° sector of a 10 foot circular deck?

$$\text{length of } 60° \text{ arc} = \frac{60}{360} \text{ of } 3.1416 \times 10 = 5.24 \text{ ft.}$$

The reason for the above operation should be apparent without any explanation.

Applying rule

tin required for 60° sector = 5.24 × ½ of 5 = 13.1 sq. ft.

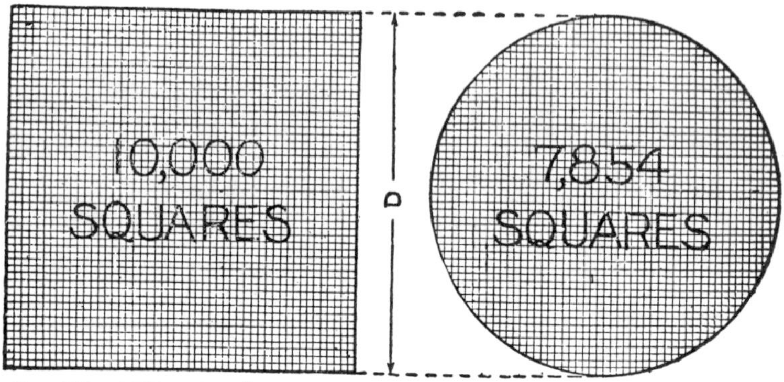

FIGS. 207 and 208.— Showing why the decimal .7854 is used to find the area of a circle. If the square be divided into 10,000 parts or small squares, a circle having a diameter D, equal to a side of the large square will contain 7,854 small squares, hence, if the area of the large square be 1 sq. in., then the area of the circle will be 7854 ÷ 10,000 or .7854 sq. ins., that is, area of the circle = .7854 × D × D = .7854 × 1 × 1 = .7854 sq. ins.

Problem 15.—To find the area of a segment of a circle.

Rule.—*Find the area of the sector which has the same arc and also the area of the triangle formed by the radii and chord; take the sum of these areas if the segment be greater than* 180°; *take the difference if less.*

Problem 16.—To find the area of a ring.

Rule.—*Take the difference between the areas of the two circles.*

Problem 17.—To find the area of an ellipse.

Rule.—*Multiply the product of the two diameters by* .7854.

Example.—What is the area of an ellipse when the minor and major axes are 6 and 10 ins. respectively?

$$10 \times 6 \times .7854 = 47.12 \text{ sq. ins.}$$

Problem 18.—To find the circular area of a cylinder.

Rule.—*Multiply* 3.1416 *by the diameter and by the height.*

Example.—How many sq. ft. of lumber are required for the sides of a cylindrical tank 8 ft. in diameter and 12 ft. high; how many pieces 4″ × 12′ will be required?

cylindrical surface 3.1416 × 8 × 12 = 302 sq. ft.

circumference of tank = 3.1416 × 8 = 25.1 ft.

Number 4″ × 12′ pieces 25.1 ÷ $^4/_{12}$ = 25.1 × 3 = 75.3, say **76.**

Problem 19.—To find the slant area of a cone.

Rule.—*Multiply* 3.1416 *by diameter of base and by one-half the slant height.*

Example.—A conical spire having a base 10 ft. diameter and altitude of 20 ft. is to be covered. Find area of surface to be covered.

In fig. 210, first find slant height, thus

$$\text{slant height} = \sqrt{5^2 + 20^2} = \sqrt{425} = 20.62 \text{ ft.}$$

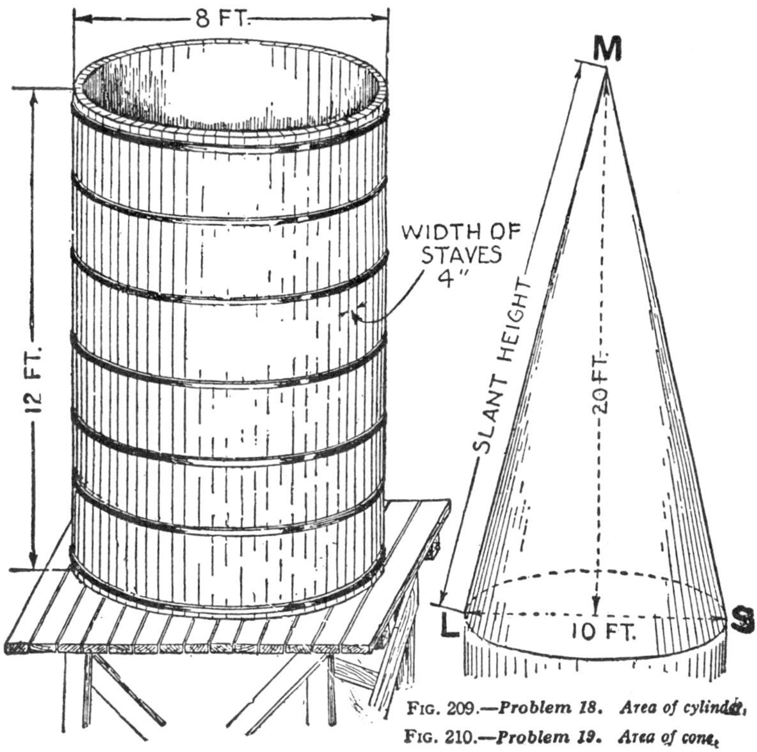

FIG. 209.—***Problem 18. Area of cylinder.***
FIG. 210.—***Problem 19. Area of cone.***

circumference of base = 3.1416 × 10 = 31.42 ft.
area of conical surface = 31.42 × ½ of 20.62 = 324 sq. ft.

Problem 20.—To find the (slant) area of the frustum of a cone.

Rule.—*Multiply half the slant height by the sum of the circumferences.*

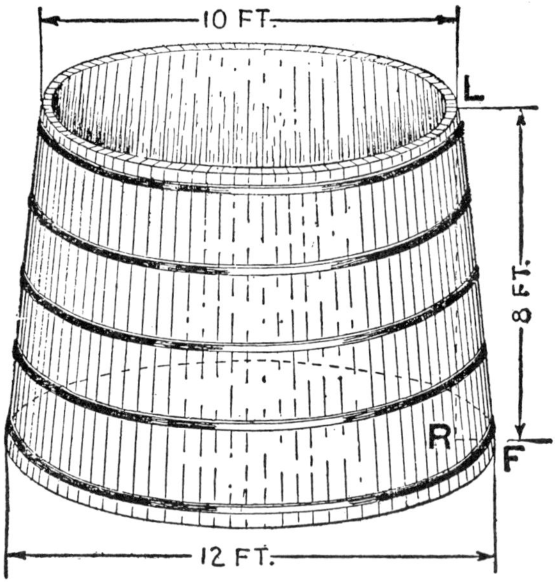

FIG. 211.—***Problem 20. Area of frustum of a cone.*** This is the shape of the ordinary wooden tank seen in wind mill towers. In the figure LR = height of tank. Since the difference between the two diameters is two feet, RF = 1 ft. Hence slant height or LF = $\sqrt{1^2 + 8^2}$ = 8.06.

Example.—A tank is 12 ft in diameter at the base, 10 ft at the top, and 8 ft. high. What is the area of the slant surface?

circumference 10 ft. circle = 3.1416 × 10 = 31.42 ft.
circumference 12 ft circle = 3.1416 × 12 = 37.7 ft
sum of circumferences = 69.1 ft.

$$\text{slant height} = \sqrt{1^2 + 8^2} = \sqrt{65} = 8.06$$

$$\text{slant surface} = \text{sum of circumferences} \times \tfrac{1}{2}\ \text{slant height}$$

$$= 69.1 \times \tfrac{1}{2} \text{ of } 8.06 = 278.5 \text{ sq. ft.}$$

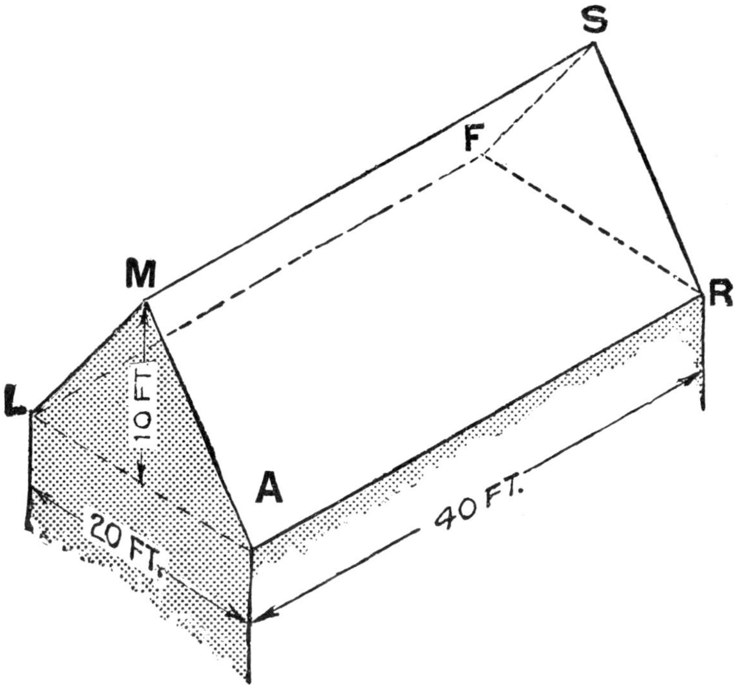

FIG. 212.—*Problem 21. Volume of rectangular wedge.*

3. Measurement of Solids

(volumes)

Problem 21.—*To find the volume of a rectangular wedge.*

Rule.—*Multiply length, breadth and one half height.*

Example.—Find the volume LARFMS of the barn shown in fig. 212.

$$40 \times 20 \times \frac{1}{2} \text{ of } 10 = 4000 \text{ cu. ft.}$$

Problem 22.—To find the volume of a cylinder.

Rule.—*Find the area of the base and multiply this by the length.*

Example.—What is the volume of a cylinder whose diameter is 4 ft. and length $7\frac{1}{2}$ ft.?

```
 4                         .7854
                              16
 4                         47124
                           7854
 —                      12.5664 = area of base in sq. ft.
16                           7.5 = length in ft.
                          628320
                         879648
            Answer,     94.24800 cu. ft.
```

Problem 23.—To find the volume of a cone.

Rule.—*Multiply the area of the base by ⅓ the altitude and the product will be the volume.*

Example.—What is the volume of a cone whose diameter is 12 ft. and altitude 10 ft.?

Area of a circle $= .7854 \times$ sq. of the diameter

Area of base $= .7854 \times 12^2 = 113.1$ sq. ft.

volume $= 113.1 \times \frac{1}{3}$ of $10 = 377$ cu. ft.

Problem 24.—To find the volume of a sphere.

Rule.—*Multiply the cube of the diameter by* .5236.

Example.—Find the volume of a sphere whose diameter is 5 ft.

```
Cube of diameter            Diam.³×.5236
      5                      .5236
      5                        125
     25                      26180
      5                     10472
    125=5³                  5236
                            65.4500 cu. ft.
```

Problem 25.—Find the volume of a segment of a sphere.

Rule 1.—*To three times the square of the radius of the segment's base, add the square of the height; then multiply this sum by the height and the product by* .5236.

Example.—How many cu. ins. in a spherical segment, having a base with diameter of 60 ins. and a height of 20 ins.?

Radius $= 60 \div 2 = 30$ ins.

Three times square of radius $= 3 \times 30 \times 30 = 2,700$

Add the square of the height

$$2,700 + (20 \times 20) = 3,100.$$

Multiply this by the height, 3100×20=62,000

Multiply by .5236 and obtain

62,000×.5236=32,463.2 cu. ins.

Rule 2.—*From three times the diameter of the sphere subtract twice the height of the segment; multiply the remainder by the square of the height, and that product by* .5236 *for the volume.*

Example.—If the diameter of a sphere be 3 ft. 6 ins. what is the volume of a segment whose height is 1 ft. 3 ins.?

$$3\times3.5=10.5$$
$$2\times1.25=2.5$$
$$\overline{\quad 8 \quad}$$
$$8\times1.25\times1.25\times.5236=6.55 \text{ cu. ft.}$$

Problem 26.—To find the volume of an irregular solid.

Rule.—1. *Divide the irregular solid into different figures; the sum of their solidities will be the solidity required.* 2. *To find the solidity of a piece of wood or stone that is craggy or uneven, put it into a tub or cistern, and pour in as much water as will just cover it; then take it out and find the contents of that part of the vessel through which the water has descended and it will be the solidity required.*

Problem 27.—To find the surface and volume of any of the five regular solids, figs. 213 to 217.

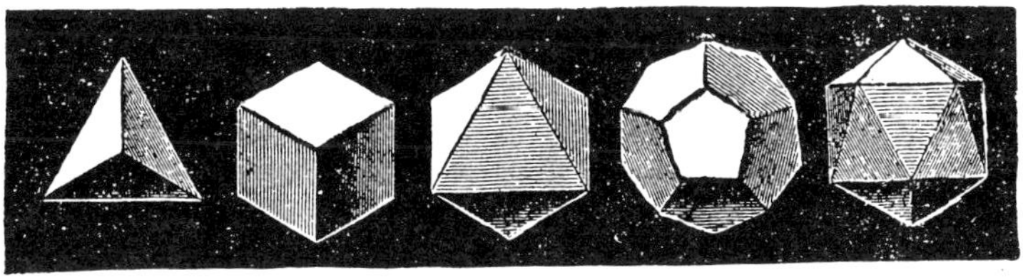

FIGS. 213 to 217.—***The five regular solids:*** Fig. 213, tetrahedron or solid, bounded by four equilateral triangles; fig. 214, hexahedron or cube, bounded by six squares; fig. 215, octahedron, bounded by eight equilateral triangles; fig. 216, dodecahedron, bounded by twelve pentagons; fig. 217, icosahedron, bounded by twenty equilateral triangles.

Rule (surface).—*Multiply the tabular area below, by the square of the edge of the solid.*

Rule (volume).—*Multiply the tabular contents below, by the cube of the given edge.*

Surfaces and Volumes of Regular Solids

Number of Sides	NAME	Area. Edge = 1	Contents. Edge = 1
4	......Tetrahedron........	1.7320	0.1178
6	......Hexahedron.......	6.0000	1.0000
8	......Octahedron.........	3.4641	0.4714
12	......Dodecahedron......	20.6458	7.6631
20	......Icosahedron.........	8.6603	2.1817

Mensuration of Surfaces and Volumes

(Summary)

Area of rectangle = length × breadth.
Area of triangle = base × ½ perpendicular height.
Diameter of circle = radius × 2.
Circumference of circle = diameter × 3.1416.
Area of circle = square of diameter × .7854.

$$\text{Area of sector of circle} = \frac{\text{area of circle} \times \text{number of degrees in arc.}}{360}$$

Area of surface of cylinder = circumference × length + area of two ends.

To find diameter of circle having given area: Divide the area by .7854, and extract the square root.

To find the volume of a cylinder: Multiply the area of the section in square inches by the length in inches = the volume in cubic inches. Cubic inches divided by 1728 = volume in cubic feet.

Surface of a sphere = square of diameter × 3.1416.
Solidity of a sphere = cube of diameter × .5236.
Side of an inscribed cube = radius of a sphere × 1.1547.

Area of the base of a pyramid or cone, whether round, square or triangular, multiplied by one-third of its height = the solidity

Diam. × .8862 = side of an equal square.
Diam. × .7071 = side of an inscribed square.
Radius × 6.2832 = circumference.

$\text{Circumference} = 3.5449 \times \sqrt{\text{Area of circle.}}$

$\text{Diameter} = 1.1283 \times \sqrt{\text{Area of circle}}$

Length of arc = No. of degrees × .017453 radius.
Degrees in arc whose length equals radius = 57°3'
Length of an arc of 1° = radius × .017453
" " " 1 Min. = radius × .0002909
" " " 1 Sec. = radius × .0000048

π = Proportion of circumference to diameter = 3.1415926.
π^2 = 9.8696044
$\sqrt{\pi}$ = 1.7724538
Log. = 0.49715
$1/\pi$ = 0.31831
$\pi/360$ = .008727
$360/\pi$ = 114.59

Lineal feet	×	.00019	= Miles.
" yards	×	.0006	= "
Square inches	×	.007	= Square feet.
Square feet	×	.111	= Square yards.
" yards	×	.0002067	= Acres.
Acres	×	4840	= Square yards.
Cubic inches	×	.00058	= Cubic feet.
" feet	×	.03704	= Cubic yards.
Circular inches	×	.00546	= Square feet.
Cyl. inches	×	.0004546	= Cubic feet.
" feet	×	.02909	= " yards.
Links	×	.22	= Yards.
"	×	.66	= Feet.
Feet	×	1.5	= Links.
Width in chains	×	8	= Acres per mile.
183346 circular in			= 1 square foot.
2200 Cylindrical in			= 1 cubic foot.
Cubic feet	×	7.48	= U. S. gallons.
" inches	×	.004329	= U. S. gallons.
U. S. gallons	×	.13368	= Cubic feet.
U. S. "	×	231	= " inches.
Cubic feet	×	.8036	= U. S. bushel.
" inches	×	.000466	= " " "
Cyl. feet of water	×	6	= U. S. gallons.
Lbs. Avoir	×	.009	= cwt. (112)
" "	×	.00045	= Tons (2240)
Cubic feet of water	×	62.5	= Lbs. Avoir.
" inch " "	×	.03617	= " "
Cyl. feet water	×	49.1	= " "
Cyl. inch water	×	.02842	= " "
13.44 U. S. gallons of water			= 1 cwt.
268.8 U. S. " " "			= 1 ton.
1.8 cubic feet of water			= 1 cwt.
35.88 cubic feet of water			= 1 ton.
Column of water, 12 inches high, and 1 inch in diameter = .341 Lbs.			
U. S. bushel	×	.0495	= Cubic yards.
" " "	×	1.2446	= " feet.
" " "	×	2150.42	= inches.

Problem 28.—To find the volume of a cylindrical ring.

Rule.—*To diameter of body of ring add inner diameter of ring; multiply sum by square of diameter of body, and product by* 2.4674.

Example.—What is volume of an anchor ring, diameter of metal, being 3 ins. and inner diameter of ring. 8 ins.?

$3+8\times3^2=99$ = product of sum of diameter and square of diameter of body of ring.

Then $99\times2.4674=244.2726$ cu. ins.

Problem 29.—To find the sectional area of a pipe.

Example.—A pipe has an external diameter of 2 ins. and an internal diameter of 1¾ ins. Find its sectional area in sq. ins.

Rule.—*From the area of the greater circle subtract that of the lesser.*

$$\text{area of } 2\ \odot = 2^2\times.7854 = 3.1416$$

$$\text{area of } 1\tfrac{3}{4}\ \odot = (1\tfrac{3}{4})^2\times.7854 = \underline{2.4053}$$

$$.7363 \text{ sq. ins.}$$

Properties of the Circle

(According to Kent)

Diameter of circle	× .88623	= side of equal square
Circumference of circle	× .28209	= side of equal square
Circumference of circle	×1.1284	= perimeter of equal square
Diameter of circle	× .7071	= side of inscribed square
Circumference of circle	× .22508	= side of inscribed square
Area of circle × .90031	÷ diameter	= side of inscribed square
Area of circle	×1.2732	= area of circumscribed square
Area of circle	× .63662	= area of inscribed square
Side of square	×1.4142	= diam. of circumscribed circle
Side of square	×4.4428	= circum.
Side of square	×1.1284	= diam. of equal circle
Side of square	×3.5449	= circum. of equal circle
Perimeter of square	× .88623	= circum. of equal circle
Square inches	×1.2732	= circular inches

TEST QUESTIONS

1. *What is mensuration?*
2. *What are the three divisions of mensuration?*
3. *What is the length of the hypotenuse of a triangle in terms of the legs?*
4. *What is the number 3.1416 used for?*
5. *What is the value of ¼π?*
6. *What length of moulding strip is required for a circular window 5 ft. in diameter?*

7. *Give rule for finding the length of an arc of a circle.*
8. *If the circumference of a circle be 6 ft. what is the length of a 60° arc?*
9. *What is the area of a rectangle 5 ft. wide and 12 ft. long?*
10. *How many sq. ft. of sheet tin are required to cover a church steeple having four triangular sides, measuring 12 ft. (base) × 30 ft. (altitude)?*
11. *Give rule for finding the area of a trapezium.*
12. *What is the method of finding the area of any irregular polygon?*
13. *What is the area of an 8 sided polygon whose sides measure 4 ft.?*
14. *Draw a diagram showing the meaning of the much used .7854.*
15. *How much sheet tin is required to cover a 60° sector of a 10 ft. circular deck?*
16. *What is the area of an ellipse whose two diameters are 10 and 6 ins.?*
17. *What is the displacement per minute of a 5×6 engine running 600 r.p.m.?*
18. *Give rule for finding the slant area of the frustum of a cone.*
19. *What is the volume of a cone whose diameter is 12 ft. and altitude 10 ft.?*
20. *Find the volume of a sphere whose diameter is 5 ft.*
21. *How many cu. ins. in a spherical segment having a base whose diameter is 60 ins. and a height of 20 ins.?*
22. *Name five regular solids.*
23. *A pipe has an external diameter of 2 ins. and an internal diameter of 1¾ ins. Find its sectional area in sq. ins.*

CHAPTER 12

The Slide Rule

The principle of the slide rule is based on the fact that *the addition of logarithms multiplies the numbers that they represent and the subtraction of logarithms divides the numbers.*

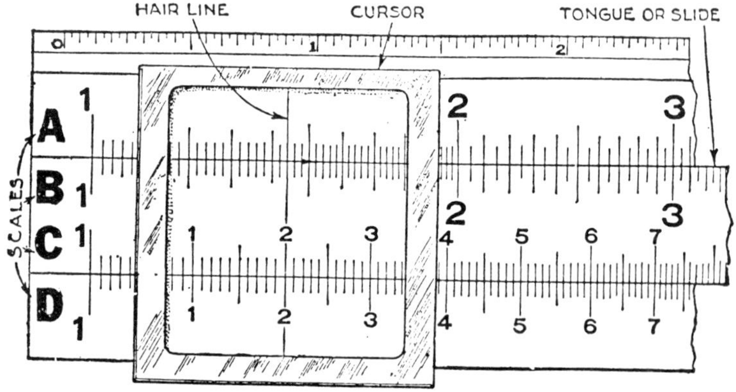

FIG. 1.—Detail of slide rule showing scales and parts. Note especially the scales as they are constantly referred to by letter, as scale **C**, scale **D**, etc.

By use of the slide rule the operations of multiplication, division, the finding of powers and the extraction of roots, may be performed rapidly and with an approximation to accuracy which is sufficient for many purposes.

With a good 10 inch rule the results obtained are usually accurate to ¼ of 1 per cent.

A slide rule has four scales as shown in fig. 1 and designated by the letters A,B,C,D. The two outer scales A and D, are stationary, but the two inner scales B and C, are on a tongue arranged to slide between the two outer scales.

The chief difficulty in learning to use the slide rule is to learn to read the graduations on the scales.

Taking first part of the D scale and beginning at the left hand end it will be seen in fig. 2, that between the large 1 and 2 there are ten divisions indicated by the small figures 1, 2, 3, etc.

Now, in fig. 3 the large 1 is read 1 and since there are 10 large divisions between large 1 and large 2, the reading at small 2 for instance is 1.2, 12,

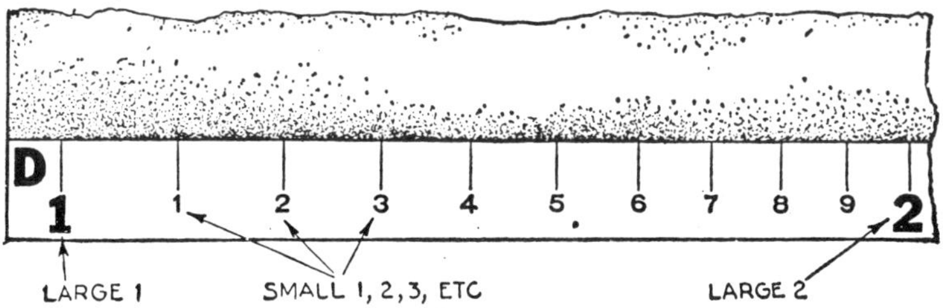

FIG. 2.—Detail of D scale showing large and small figures.

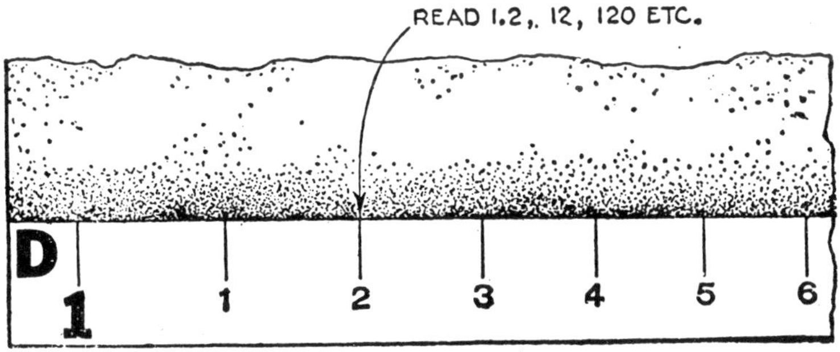

FIG. 3.—Detail of D scale showing how to read 1.2, 12, 120, etc.

120, etc. Hence it will be noted that the slide rule does not place the decimal point.

It will be further noted that between each of the divisions designated by small 1, 2, etc., there are 10 smaller divisions (fig. 1). Hence each one of these is one hundredth of the large division from large 1 to large 2. This permits reading accurately a number of three figures and by "guess" a number of four figures. Thus to read 1 1 3, read the big 1 (for 100), then the small 1 (for 10) and finally the third hundredths division (for 3) counting to the right from the little 1 division as indicated in fig. 4.

To read a number such as 104, remember that the big 1 can be read either 1 or 10. Hence call the big 1 ten and count 4 of the little divisions to the right as in fig. 5.

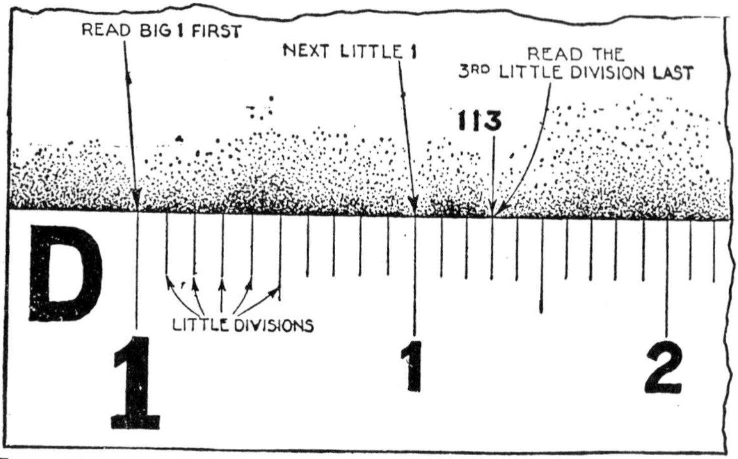

FIG. 4.—Detail of **C** scale showing how to read numbers such as 113.

The accuracy with which a number of four figures can be read depends upon the ability of the eye to gauge the position of the 4th figure between two of the small divisions. Thus, in fig. 6, to read the number 1,057 take big 1 for the first two figures, that is 10; count to right from big 1 five of the little divisions indicated by *a*, for the third figure.

Now the fourth figure is found between the two adjacent little divisions *a* and *b*. The illustration which is greatly magnified shows *a*,*b*, divided into 10 divisions, but these divisions are not on the rule itself and must be judged or guessed at "by the eye" and upon the precision with which this is done depends the accuracy of the fourth figure or in this case the 7 of the number 1057.

To read numbers beginning with 2, 3, etc., start with the big 2, big 3, etc., and proceed similarly as in the foregoing explanations.

Thus to read the number 275, read first in fig. 7, the big 2; count off seven small divisions to right, that is to division *a*, and then gauge by eye half the distance or 5 invisible divisions between *a* and *b* obtaining the number 275.

Example.—Multiply 2×7.

In fig. 8 slide last division or "index" at right end of C scale opposite big 2 on D scale. Move "cursor" (the sliding glass with hair line) opposite big 7 on C scale and read 14 on D scale indicated by the hair line.

The "index" at either end may be used, but if the second factor come "off the scale" take the index at the other end of the tongue.

Example.—Divide 4 by 2.

In fig. 9 slide cursor till hair line registers with large 4 on **D** scale; slide tongue to register large 2 (on **C** scale) with hair line. At large 1 on the end of tongue (scale **C**) read 2 on scale **D**, which is the answer.

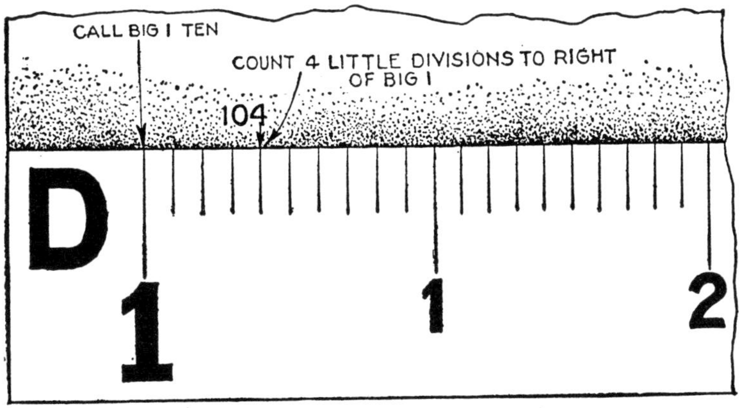

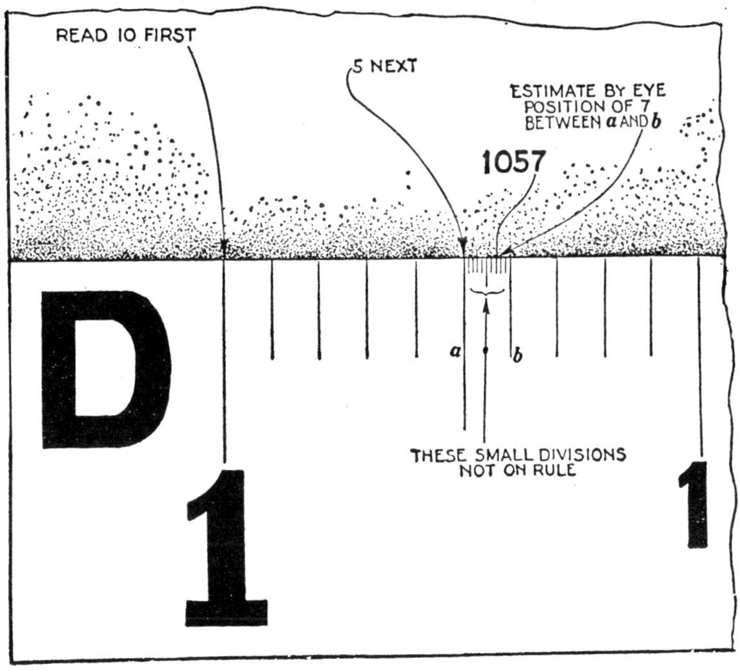

FIG. 5.—Detail of **D** scale showing how to read numbers containing a cipher (or ciphers) such as 104.

FIG. 6.—Detail of **D** scale showing how to read a number of four figures.

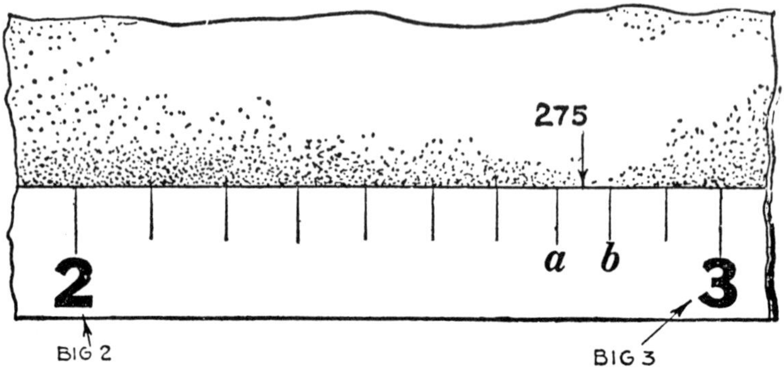

FIG. 7.—Detail of **D** scale showing how to read numbers of three figures beginning with **2**, 3, etc.

Example.—Divide 475 by 23.

In fig. 10 slide cursor till hair line registers with 475 on scale **D**; slide tongue until 23 on scale **C**, registers with 475 on scale **D**. At large 1 on end of tongue read 2,065 (scale **D**) and place the decimal point to read 20.65. As the slide rule does not give the decimal point this has to be done independently. This example illustrates how to read a number of four figures as explained under figs. 10 and 11.

Continued Multiplication.—When there are more than two factors proceed as follows:

Rule.—*Multiply the first two factors in the regular way; set hair line to register first product and continue by placing index to register with hair line.*

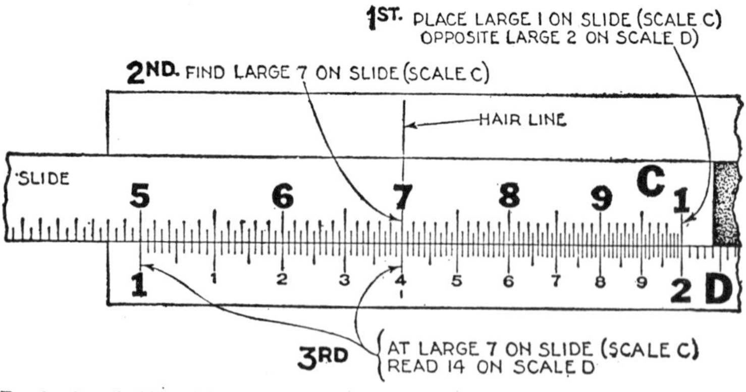

FIG. 8.—Detail of **D** and **C** scales showing how to multiply.

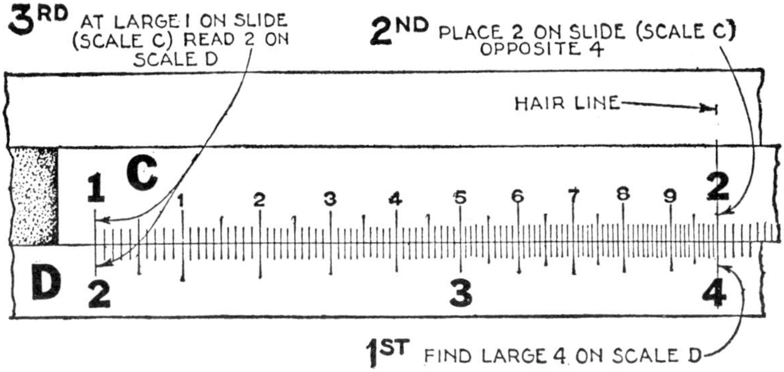

FIG. 9.—Detail of **D** and **C** scales showing how to divide numbers of one figure each.

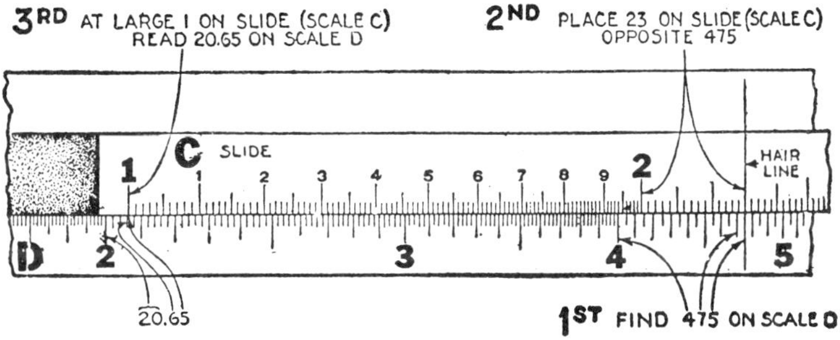

FIG. 10.—Detail of **D** and **C** scales showing how to divide numbers of three figures each.

Squares and Square Roots.—These are found by using the A and D scales. Thus:

Rule.—*Set the cursor so that the hair line stands over the number on the D scale and the square of that number will be found on the A scale under the hair line.*

In finding the square root of a number, note that the number is located on the A scale and the root is found directly below, on the D scale, the hair line being used to make the readings accurately.

Cubes and Cube Roots.—Finding the cube of a number requires the use of all four scales. The index of the C scale is set to the given number on the D scale and opposite the same number on the *left hand* B scale, the cube is read from the A scale. Three cases may arise, as follows:

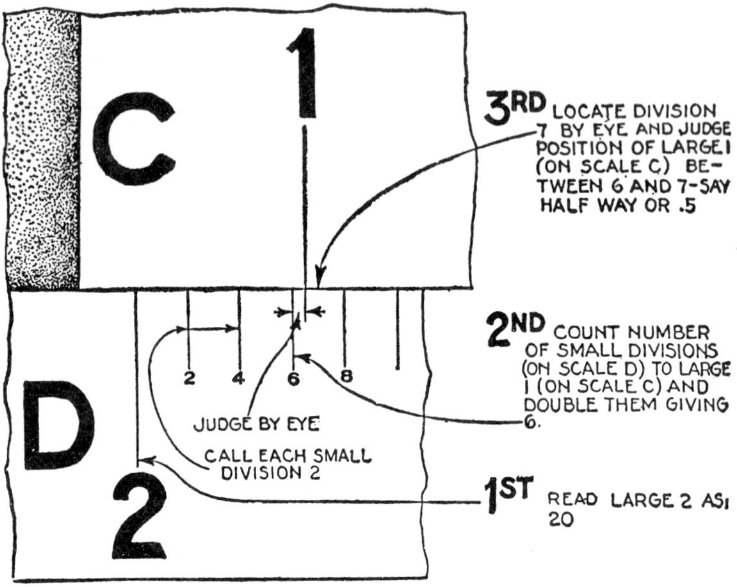

FIG. 11.—Magnified detail of **D** and **C** scales showing how to read the answer in the example as given in fig. 10.

Case 1.—*The slide may project to the left and the cube be found on the left hand A scale.*

Case 2.—*The slide may project to the right and the cube be found on the right hand A scale.*

Case 3.—*The slide may project to the right and the cube be found on the left hand A scale.*

Cube Root.—The operation of finding the cube root is *the reverse of finding the cube*, and the method of determining the number of figures in the root, preceding or following the decimal point, is the same as that used in arithmetic; that is, the number is pointed off into periods of three figures each, beginning at the decimal point, and the root then contains one figure for each period or part of a period.

If the number contain a decimal, add ciphers to the right of it, if necessary, so that the number of figures in the decimal part will be exactly divisible by 3, without a remainder.

The scales and indexes to be used in finding the cube root depend upon the number of figures in the first period at the left, not counting ciphers when they *immediately* follow the decimal point.

The rules to be followed are:

Case 1.—*If there are three figures in the first period, use the left-hand A scale and the right-hand index of the C scale.*

Case 2.—*If there are two figures in the first period, use the right-hand A scale and the left-hand index of the C scale.*

Case 3.—*If there is only one figure in the first period, use the left-hand A scale and the left-hand index of the C scale.*

Tables and Data

Metric and English Equivalent Measures

MEASURES OF LENGTH

Metric		*English*
1 meter	=	39.37 inches, or 3.28083 feet, or 1.09361 yards
.3048 meter	=	1 foot
1 centimeter	=	.3937 inch
2.54 centimeters	=	1 inch
1 millimeter	=	.03937 inch, or nearly 1-25 inch
25.4 millimeters	=	1 inch
1 kilometer	=	1093.61 yards, or 0.62137 mile

MEASURES OF WEIGHT

Metric		*English*
1 gram	=	15.432 grains
.0648 gram	=	1 grain
28.35 grams	=	1 ounce avoirdupois
1 kilogram	=	2.2046 pounds
.4536 kilogram	=	1 pound
1 metric ton 1000 kilograms	=	.9842 ton of 2240 pounds 19.68 cwt. 2204.6 pounds
1.016 metric tons 1016 kilograms	=	1 ton of 2240 pounds

MEASURES OF CAPACITY

Metric		*English*
1 liter (= 1 cubic decimeter)	=	61.023 cubic inches .03531 cubic foot .2642 gal. (American) 2.202 lbs. of water at 62° F.
28.317 liters	=	1 cubic foot
3.785 liters	=	1 gallon (American)
4.543 liters	=	1 gallon (Imperial)

English Conversion Table

Length

Inches	×	.0833	= feet
Inches	×	.02778	= yards
Inches	×	.00001578	= miles
Feet	×	.3333	= yards
Feet	×	.0001894	= miles
Yards	×	36.00	= inches
Yards	×	3.00	= feet

Yards	×	.0005681	= miles
Miles	×	63360.00	= inches
Miles	×	5280.00	= feet
Miles	×	1760.00	= yards
Circumference of circle	×	.3188	= diameter
Diameter of circle	×	3.1416	= circumference
Area			
Square inches	×	.00694	= square feet
Square inches	×	.0007716	= square yards
Square feet	×	144.00	= square inches
Square feet	×	.11111	= square yards
Square yards	×	1296.00	= square inches
Square yards	×	9.00	= square feet
Dia. of circle squared	×	.7854	= area
Dia. of sphere squared	×	3.1416	= surface
Volume			
Cubic inches	×	.0005787	= cubic feet
Cubic inches	×	.00002143	= cubic yards
Cubic inches	×	.004329	= U. S. gallons
Cubic feet	×	1728.00	= cubic inches
Cubic feet	×	.03704	= cubic yards
Cubic feet	×	7.4805	= U. S. gallons
Cubic yards	×	46656.00	= cubic inches
Cubic yards	×	27.00	= cubic feet
Dia. of sphere cubed	×	.5236	= volume
Weight			
Grains (avoirdupois)	×	.002286	= ounces
Ounces (avoirdupois)	×	.0625	= pounds
Ounces (avoirdupois)	×	.00003125	= tons
Pounds (avoirdupois)	×	16.00	= ounces
Pounds (avoirdupois)	×	.01	= hundredweight
Pounds (avoirdupois)	×	.0005	= tons
Tons (avoirdupois)	×	32000.00	= ounces
Tons (avoirdupois)	×	2000.00	= pounds

English Conversion Table

Energy			
Horsepower	×	33000.	=ft.-lbs. per min.
B. t. u.	×	778.26	=ft.-lbs.
Ton of refrigeration	×	200.	=B. t. u. per min.
Pressure			
Lbs. per sq. in.	×	2.31	= ft. of water (60°F.)
Ft .of water (60°F.)	×	.433	= lbs. per sq. in.
Ins. of water (60°F.)	×	.0361	= lbs. per sq. in.
Lbs. per sq. in.	×	27.70	= ins. of water (60°F.)
Lbs. per sq. in.	×	2.041	= ins. of Hg. (60°F.)
Ins. of Hg (60°F.)	×	.490	= lbs. per sq. in.
Power			
Horsepower	×	746.	= watts
Watts	×	.001341	= horsepower
Horsepower	×	42.4	= B. t. u. per min.

Water Factors (at point of greatest density—39.2°F)

Miners inch (of water)	×	8.976	= U. S .gals. per min.
Cubic inches (of water)	×	.57798	= ounces
Cubic inches (of water)	×	.036124	= pounds
Cubic inches (of water)	×	.004329	= U. S. gallons
Cubic inches (of water)	×	.003607	= English gallons
Cubic feet (of water)	×	62.425	= pounds
Cubic feet (of water)	×	.03121	= tons
Cubic feet (of water)	×	7.4805	= U. S. gallons
Cubic feet (of water)	×	6.232	= English gallons
Cubic foot of ice	×	57.2	= pounds
Ounces (of water)	×	1.73	= cubic inches
Pounds (of water)	×	26.68	= cubic inches
Pounds (of water)	×	.01602	= cubic feet
Pounds (of water)	×	.1198	= U. S. gallons
Pounds (of water)	×	.0998	= English gallons
Tons (of water)	×	32.04	= cubic feet
Tons (of water)	×	239.6	= U. S. gallons
Tons (of water)	×	199.6	= English gallons
U. S. gallons	×	231.00	= cubic inches
U. S. gallons	×	.13368	= cubic feet
U. S. gallons	×	8.345	= pounds
U. S. gallons	×	.8327	= English gallons
U. S. gallons	×	3.785	= liters
English gallons (Imperial)	×	277.41	= cubic inches
English gallons (Imperial)	×	.1605	= cubic feet
English gallons (Imperial)	×	10.02	= pounds
English gallons (Imperial)	×	1.201	= U. S. gallons
English gallons (Imperial)	×	4.546	= liters

Metric Conversion Table

Length

Millimeters	×	.03937	= inches
Millimeters	÷	25.4	= inches
Centimeters	×	.3937	= inches
Centimeters	÷	2.54	= inches
Meters	×	39.37	= inches (Act. Cong.)
Meters	×	3.281	= feet
Meters	×	1.0936	= yards
Kilometers	×	.6214	= miles
Kilometers	÷	1.6093	= miles
Kilometers	×	.3280.8	= feet

Area

Sq. Millimeters	×	.00155	= sq. in.
Sq. Millimeters	÷	645.2	= sq. in.
Sq. Centimeters	×	.155	= sq. in.
Sq. Centimeters	÷	6.452	= sq. in.
Sq. Meters	×	10.764	= sq. ft.
Sq. Kilometers	×	247.1	= acres
Hectares	×	2.471	= acres

Volume

Cu. Centimeters	÷	16.387	= cu. in.
Cu. Centimeters	÷	3.69	= fl. drs. (U.S.P.)
Cu. Centimeters	÷	29.57	= fl. oz. (U.S.P.)
Cu. Meters	×	35.314	= cu. ft.

Metric Conversion Table (Cont.)

Cu. Meters	×	1.308	= cu. yards
Cu. Meters	×	264.2	= gals. (231 cu. in.)
Litres	×	61.023	= cu. in. (Act. Cong.)
Litres	×	33.82	= fl. oz. (U.S.P.)
Litres	×	.2642	= gals. (231 cu. in.)
Litres	÷	3.785	= gals. (231 cu. in.)
Litres	÷	28.317	= cu. ft.
Hectolitres	×	3.531	= cu. ft.
Hectolitres	×	2.838	= bu. (2150.42 cu. in.)
Hectolitres	×	.1308	= cu. yds.
Hectolitres	×	26.42	= gals. (231 cu. in.)
Weight			
Grams	×	15.432	= grains (Act. Cong.)
Grams	÷	981.	= dynes
Grams (water)	÷	29.57	= fl. oz.
Grams	÷	28.35	= oz. avoirdupois
Kilo-grams	×	2.2046	= lbs.
Weight			
Kilo-grams	×	35.27	= oz. avoirdupois
Kilo-grams	×	.0011023	= tons (2000 lbs.)
Tonneau (Metric ton)	×	1.1023	= tons (2000 lbs.)
Tonneau (Metric ton)	×	2204.6	= lbs.
Unit Weight			
Grams per cu. cent.	+	27.68	= lbs. per cu. in.
Kilo per meter	×	.672	= lbs. per ft.
Kilo per cu. meter	×	.06243	= lbs. per cu. ft.
Kilo per Cheval	×	2.235	= lbs. per h. p.
Grams per liter	×	.06243	= lbs. per cu. ft.
Pressure			
Kilo-grams per sq. cm.	×	14.223	= lbs. per sq. in.
Kilo-grams per sq. cm.	×	32.843	= ft. of water (60°F.)
Atmospheres (international)	×	14.696	= lbs. per sq. in
Energy			
Joule	×	.7376	= ft. lbs.
Kilo-gram meters	×	7.233	= ft. lbs.
Power			
Cheval vapeur	×	.9863	= h. p.
Kilo-watts	×	1.341	= h. p.
Watts	+	746.	= h. p.
Watts	×	.7373	= ft. lbs. per sec.

Miscellaneous

Kilogram calorie	×	3.968	= B. t. u.
Standard gravity (Sea level 45° lat.)	÷	980.665	= centimeters per sec. per sec.
Frigories/hr. (French)	÷	3023.9	= Tons refrigeration

Decimal and Millimeter Equivalents of Fractional Parts of an Inch

Inches	Inches	mm	Inches	Inches	mm
1-64	.01563	.397	33-64	.51563	13.097
1-32	.03125	.794	17-32	.53125	13.494
3-64	.04688	1.191	35-64	.54688	13.890
1-16	.0625	1.587	9-16	.5625	14.287
5-64	.07813	1.984	37-64	.57813	14.684
3-32	.09375	2.381	19-32	.59375	15.081
7-64	.10938	2.778	39-64	.60938	15.478
1-8	.125	3.175	5-8	.625	15.875
9-64	.14063	3.572	41-64	.64063	16.272
5-32	.15625	3.969	21-32	.65625	16.669
11-64	.17188	4.366	43-64	.67188	17.065
3-16	.1875	4.762	11-16	.6875	17.462
13-64	.20313	5.159	45-64	.70313	17.859
7-32	.21875	5.556	23-32	.71875	18.256
15-64	.23438	5.953	47-64	.73438	18.653
1-4	.25	6.350	3-4	.75	19.050
17-64	.26563	6.747	49-64	.76563	19.447
9-32	.28125	7.144	25-32	.78125	19.844
19-64	.29688	7.541	51-64	.79688	20.240
5-16	.3125	7.937	13-16	.8125	20.637
21-64	.32813	8.334	53-64	.82813	21.034
11-32	.34375	8.731	27-32	.84375	21.431
23-64	.35938	9.128	55-64	.85938	21.828
3-8	.375	9.525	7-8	.875	22.225
25-64	.39063	9.922	57-64	.89063	22.622
13-32	.40625	10.319	29-32	.90625	23.019
27-64	.42188	10.716	59-64	.92188	23.415
7-16	.4375	11.113	15-16	.9375	23.812
29-64	.45313	11.509	61-64	.95313	24.209
15-32	.46875	11.906	31-32	.96875	24.606
31-64	.48438	12.303	63-64	.98408	25.003
1-2	.5	12.700	1	1.00000	25.400

Diesel Engine Manual (23199)

A practical treatise on the theory, operation and maintenance of modern Diesel engines. Explains Diesel principles—valves—timing—fuel pumps—pistons and rings—cylinders—lubrication—cooling system—fuel oil—engine indicator—governors—engine reversing—answers on operation—calculations. AN IMPORTANT GUIDE FOR ENGINEERS, OPERATORS, STUDENTS.

Gas Engine Manual (23061)

A completely practical book covering the construction, operation and repair of all types of modern gas engines. Part I covers gas-engine principles; engine parts; auxiliaries; timing methods; ignition systems. Part II covers troubleshootng, adjustment and repairs.

BUILDING AND MAINTENANCE

Answers on Blueprint Reading (23041)

Covers all types of blueprint reading for mechanics and builders. The man who can read blueprints is in line for a better job. This book gives you the secret language, step by step in easy stages. NO OTHER TRADE BOOK LIKE IT.

Building Construction and Design (23180)

A completely revised and rewritten version of Audel's **Architects and Builders Guide.** New illustrations and extended coverage of material makes this treatment of the subject more valuable than ever. Anyone connected in any way with the building industry will profit from the information contained in this book.

Building Maintenance (23140)

A comprehensive book on the practical aspects of building maintenance. Chapters are included on: painting and decorating; plumbing and pipe fitting; carpentry; calking and glazing; concrete and masonry; roofing; sheet metal; electrical maintenance; air conditioning and refrigeration; insect and rodent control; heating maintenance management; cutodial practices: A BOOK FOR BUILDING OWNERS, MANAGERS, AND MAINTENANCE PERSONNEL.

Gardening & Landscaping (23229)

A comprehensive guide for the homeowner, industrial, municipal, and estate groundskeepers. Information on proper care of annual and perennial flowers; various house plants; greenhouse design and construction; insect and rodent control; complete lawn care; shrubs and trees; and maintenance of walks, roads, and traffic areas. Various types of maintenance equipment are also discussed.

Carpenters & Builders Library—4 Vols. (23169)

A practical illustrated trade assistant on modern construction for carpenters, builders, and all woodworkers. Explains in practical, concise language and illustrations all the principles, advances and short cuts based on modern practice. How to calculate various jobs.

Vol. 1—(23170)—Tools, steel square, saw filing, joinery, cabinets.
Vol. 2—(23171)—Mathematics, plans, specifications, estimates.
Vol. 3—(23172)—House and roof framing, laying out, foundations.
Vol. 4—(23173)—Doors, windows, stairs, millwork, painting.

Carpentry and Building (23142)

Answers to the problems encountered in today's building trades. The actual questions asked of an architect by carpenters and builders are answered in this book. No apprentice or journeyman carpenter should be without the help this book can offer.

Do-It-Yourself Encyclopedia (23207)

An all-in-one home repair and project guide for all do-it-yourselfers. Packed with step-by-step plans, thousands of photos, helpful charts. A really authentic, truly monumental, home-repair and home-project guide.

Home Workshop & Tool Handy Book (23208)

The most modern, up-to-date manual ever designed for home craftsmen and do-it-yourselfers. Tells how to set up your own home workshop, (basement, garage, or spare room), all about the various hand and power tools (when, where, and how to use them, etc.). Covers both wood- and metal-working principles and practices. An all-in-one workshop guide for handy men, professionals and students.

Plumbers and Pipe Fitters Library—3 Vols. (23155)

A practical illustrated trade assistant and reference for master plumbers, journeyman and apprentice pipe fitters, gas fitters and helpers, builders, contractors, and engineers. Explains in simple language, illustrations, diagrams, charts, graphs and pictures, the principles of modern plumbing and pipe-fitting practices.

Vol. 1—(23152)—Materials, tools, calculations.

Vol. 2—(23153)—Drainage, fittings, fixtures.

Vol. 3—(23154)—Installation, heating, welding.

Masons and Builders Library—2 Vols. (23185)

A practical illustrated trade assistant on modern construction for bricklayers, stonemasons, cement workers, plasterers, and tile setters. Explains in clear language and with detailed illustrations all the principles, advances, and shortcuts based on modern practice—including how to figure and calculate various jobs.

Vol. 1—(23182)—Concrete, Block, Tile, Terrazzo.

Vol. 2—(23183)—Bricklaying, Plastering, Rock Masonry, Clay Tile.

Upholstering (23189)

Upholstering is explained for the average householder and apprentice upholsterer in this Audel text. Selection of coverings, stuffings, springs, and other upholstering material is made simple. From repairing and reglueing of the bare frame, to the final sewing or tacking, for antiques and most modern pieces, this book gives complete and clearly written instructions and numerous illustrations.

Questions and Answers for Plumbers Examinations (23206)

Many questions are answered as to types of fixtures to use, size of pipe to install, design of systems, size and location of septic tank systems, and procedures used in installing material. Subjects such as traps, cleanouts, drainage, vents, water supply and distribution, sewage treatment, plastic pipe, steam and hot water fittings, just to name a few.

Wood Furniture, Finishing, Refinishing Repairing (23216)

The basic technical and practical information needed for complete wood finishing. This book presents the fundamentals of furniture repair, both veneer and solid wood and complete refinishing procedures, which includes stripping the old finish, sanding, selecting the finish and using wood fillers. Complete step-by-step procedure for antiquing, painting, staining, flocking, inlay patterns and gold- and silver-leaf finishing. Various woods, along with actual grain photos, are discussed as to their characteristics and their reactions to various finishes.

Building a Vacation Home: Step-by-Step (23222)

Step-by-step procedure in the construction of this vacation home includes fifteen chapters, numerous illustrations, actual photographs, and a complete set of blueprints, showing cabinet construction as well as building and foundation structure. It points out difficulties which might be avoided by informing you of the proper procedures for obtaining property lines, building and sewage permits, and various hidden costs which are bound to occur.

ELECTRICITY-ELECTRONICS

Wiring Diagrams for Light & Power (23028)

Electricians, wiremen, linemen, plant superintendents, construction engineers, electrical contractors, and students will find these diagrams a valuable source of practical help. Each diagram is complete and self-explaining. A PRACTICAL HANDY BOOK OF ELECTRICAL HOOK-UPS.

Electric Motors (23150)

Covers the construction, theory of operation, connection, control, maintenance, and troubleshooting of all types of electric motors. A HANDY GUIDE FOR ELECTRICIANS AND ALL ELECTRICAL WORKERS.

Practical Electricity (23218)

This updated version is a ready reference book, giving complete instruction and practical information on the rules and laws of electricity—maintenance of electrical machinery—AC and DC motors—wiring diagrams—lighting—house and power wiring—meter and instrument connections—transformer connectors—circuit breakers—power stations—automatic substations. THE KEY TO A PRACTICAL UNDERSTANDING OF ELECTRICITY.

House Wiring—3rd Edition (23224)

Answers many questions in plain simple language concerning all phases of house wiring. A ready reference book with over 100 illustrations and concise interpretations of many rulings contained in the National Electrical Code. Electrical contractors, wiremen, and electricians will find this book invaluable as a tool in the electrical field

Guide to the 1975 National Electric Code (23223)

This important and informative book is now revised to conform to the 1975 National Electrical Code. Offers an interpretation and simplification of the rulings contained in the Code. Electrical contractors, wiremen, and electricians will find this book invaluable for a more complete understanding of the NEC.

Questions and Answers for Electricians Examinations (23225)

Newly revised to conform to the 1975 National Electrical Code. A practical book to help you prepare for all grades of electricians examinations. A helpful review of fundamental principles underlying each question and answer needed to prepare you to solve any new or similar problem. Covers the NEC; questions and answers for license tests; Ohm's law with applied examples, hook-ups for motors, lighting, and instruments. A COMPLETE REVIEW FOR ALL ELECTRICAL WORKERS.

Electrical Library—6 Vols. (23234)

Electric Generating Systems (23179)

Answers many questions concerning the selection, installation, operation, and maintenance of engine-driven electric generating systems for emergency, standby, and away-from-the-power-line applications. Private homes, hospitals, radio and television stations, and pleasure boats are only a few of the installations that owners either desire or require for primary power or for standby use in case of commercial power failure. THE MOST COMPREHENSIVE COVERAGE OF THIS SUBJECT TO BE FOUND TODAY.

Electrical Power Calculations (23050)

275 TYPICAL PROBLEMS WORKED OUT. Presents and explains the mathematical formulas and the fundamental electrical laws for all the everday, practical problems in both AC and DC electricity. EVERY ELECTRICAL WORKER AND STUDENT NEEDS THIS MODERN MATHEMATICAL TOOL.

Electrical Course for Apprentices and Journeymen (23209)

A basic study course for apprentice or journeyman electricians which may be used as a classroom or self-taught program. This book can be utilized without any other books on electrical theory. Review questions included.

Electronic Security Systems (23205)

Protect your home and business. A basic and practical text written for the electrician, electronic technician, security director and do-it-yourself householder or businessman. Such subjects as sensors and encoders, indicators and alarms, electrical control and alarm circuits, security communications, closed circuit television, and security system installations are covered.

New Electric Library

For engineers, electricians, electrical workers, mechanics and students. Presenting in simple, concise form the fundamental principles, rules and applications of applied electricity. Fully illustrated with diagrams and sketches, also calculations and tables for ready reference. Based on the best knowledge and experience of applied electricity.

Vol. 1 (23031)—Electricity, magnetism, armature winding, repairs.

Vol. 3 (23033)—Electrical testing instruments, storage battery construction and repairs.

ENGINEERS-MECHANICS-MACHINISTS

Machinists Library (23174)

Covers modern machine-shop practice. Tells how to set up and operate lathes, screw and milling machines, shapers, drill presses and all other machine tools. A complete reference library. A SHOP COMPANION THAT ANSWERS YOUR QUESTIONS.

Vol. 1—(23175)—Basic Machine Shop.
Vol. 2—(23176)—Machine Shop.
Vol. 3—(23177)—Toolmakers Handy Book.

Millwrights and Mechanics Guide— 2nd Edition (23201)

Practical information on plant installation, operation, and maintenance for millwrights, mechanics, maintenance men, erectors, riggers, foremen, inspectors, and superintendents. Partial contents: • Drawing and Sketching • Machinery Installation • Power-Transmission Equipment • Couplings • Packing and Seals • Bearings • Structural Steel • Mechanical Fasteners • Pipe Fittings and Valves • Carpentry • Sheet-Metal Work • Blacksmithing • Rigging • Electricity • Welding • Mathematics and much more.

Practical Guide to Mechanics (23102)

A convenient reference book valuable for its practical and concise explanations of the applicable laws of physics. Presents all the basics of mechanics in everyday language, illustrated with practical examples of their applications in various fields.

Power Plant Engineers Guide—2nd Edition (23220)

A complete steam or diesel power plant engineers library. This book covers (in question-and-answer form) facts for all engineers, fireman, water tenders, oilers, and operators of steam and diesel power systems. A valuable book to the applicant for engineer's and fireman's licenses.

Questions & Answers for Engineers and Firemans Examinations (23217)

An aid for stationary, marine, diesel & hoisting engineers' examinations for all grades of licenses. A new concise review explaining in detail the principles, facts and figures of practical engineering. Questions & Answers.

Welders Guide—2nd Edition (23202)

New revised edition. Covers principles of electric, oxyacetylene, thermit, unionmelt welding for sheet metal; spot and pipe welds; pressure vessels; aluminum, copper brass, bronze, plastics, and other metals; airplane work; surface hardening and hard facing; cutting brazing; underwater welding; eye protection. EVERY WELDER SHOULD OWN THIS GUIDE.

Mechanical Trades Pocket Manual (23215)

This "paper back—pocket size manual" provides reference material for mechanical tradesman. The handbook covers methods, tools, equipment, procedures, etc. in convenient form and plain language to aid the mechanic in performance of day-to-day tasks concerned with installation, maintenance, and repair of machinery and equipment.

FLUID POWER

Pumps (23167)

A detailed book on all types of pumps from the old-fashioned kitchen variety to the most modern types. Covers construction, application, installation, and troubleshooting.

MATHEMATICS

Practical Mathematics for Everyone—2 Vols. (23112)

A concise and reliable guide to the understanding of practical mathematics. People from all walks of life, young and old alike, will find the information contained in these two books just what they have been looking for. The mathematics discussed is for the everyday problems that arise in every household and business.
Vol. 1—(23110)—Basic Mathematics.
Vol. 2—(23111)—Financial Mathematics.

OUTBOARD MOTORS

Outboard Motors & Boating (23168)

Provides the information necessary to adjust, repair, and maintain all types of outboard motors. Valuable information concerning boating rules and regulations is also included.

RADIO-TELEVISION-AUDIO

Handbook of Commercial Sound Installations (23126)

A practical complete guide to planning commercial systems, selecting the most suitable equipment, and following through with the most proficient servicing methods. For technicians and the professional and businessman interested in installing a sound system.

Practical Guide to Auto Radio Repair (23128)

A complete servicing guide for all types of auto radios, including hybrid, all-transistor, and FM . . . PLUS removal instructions for all late model radios. Fully illustrated.

Practical Guide to Servicing Electronic Organs (23132)

Detailed, illustrated discussions of the operation and servicing of electronic organs. Including models by Allen, Baldwin, Conn, Hammond, Kinsman, Lowrey, Magnavox, Thomas, and Wurlitzer.

Radioman's Guide (23163)

Audel best-seller, containing the latest information on radio and electronics from the basics through transistors. Covers radio fundamentals—Ohm's law—physics of sound as related to radio—radio-wave transmission—test equipment—power supplies—resistors, inductors, and capacitors—transformers—vacuum tubes—transistors—speakers—antennas—troubleshooting. A complete guide and a perfect preliminary to the study of television servicing.

Television Service Manual (23162)

Includes the latest designs and information. Thoroughly covers television with transmitter theory, antenna designs, receiver circuit operation and the picture tube. Provides the practical information necessary for accurate diagnosis and repair of both black-and-white and color television receivers. A MUST BOOK FOR ANYONE IN TELEVISION.

Radio-TV Library—2 Vol. (23161)

TO ORDER AUDEL BOOKS mail this handy form to

Theo. Audel & Co., 4300 W. 62nd
Indianapolis, Indiana 46268

Please send me for FREE EXAMINATION books marked (x) below. If I decide to keep them I agree to mail $3 in 10 days on each book or set ordered and further mail 1/3 of the total purchase price 30 days later, with the balance plus shipping costs to be mailed within another 30 days. Otherwise, I will return them for refund.

APPLIANCES

- ☐ (23196) Air Conditioning & Refrigeration Library (2 Vols.) 13.75
  - ☐ (23159) Air Conditioning 6.95
  - ☐ (23195) Commercial Refrigeration 7.50
- ☐ (23214) Home Appliance Servicing 7.95
- ☐ (23133) Home Refrigeration and Air Conditioning 8.95
- ☐ (23151) Oil Burners 6.25

AUTOMOTIVE

- ☐ (23198) Automotive Library (2 Vols.) 15.95
  - ☐ (23192) Automobile Guide 10.25
  - ☐ (23181) Auto Engine Tune-Up 6.75
- ☐ (23199) Diesel Engine Manual 8.50
- ☐ (23061) Gas Engine Manual 5.50

BUILDING AND MAINTENANCE

- ☐ (23041) Answers on Blueprint Reading 6.50
- ☐ (23180) Building Construction and Design 6.75
- ☐ (23140) Building Maintenance 6.75
- ☐ (23169) Carpenters and Builders Library (4 Vols.) 21.25
  - ☐ Single Volumes sold separately .... ea. 5.50
- ☐ (23142) Carpentry and Building 7.50
- ☐ (23207) Do-It-Yourself Encyclopedia 13.50
- ☐ (23229) Gardening & Landscaping 7.95
- ☐ (23208) Home Workshop & Tool Handy Book 6.50
- ☐ (23185) Masons & Builders Library (2 Vols.) 12.95
  - ☐ Single Volumes sold separately .... ea. 6.95
- ☐ (23189) Upholstering 6.75
- ☐ (23155) Plumbers and Pipe Fitters Library (3 Vols.) 15.50
  - ☐ Single Volumes sold separately .... ea. 5.50
- ☐ (23206) Q & A for Plumbers Exam 6.95
- ☐ (23216) Wood Furniture Finishing, Repair 6.95
- ☐ (23222) Building a Vacation Home 7.95

ELECTRICITY-ELECTRONICS

- ☐ (23179) Electric Generating Systems 6.75
- ☐ (23234) Electrical Library (6 Vols.) 38.00
  - ☐ (23232) Wiring Diagrams for Light and Power 6.95
  - ☐ (23150) Electric Motors 6.95
  - ☐ (23218) Practical Electricity 6.95
  - ☐ (23224) House Wiring (3rd Edition) 6.50
  - ☐ (23223) Guide to the 1975 National Electrical Code 8.50
  - ☐ (23225) Questions & Answers for Electrician's Exams 6.50
- ☐ (23205) Electronic Security Systems 6.95
- ☐ (23209) Electrical Course for Apprentices & Journeymen 6.95

NEW ELECTRIC LIBRARY

- ☐ Single Volumes sold separately .... ea. 4.00

ENGINEERS-MECHANICS-MACHINISTS

- ☐ (23174) Machinists Library (3 Vols.) 19.50
  - ☐ Single Volumes sold separately .... ea. 6.75
- ☐ (23202) Welders Guide (2nd Edition) 10.95
- ☐ (23215) Mechanical Trades Pocket Manual 3.95
- ☐ (23201) Millwrights and Mechanics Guide (2nd Edition) 10.95
- ☐ (23220) Power Plant Engineers Guide 9.95
- ☐ (23102) Practical Guide to Mechanics 5.50
- ☐ (23217) Q&A for Engineers and Firemans Exams 7.95

FLUID POWER

- ☐ (23167) Pumps 8.95

MATHEMATICS

- ☐ (23112) Practical Math for Everyone (2 Vols.) 10.25
  - ☐ Single Volumes sold separately .... ea. 5.50

OUTBOARD MOTORS

- ☐ (23168) Outboard Motors and Boating 5.50

RADIO-TELEVISION-AUDIO

- ☐ (23126) Handbook of Commercial Sound Installations 5.95
- ☐ (23128) Practical Guide to Auto Radio Repair 4.95
- ☐ (23132) Practical Guide to Servicing Electronic Organs 5.50
- ☐ (23161) Radio & Television Library (2 Vols.) 13.75
  - ☐ (23163) Radiomans Guide 6.75
  - ☐ (23162) Television Service Manual 7.50

Prices Subject to Change Without Notice

Name ______________________________

Address ______________________________

City ____________________ State ____________ Zip ______

Occupation ____________________ Employed by ____________________

☐ **SAVE SHIPPING CHARGES! Enclose Full Payment With Coupon and We Pay Shipping Charges.**

PRINTED IN USA